Advance Praise

Dr. Spiekermann's book is nothing short of a watershed volume, one that is extraordinary in several ways. First, it develops an exceptionally practical approach to ethics in IT design. Central to this practical approach is its solid pedagogy: it is carefully crafted to help students as well as interested professionals in the IT industries develop a better understanding of ethics and how ethics can be usefully interwoven in the design and development of information and communications technology (ICTs). Indeed, a primary virtue of the book is that it is clear and accessible to these audiences as it is written by someone with long experience in and deep understanding of IT in business and professional terms. At the same time, the work is exceptionally well informed in philosophical terms. Indeed, both beginning students and advanced scholars in Information and Computing Ethics (ICE) will profit from reading this text. In these ways, the book thus stands as an exceptional synthesis and bridge between two sets of disciplines and cultures that are otherwise, to put it mildly, difficult to bring together—both theoretically, and certainly in the most practical terms, i.e., of integrating philosophical ethics in the design and development of ICTs.

Perhaps best of all (certainly from my perspective), Dr. Spiekermann not only takes up the requisite and usual attention to both utilitarian and deontological ethics; more fundamentally, the book stands solidly in the tradition of virtue ethics as applied to ICTs from the very beginning of computer ethics in the work of Norbert Wiener. Even more importantly, as Dr. Spiekermann makes clear, virtue ethics is becoming ever more useful and important in our efforts to ethically evaluate and guide the design of digital technologies—i.e., in hopes of developing these technologies so that they indeed contribute to flourishing lives. The book incorporates much of the most recent and important research and scholarship in virtue ethics and IT, and at the same time, Dr. Spiekermann significantly expands upon this work through her more extensive syntheses between IT design and philosophical ethics, and so establishes a new level of accomplishment and contribution within both domains.

The book is thus a breakthrough volume; it will stand as essential reading and a primary reference in the further development of ethics and IT design, most especially as informed by virtue ethics approaches.

Charles Ess, Professor
University of Oslo

Ethical IT Innovation

A Value-Based System Design Approach

Ethical IT Innovation

A Value-Based System Design Approach

Sarah Spiekermann

Vienna University of Economics and Business
Institute for Management Information Systems

CRC Press
Taylor & Francis Group
Boca Raton   London   New York

CRC Press is an imprint of the
Taylor & Francis Group, an **informa** business

AN AUERBACH BOOK

This book was language edited by Julian Cantella.

CRC Press
Taylor & Francis Group
6000 Broken Sound Parkway NW, Suite 300
Boca Raton, FL 33487-2742

© 2016 by Taylor & Francis Group, LLC
CRC Press is an imprint of Taylor & Francis Group, an Informa business

No claim to original U.S. Government works

Printed on acid-free paper
Version Date: 20151007

International Standard Book Number-13: 978-1-4822-2635-5 (Hardback)

Visit the Taylor & Francis Web site at
http://www.taylorandfrancis.com

and the CRC Press Web site at
http://www.crcpress.com

Contents

List of Figures

List of Tables

List of Abbreviations

AI	Artificial intelligence		**IT**	Information technology
AR	Augmented reality		**NPD**	New product development
AU	Act utilitarism		**NPV**	Net present value
CIA	Confidentiality, integrity, availability		**PET**	Privacy enhancing technology
CRM	Customer relationship management		**PIU**	Problematic Internet use
CSR	Corporate social responsibility		**POS**	Point of sale
CST	Critical system thinking		**RFID**	Radio-frequency identification
e-2-e	End-to-end		**ROI**	Return on investment
ERP	Enterprise resource planning systems		**RU**	Rule utilitarianism
E-SDLC	Ethical system development life cycle		**SAR**	Specific absorption rate
EUD	End-user development		**SDLC**	System development life cycle
GU	General utilitarianism		**SME**	Small and medium enterprise
HBR	*Harvard Business Review*		**SOX**	Sarbanes–Oxley Act
HCI	Human–computer interaction		**UML**	Unified modeling language
IP	Intellectual property		**UX**	User experience
IS	Information systems		**VR**	Virtual reality

Preface

"To lift a weight so heavy, Sisyphus
Would take your courage,
Although one's heart is in the work
Art is long and Time is short."

Charles Baudelaire (1857)

Hardly a day passes without news of a large-scale privacy or security breach: incidents where personal data or entire identities are stolen or misused, where critical systems are hacked or destroyed by malicious attackers. Edward Snowden revealed how governments and companies abuse our information technology (IT), recording every step or social media post and using this knowledge to increase their power and profit. But this is just the tip of the iceberg. Economies now lose billions in productivity from knowledge workers who are constantly distracted, interrupted, and tempted by IT systems. Family lives, friendship, and intimacy suffer as conversations are constantly interrupted and people become addicted to their IT devices. Millions of people escape into the fantasy of virtual worlds and social networks, living out dreams that reality doesn't hold ready for them. The price of escape is often physical and mental health problems, as the body is forgotten by the straying mind.

Another textbook on computer ethics was titled *Goblet of Fire*, referring to the Greek myth where Prometheus brought fire to humanity. I agree with this analogy; IT is like fire. IT enables enormous leaps in progress, leading humanity into a new, potentially better age. But like fire, it is also dangerous in that IT can harm human values. As we play with fire today, we are often burned. This textbook is written for those who understand that our current way of developing IT systems and investing in them is a bit like playing with fire and has problems. It is for those who want to construct IT as a benevolent power rather than an unregulated force of destruction.

To bring order and control to IT, we must first understand its goals. We must understand its desired properties and then choose the right methods to cater IT to those properties. This is what this textbook is about. After introducing concepts of IT innovation and the future of IT, I describe the ethical values that matter to humans and human societies across the world. I detail what these values stand for and how they relate to IT. I explain how to build IT so that it accommodates the things we all care about: intrinsic values like freedom, knowledge, health, safety and security, trust, belonging and friendship, dignity and respect, and qualities that support these intrinsic values, such as privacy, transparency, control, truth, fairness, accessibility, objectivity, authenticity, accuracy, accountability, empathy, reciprocity, and politeness.

Why is the concept of Sisyphus' rock so important that I quote it above? First, building IT is a Sisyphean task. You engineers have to pay attention to the implications, uses, and abuses of your systems, their true requirements and to the details of your code. If you do not, you will never reach the mountaintop. Once a program runs well, it feels like sitting on the mountaintop. But Baudelaire warned you, warned us: *L'Art est long et le Temps est court* ("Art is long and time is short"). To create real art, which I equate here with good IT systems, you need to invest time—time that is usually too short. If you do not invest the time, then you will not produce anything that deserves to be called art; the IT will not be good. This book propagates the idea that thinking and crafting IT for value need time. They need the patience of Sisyphus. Those who are not willing to invest time build second-class objects that create more entropy than good.

Sisyphus had to meticulously roll his rock up the hill, and he could not stand still until it was at the top. Once it was there, it rolled down again and he had to start anew. At times, a message of this nature and size feels like a big rock. This book is large in ambition and scope, and its knowledge spans four disciplines: philosophy, computer science, psychology, and management. With four disciplines embedded in this work, it has become a rich and dense piece. When four disciplines are forming a mosaic, a beautiful new landscape emerges—a landscape in this case of an artful future IT environment. That said, I am aware that this work cannot stop here. The rock will role down the hill again. Many scholars will probably turn to me and outline more insights that I could have or should cover. Future editions might therefore become necessary.

Discussing the Sisyphus myth, Nobel laureate Günther Grass once said:

"Sisyphus is nothing else than the insight that the rock will not stay on the mountaintop—and to consent to this. Ideology always promises mankind that the rock will stay on the mountaintop one day. This would be paradise. All utopia work with this promise."

I must confess that writing about ethical IT reflects my ideology; it is an "evangelizing" exercise. Of course, I would like to promise that if you engineers and managers built IT systems with the ideology of this book in mind, the world would become a better place (the rock would stay on the mountaintop). And I really believe that! But I must acknowledge that much more exists than I have written about here: many more views, many more values, and a history of values that will progress. So, humbly, the only thing I can do is urge you to face the rock, take the time and use your art and virtuousness to create value instead of entropy!

Acknowledgments

Writing this book forced me into some isolation. But unlike Sisyphus, I was not truly alone. I would therefore like to thank the 20 people who helped me in this endeavor. Most important, I thank my husband, Johannes, who lightened these days of hard labor through his presence. Without him, I do not know how I could have made it. But emotional support also came from my team at Vienna University of Economics and Business, who constantly interacted with me. My greatest thanks go to Alexander Novotny, my PhD student at the time, who gave me in-depth feedback for every chapter. Later, Domink Berger, Peter Blank, and Roswitha Kreitner took over and helped me to round out all the chapters. Sandra Paulhart did an enormous job digging out the computer ethics literature on relevant values in IT. My PhD students Jana Korunovska and Olja Drozd recommended changes to Chapters 5, 8, and 10. Very important to this book is Julian Cantella. He did all of the English language editing and thoroughly revised everything I wrote. Finally, Marthe Kerkwijk found the original citations for all the philosophical works cited in this book.

In addition to this comprehensive support, many experts took the time to read those chapters where they are recognized specialists. Professor Johannes Hoff (Heythrop College, University of London) revised all the chapters that address philosophy (Chapter 4 on value ethics, Chapters 5 and 6 on knowledge and freedom, Chapter 11 on dignity and respect, and Chapter 15 on wise leadership). Professor Thomas Gross (University of Newcastle) and Magister Stefan Strauß (Institute of Technology Assessment, Vienna) revised Chapter 8 on IT security and safety. Professor Katie Shilton (University of Maryland) gave feedback on Chapters 4 and 5 on value ethics and ethical knowledge. Dr. Christof Miska (Vienna University of Economics and Business) gave feedback on Chapter 15 on leadership. And Ulrike Rauer (PhD student at the Oxford Internet Institute) helped to improve Chapter 7 on health.

Besides these academic experts, I also received in-depth comments and revisions from industry. In particular, Dr. Peter Lasinger (investment manager at Austria's premier seed-funding investment company) provided great feedback to Chapter 2 on innovation, Chapter 13 on system design, and Chapter 15 on leadership. Stephane Chaudron (from the Digital Citizen Security Unit of the European Commission's research unit JRC) provided insightful comments and ideas for my future scenarios described in Chapter 3. Peter Bubestinger from the Free Software Foundation revised Chapter 6 on freedom, Chapter 11 on dignity, and Chapter 13 on system design. Finally, IT thought leaders and senior managers took the time to read and provided important comments, in particular, Magister Gregor Herzog and Jakob Haesler. I am deeply indebted to all of these wonderful supporters.

Author

Sarah Spiekermann chairs the Institute for Management Information Systems at Vienna University of Economics and Business (WU Vienna). She has published more than 70 articles on the social and ethical implications of computer systems and given more than a hundred presentations and talks about her work throughout the world. Her main expertise is electronic privacy, disclosure behavior, and ethical computing. Spiekermann has coauthored US/EU privacy regulation for radio-frequency identification (RFID) technology and regularly works as an expert and adviser to companies and governmental institutions, including the EU Commission and the Organisation for Economic Co-operation and Development (OECD). She is on the supervisory board of the Foundation of Data Protection of the German Parliament.

She also maintains a blog on "The Ethical Machine" at Austria's leading daily newspaper *Standard.at* and is on the board of the Austrian art and science think-tank GlobArt. Before being tenured in Vienna in 2009, Spiekermann was assistant professor at the Institute of Information Systems at Humboldt University Berlin (Germany), where she headed the Berlin Research Centre on Internet Economics (2003–2009), was adjunct visiting research professor with the Heinz College of Public Policy and Management at Carnegie Mellon University (Pittsburgh, Pennsylvania, 2006–2009), founded and shut down the company Skillmap (visualizing social networks; 2008–2011), and worked as a management consultant and marketing manager with A.T. Kearney and Openwave Systems.

Chapter 1

Introduction

"It takes only a tiny group of engineers to create technology that can shape the entire future of human experience with incredible speed."

Jaron Lanier (2011)

In late 2010, Antonio Krüger invited me to speak at the German Research Center for Artificial Intelligence (DFKI). Antonio is a professor of computer science who heads the Ubiquitous Media Technology Lab at the University of Saarbrücken and the Innovative Retail Group at DFKI. He and his team are bright, gifted, and playful; deeply involved in how to get interesting new technology to work. They try it out, test its technical limits, and constantly invent new, challenging applications. Antonio explores a lot of different things we could do with technology, not always questioning the applicability of everything he builds. He is perhaps a bit like Heinrich Hertz, who was once asked what the "radio waves" he discovered were good for. Hertz said that he has no idea, but he hoped that somebody would find use for them some day.

Antonio knows that some of his technology could be misused in the future, and he cares about that. But this concern is purely theoretical. Avoiding technology harms is not *his* business, he thinks. He was trained to see his job as something else: he wants to build an exciting new future in which machines become a positive part of our lives. Thinking about all the negatives that could happen with the technology is a nuisance, not only for Antonio but for anyone who loves to create things. It is like criticizing someone's baby. So, like many engineers, Antonio feels more comfortable viewing technology as neutral; if people abuse it later, that is *their* problem and *their* moral responsibility. Antonio is a typical engineer as I have come to know them. As I wrote this book, it was Antonio and his team that I thought about whenever I used the term *engineer*.

The topic of the talk that the DFKI invited me to give was on the adoption dynamics of ubiquitous computing technologies such as self-driving cars and intelligent fridges. Because most of these technologies rely on peoples' data to function, a primary adoption factor is to what extent the technologies protect privacy. Privacy is recognized as a human right, but the machines we build often undermine this right. Worldwide, 88% of people are concerned about who has access to their data (Fujitsu 2010), and international regulation is increasingly sanctioning privacy breaches. As a result, researchers like Antonio and his team must think about privacy issues more often than they used to. Consequently, they wanted to learn from me about the relevance of privacy concerns for the adoption of their technologies. They invited me because I am a privacy expert and had done experiments on the adoption of ubiquitous computing.

My research yielded a surprising answer for them: Privacy issues did not significantly influence people's initial intent to use a technology. In fact, privacy concerns had a surprisingly small influence, while other values such as user control over technology played a huge role for study participants' emotions and adoption plans (Spiekermann 2008). When devices become autonomous, people can feel helpless and deprived of their sense of being and their freedom (Langer 1983; Guenther and Spiekermann 2005).

The study confirmed what I have always believed: privacy and security concerns are not currently the most important values for initial technology *adoption*, and they will not be until people have really *experienced* the negative effects of privacy breaches. Perhaps now, with Edward Snowden having revealed the degree of state surveillance, privacy might become more important. But privacy and security concerns are still just the tip of the iceberg of many other values that count for information technology (IT) users. Values such as user trust, autonomy and control, transparency, attention sensitivity, and safety are all popping up as relevant IT design issues. And so, when I spoke to Antonio and his team, I recommended that somehow their IT system designs need to accommodate for these values or "nonfunctional requirements." IT designs must cater to these values, protect them, even foster them.

My second recommendation to the DFKI team did concern privacy. I wondered: "If people state that they care about privacy so much but do not act on their concerns, perhaps systems need to come with built-in privacy to protect them proactively." We know that people underestimate their long-term privacy risks when they make decisions and use technologies (Acquisti and Grossklags 2005). Can engineers engage in wise foresight for their users? Should engineers build technology that minimizes the potential for later abuse?

These questions about a proactive privacy by design opened a debate between myself and the DFK team. Although Antonio and his team agreed in principle that they might need to think more about privacy and other values, they asked me *when* and *how* they should actually think about privacy in the development process. I was totally surprised by this question. I thought that the answer to this question should be perfectly clear to any software engineer. Did they not learn when and how to think about ethical requirements in system design? Did they not practice this extensively in their undergraduate computer science education? I soon found that, even among high-end IT professionals, this is not the case. Computer science education and the IT industry have so far mainly focused on *functional* system requirements engineering. Engineers are taught to understand technical phenomena, use them, and make things work. Nonfunctional requirements in contrast have been treated as stepchildren, second-order problems.

However, a massive change has captured the IT industry in the last 10 years; technologists needed to embrace usability and aesthetics as key nonfunctional requirements for technology design. The field of human–computer interaction (HCI) has gained in importance, and many methods for participatory design have been developed to enable engineers to build systems people are more comfortable using. But unresolved issues remain: How do we go beyond aesthetics? How can we recognize human values in system design, engage in privacy by design, design for autonomy or trust or attention sensitivity? When Antonio asked me these questions in 2010, I realized I did not know the answer. I knew many privacy-enhancing technologies (PETS), individual pieces of software that would support privacy-sensitive machine behavior. I could have reported on similar technologies that supported security, attention sensitivity, or user control. I knew about various languages for modeling processes, logic, and data. But I was not able to answer Antonio's question, which looked for a methodology to support and guide engineers so that they could systematically consider human values as part of their creative IT construction process. This is how the journey of this book began, a book that tries to answer Antonio's question.

1.1 How This Book Frames Computer Ethics

When I searched for value-sensitive IT system design methodologies, the first work I found was that of Batya Friedman and her group. In the late '90s and early 2000s, she started to think about human values in IT. She developed some central puzzle pieces for an overall methodology to use for value-based design. The most important insight in her work was that we need to understand values first; we need to conceptually investigate a value and break it into small pieces that are so concrete that engineers can start actually working on them. For instance, a term like "privacy" is so generic that an engineer cannot possibly know how to build a privacy-sensitive system. Such a value needs to be broken down to component concepts, such as "informed consent to data collection" or "confidential information transfers," in order for engineers to start working.

For engineering teams like those at DFKI, it is not easy to conceptually investigate a value. Understanding a human value in detail is not something engineers have learned to do, nor is it only a matter of gut feeling and stakeholder discussion. Major values such as freedom, safety, trust, friendship, and dignity are age-old concepts. Their meaning has been discussed among philosophers and political scientists for more than 2000 years. To understand the values that matter to us in IT, we need to dig deeply into the humanities. Only when we understand a concept such as freedom can we think about building systems that set us free (or at least do not undermine our freedom).

A major concern in this book is therefore—first—to make the humanities accessible to engineers and IT innovation teams. Almost half of this book (Chapters 3 to 12) is dedicated to conceptually analyzing major human values and breaking them down for engineering. These chapters give engineers and IT innovation teams a flavor of what a respective value (like privacy) stands for, where it comes from, and how it is relevant to IT system design. In this endeavor, I drew not only from original philosophical sources and legal studies, but also the last two decades of academic work in the field of computer ethics. Computer ethicists, whose academic background is to a large extent rooted in the humanities, have done an incredible job of accumulating and deepening our understanding of human values relevant to IT.

When I researched computer ethics textbooks for university education, I was surprised. Against my expectations, I could hardly find any material about value-ethics. Where could students learn about those values IT users and designers face, such as trust, autonomy, health, dignity, etc.? Although the textbooks cover some values, such as privacy values for IT were not the focus. Instead, the books teach normative ethical theory: they reflect philosophies such as deontology, utilitarianism, virtue ethics, and justice. Often, the examples used to explain these philosophical theories do not even refer to technology use at all.

I wondered why are these normative ethical theories relevant for engineers at all? What use can IT teams make of knowing about Kant's categorical imperative?

I found one useful IT design area where normative ethical knowledge is helpful. We can apply Kant or utilitarianism to what experts now call "machine ethics" (see Chapter 19). Machine ethics is mainly about artificial intelligence (AI) for ethical reasoning, algorithms that integrate ethical decision theory (Anderson and Anderson 2011). AI scholars have started to work on the grand challenge to build machines with morals, machines that promise to be similar to humans or "full ethical agents" (Moor 2011). Many scholars now engage in discussions about how presumably autonomous systems, such as drones or robots, could take ethical decisions. A good example to illustrate their struggle is given in Box 19.1 where a self-driving car needs to take the ethical decision of whether to bump into a Volvo or a Mini-Cooper. A lot of attention is also given to the logic of killing-algorithms of drones used in war zones.

Such reflections on ethical AI are not the goal of this textbook. I am not interested here in ethical theory for algorithm design. But still I think that the grand theories for ethical decision-making such as virtue ethics, deontology, and utilitarianism are crucial for the IT world for a second reason. The theories can be used to determine whether we want certain technologies in the first place. They are relevant when we take investment decisions, set value priorities for our IT systems, and investigate their ethical and political feasibility. At the core of this effort is IT leadership. Senior executives and IT investors must try to think more in normative ethical terms when they judge technology investments and define requirements. Chapter 15 introduces how wise IT leaders could think about technology with the help of normative ethical theory.

Of course, making managers think about Kant or Aristotle when they invest in technology seems to be a naïve endeavor at first sight. For the most part, managers do not even know what these great thinkers have taught us, let alone how their thoughts relate to IT innovation management. Many companies today are effectively driven by short-term shareholder value. Despite a call for more responsible leadership and many studies about the positive effects of genuine corporate social responsibility, most managers are still mandated to pursue only profit and efficiency. Managers are not hired to lead or to ethically question, but to execute. In this climate, I realize it is very difficult to act as a responsible leader, who reflects on his decisions with the help of normative ethical instruments. That said, this book is not written for those who merely "execute." It is written for students of computer science, engineering, information systems (IS), and innovation management who want to make a difference. It is also written for CIOs and

IT innovators who are willing to lead. For those people, it is important to understand what societies value and how we can make decisions for ethical IT investment and design. As Jaron Lanier states in the quote above: "It takes only a tiny group of engineers to create technology that can shape the entire future of human experience with incredible speed." This book is written for those who are aware of this.

1.2 The Ethical System Development Life Cycle (E-SDLC)

Taking the right IT investment decisions and prioritizing the values we want to see it is not enough to produce "good" IT systems. After IT innovation teams conceptually investigate human values, prioritize them, and break them down, they must next determine how to actually build them into IT systems.

In Chapters 13 to 18 of this book, I provide a methodology that I call the "E-SDLC." "E" stands for "ethical" and "SDLC" is the standard abbreviation for "system development life cycle." From many years of teaching courses on system analysis and design and innovation management, I know that the academic field of IS has developed a rich body of knowledge that systematizes IT design processes. IS scholars have developed not only methodologies for system design but also modeling tools to support systematic planning of IT infrastructures. The only missing piece is how to systematically bring ethics into these methodologies: "folding human values into the research and design cycle" (Sellen et al. 2009, p. 64).

When I started work on privacy and security risk assessment methods in 2010, I realized how this folding exercise could be done. Security risk assessment methodologies have appeared in recent years, designed to identify where valued assets are threatened and how to systematically build IT systems to mitigate these threats. Transferring this "asset-protection-thinking" to values beyond security, I developed a methodology for building privacy-sensitive systems by using a privacy impact assessment (Oetzel and Spiekermann 2013). In doing so, I reckoned that the methodological steps that we use in security and privacy risk analysis can be applied to other human values as well. Chapters 17 and 18 explain this process in detail, providing a tool-supported methodology to work with.

1.3 Value-Based IT Design: A Positive Message for IT Markets

Risk analysis always takes a negative perspective on IT systems. It looks at all the bad things that could happen: how the new system could be attacked by a malicious entity, fall victim to abuses and environmental catastrophes, or cause customer backlash when values are destroyed. But this book does not take the negative perspective. Even though the methodology stemming from risk analysis helps us to systematically think about and address values in IT design, values as such are positive. The term "value" comes from the Latin word *valere*, which means "to be strong" or "to be worth." Something that is valued is treasured and therefore perceived as "good." As a result, this book frames ethical IT innovation in a positive light. "What is desirable, good or worthwhile in life? … What values should we pursue for ourselves and others?" (Frankena 1973, p. 79). When we identify what values we share and want to foster through information technology, then we can start building the technology with values in mind. This point was recently emphasized by Michael Porter, one of today's leading management scholars: "Shared value is not social responsibility, philanthropy, or even sustainability, but a new way to achieve economic success. It is not on the margin of what companies do, but at the center" (Porter and Kramer 2011).

1.4 Engineers, Managers, and "Ethical Machines"

Hans Jonas once wrote that "the future is not represented in any committee … the non-existent has no lobby and the unborn are powerless" (1979/2003, p. 55). I would like to contradict this view: I think that the future is represented by the powerful people who design and build our IT systems. This book shows those people how to begin. So, let's return to Antonio and his team. After reading this book, will they be able to work differently than they do now? What will change for them if they take its content seriously?

For one thing, I hope to change the mindset of many engineers by refuting, once and for all, the "technology is neutral" argument. The future scenarios in Chapter 3 and later chapters show how technology can be built in completely different ways when value ethics is considered. I give many examples of what I call a "value-based *alternative* approach to computer ethics" (Chapter 13). With this approach, an IT team can come up with ethical alternatives for every initial technical design they had in mind. Real magic is realized when IT teams find design alternatives that have fewer value problems and more potential for shared positive value. From Chapter 4 onward, this book demonstrates that "good" machine behavior is deeply rooted in how it is engineered. Engineers like Antonio and IT innovation teams now receive guidance on what various values mean conceptually and at a level of detail where they can start constructing technology. The methodological chapters show them when and how to systematically tackle the value details in an IT project. Only

one pitfall remains: IT management must provide enough time to think through ethical alternatives.

Of course, one might argue that every technology has drawbacks. A value-based alternative will create new problems of its own. A value-based machine is not a machine without flaws. But there is an important difference in its setup: it is a technology created by an innovation and engineering team that has done everything possible to implement a technology with ethics in mind. I would like to argue that building ethical machines is like raising children—"mind children" (Moravec 1988). You cannot guarantee that your children will be good people and succeed in life, but you can do your best to influence them. Everyone who has raised children will agree on this point. Good education sets boundaries, teaching children what they are allowed to do and what they are not. We must do the same thing with our machines. Engineers are responsible for building machines as implicit ethical agents: they have to build them so that they cannot be abused at the large scale, as many are today.

Furthering the educational analogy, everyone will agree that parents' attitude toward the world and other people influences the attitude of their children. So what attitude dominates in engineering circles today? In their seminal book *The Race Against the Machine*, Eric Brynjolfsson and Andrew McAfee cite a NASA report that suggests one dominant attitude: "Man is the lowest-cost, 150-pound, nonlinear, all-purpose computer system which can be mass-produced by unskilled labor" (Brynjolfsson and McAfee 2012, p. 25). This view on mankind is reflected in the "resource-based" view of the firm that is preached in management literature. By talking about "resources" instead of "humans," management students are trained to regard employees as assets with an exploitable lifecycle instead of beings with a social existence. If managers and engineers continue to share such attitudes, the technologies they create might not be what we as human societies want them to be.

Note that I do not address only engineers in this book. It is not only Antonio and his team who should be at the focus of this discussion. Engineers are normally embedded in organizations, sometimes universities, but mostly industrial research labs or companies' software and hardware production units. Here, they do not work freely. They largely follow what their organizational environment and their managers expect from them and pay them for. They team up with IT innovation managers who often tell them what to build or jointly explore the IT innovation space with them. As becomes clear in Chapters 14 to 18, IT managers (as opposed to engineers) often choose how to make IT investments. So the organization, its management, its processes, and the attitude reigning there are equally important for ethical system design.

The influence of organizational management on IT's privacy design became apparent in a recent study I did with Marc Langheinrich. We surveyed over a hundred software engineers from around the world and conducted face-to-face interviews. We found that 48% of the engineers were not sure or did not think that their organizations wanted them to respect privacy issues in the systems they build. Also, 49% thought that it is not mostly up to them to incorporate privacy in a system. Almost half of those we questioned (45%) think that the time allotted to incorporate privacy mechanisms makes it difficult for them (or prevents them from) implementing privacy requirements into their systems. These findings show that even for a well-known IT value that has been discussed for years in the public media, many companies lack awareness, willingness, or knowledge to design their systems in a more sensitive way.

References

Acquisti, A. and J. Grossklags. 2005. "Privacy and Rationality in Individual Decision Making." *IEEE Security & Privacy*, January/February, 26–33.

Anderson, M. and S. L. Anderson. 2011. *Machine Ethics*. New York: Cambridge University Press.

Brynjolfsson, E. and A. McAfee. 2012. *Race Against the Machine: How the Digital Revolution is Accelerating Innovation, Driving Productivity, and Irreversibly Transforming Employment and the Economy*. Lexington, MA: Digital Frontier Press.

Frankena, W. 1973. *Ethics*. 2nd ed. Englewood Cliffs, NJ: Prentice-Hall.

Fujitsu. 2010. "Personal Data in the Cloud: A Global Survey of Consumer Attitudes." Tokyo, Japan.

Guenther, O. and S. Spiekermann. 2005. "RFID and Perceived Control—The Consumer's View." *Communications of the ACM* 48(9):73–76.

Jonas, H. 1979/2003. *Das Prinzip Verantwortung—Versuch einer Ethik für die technologische Zivilisation*. Vol. 3492. Frankfurt am Main: Suhrkamp Taschenbuch Verlag.

Langer, E. 1983. *The Psychology of Control*. Beverly Hills, CA: Sage Publications.

Lanier, J. 2011. *You Are Not a Gadget*. London: Penguin Books.

Moor, J. 2011. "The Nature, Importance, and Difficulty of Machine Ethics." In *Machine Ethics*, edited by Michael Anderson and Susan Leigh Anderson. New York: Cambridge University Press.

Moravec, H. 1988. *Mind Children: The Future of Robot and Human Intelligence*. Cambridge, MA: Harvard University Press.

Oetzel, M. and S. Spiekermann. 2013. "Privacy-By-Design through Systematic Privacy Impact Assessment—Presentation of a Methodology." *European Journal of Information Systems* 23(2):126–150.

Porter, M. and M. R. Kramer. 2011. "Creating Shared Value." *Harvard Business Review* 89(1).

Sellen, A., Y. Rogers, R. Harper, and T. Rodden. 2009. "Reflecting Human Values in the Digital Age." *Communications of the ACM* 52(3):58–66.

Spiekermann, S. 2008. *User Control in Ubiquitous Computing: Design Alternatives and User Acceptance*. Aachen, Germany: Shaker Verlag.

Chapter 2

IT Innovation

"If a man does not know to what port he is steering, no wind is favorable to him."

Seneca (1 BCE–CE 65)

Technology is rapidly evolving around us, and if we believe in the theory of accelerating returns, the pace of evolution will only increase. Ray Kurzweil, a serious contemporary futurist, expects that "measured by today's rate of progress … we won't experience 100 years of technological advance in the twenty-first century; we will witness on the order of 20,000 years of progress" (2006, p. 11). His prediction, which most experts conceptually acknowledge, is based on a key observation: Information technology (IT) is not only powerful in itself; it also facilitates the success and convergence of other powerful technological domains such as nanotechnology, genetic engineering, and robotics. Technological change is like a raging river into which individuals, companies, and societies dive enthusiastically, but then have trouble staying afloat.

This book aims to help managers, entrepreneurs, and engineers to successfully maneuver through the explosion of IT innovation by embracing a new vision for IT design: "ethical IT innovation" driven by values. In this case, *value* does not mean short-term financial profit. Value implies a systematic shaping of future machines to support the growth and welfare of people or, what Aristotle (384–322 BC) called "eudemonia" (Greek: εὐδαιμονία stands for "happiness, welfare"). Function-centered engineering is replaced by human-centered engineering.

Aiming for eudemonia through machines is not easy. We currently prioritize machine functionality over ethical requirements such as the privacy or security of systems, not to mention other human values such as trust, transparency, attention-sensitivity, objectivity, accessibility, reliability, autonomy, or property. At the core of current technological change is a current software engineering practice that lives an extremely rapid pace and has the tendency to minimize requirements engineering. "Release early, release often" is a common mantra, or "rapid application development." As a result of this approach, incredibly powerful machines are created in a mostly unregulated space without any ethical considerations.

In this creative torrent, we risk creating machines that do not serve human uses and moving toward a future in which humans serve machines. Instead of helping people grow and advance—a great potential of information technology!—innovation could turn against us and wash us onto the shores of a machine age that throws humanity back into what Kant should call "immaturity" (Immanuel Kant, 1724–1804).

2.1 Human Growth and Eudemonia through IT

Eudemonia has not only been recognized by Aristotle as the core goal of human life, but many more current authors in the field of cybernetics and computer ethics, such as Norbert Wiener (1954), James Moor (1999), and Terrell Ward Bynum (2006), have embraced human flourishing as a core concept in their ethical considerations vis-à-vis IT. Using the word "happiness," economists have started to embrace eudemonia as an additional variable besides economic return and productivity (Frey and Stutzer 2001; Bruni and Porta 2007). Economists work is informed by now decades of psychological research into the drivers of human motivation and well-being (Diener 2009). Peter Drucker spoke of companies' "inescapable accountability to the quality of life" (Smith and Lenssen 2009, p. 375). And in 2011 Michael Porter laid the foundations for a "shared value" initiative at Harvard Business School in which he calls for the creation of economic value in a way that also creates value for society.

Eudemonia is not something that comes easily. Its first requirement is an optimal environment. In many respects, information technology has helped us to create a good

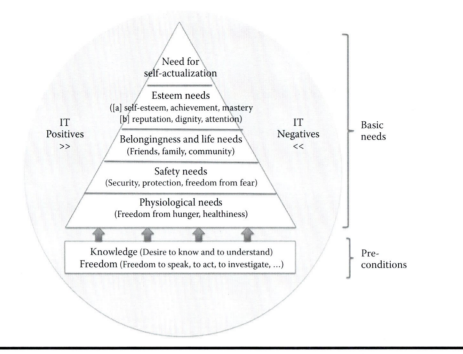

Figure 2.1 Maslow's hierarchy of human needs.

environment already. For the first time in human history, we have reached—at least in the wealthy parts of the world—a point where we start to understand important pieces of the natural forces around us. IT helps to model and unravel parts of their magic. We have started to influence some of these forces and experiment with them; also in great parts helped by the simulation and processing power of our machines. Some scholars have even called out the "age of the anthropocene."* In rich nations a majority of the population does not have to struggle any more for daily survival. Globally integrated, IT-controlled supply chains provide us with the food, shelter, and luxuries we desire. Efforts are being made to extend some of this wealth to the Third World. In fact, IT-driven manufacturing provides goods and services in quantities that often surpass demand. A majority of people is employed and earn enough income to live. This income, as well as remaining faith in the democratic balance of power between citizens, corporations, and elected governments, has led to a perception of personal safety and security in large parts of First World societies.

Against this background, the majority of readers of this book currently benefit from the fulfillment of what Abraham Maslow would consider to be the lower physiological and safety needs of people. Maslow is the founding father of humanistic psychology, a school of psychology that aims to

support human eudemonia. Maslow's list of human growth needs has been illustrated as a pyramid. The pyramid symbolizes that once people have fulfilled a lower need, they are motivated by the outlook of achieving the next higher one (Maslow 1970; Figure 2.1). Physiological needs are the lowest needs that must be fulfilled before any higher needs can actually materialize. If we are hungry or sick, we first have to overcome this state before we can truly benefit from higher needs, such as love, reputation, or creativity. The same is true for safety. If we feel insecure or threatened, our only urge is to first reestablish order before we can turn to higher goals in life. Once our physiological needs and safety needs are fulfilled we strive for the satisfaction of the higher social and individual needs, such as love, belongingness, and self-esteem. The highest level of the hierarchy is self-actualization, in which people aim to leverage their potential and talents to the fullest degree. "What a man can be he must be," wrote Maslow (1970, p. 46). None of these needs can be realized if not two preconditions are met. These are—according to Maslow—first the freedom to speak, to act, to investigate, and to defend oneself. Second, he discusses the importance of knowledge. Humans' desire to know and to understand enables them to build the competencies to fulfill their needs.

Over the last decades, IT has helped to not only secure our basic physiological and safety needs, but also provided us with many opportunities to address our higher needs. For example, IT helps us address our *need for belonging* by enabling us to stay in touch with friends and family more closely and over greater distances through e-mail, chats, and

---

* "The Anthropocene—A Man-Made World," *The Economist*, May 26, 2011, accessed October 9, 2014, http://www.economist.com /node/18741749.

videotelephony. We can easily share precious moments by exchanging digital photos and videos, sending each other emojis, or jointly visiting a museum while on two separate continents. Individual needs for *esteem* is represented by the next step in the ladder. It includes on one side the desire for self-esteem through feelings of personal achievement and competence. On the other side we desire reputation and appreciation from others. Shared online resources help us to be more informed. For example: In situations where we feel insecure because we do not know something by heart, a quick lookup on Wikipedia feels really good. Many of us have started to write blogs, share what we find important, and publish our own videos and photos. Being part of an online group again has potential to gratify us with esteem from others. We get likes (recognition) on Facebook. We are retweeted on Twitter. "In every person there is an artist," said Joseph Beuys, himself one of the most famous twentieth-century artists. And our creative potential is fostered not only through the use of IT applications but also by mastering technology itself. Programming one's own applications, being a hacker, or even engaging in simple end-user programming tasks has become something of a youth movement. Summing up, IT influences human flourishing enormously and at most levels of Maslow's hierarchy of needs.

2.2 Human Nightmares Caused by IT and a Short Reflection on Good and Bad

The peaceful beauty of technology and its potential for human eudemonia is not, however, the full story. Technological innovation is rushing ahead, and individuals, companies, and open source projects must try to navigate the accelerated flow of innovation in a way that feels safe and predictable. We do not want the benefits and enjoyment of technology flow to be a beautiful intermezzo that is soon replaced by an experience of drowning in the rapids. Today, we already feel some bumpy warnings of what may lie ahead: "Security nightmares" are increasingly receiving attention from upper management, as ill-defined IT security architectures have turned our most critical infrastructures into vulnerable targets for criminals. Every year millions of people see their electronic identities stolen and misused. Public surveillance by governments and secret services is taking place at a scale where our democratic freedoms and citizenship is openly questioned. Individual privacy and corporate secrets have been compromised. The global financial market has turned into an automated algorithm-driven zombie that hardly anyone understands and that destroys property and pensions without remorse. Millions of people suffer from burnout, partly due to information overload; addiction to their IT

devices; and opaque, uncontrollable, and undocumented IT systems. Over a billion people have joined virtual worlds, many of them using their artificial lives as a replacement for natural environments. These are the downsides of our information technology riches, and they are perhaps only warning signs of what lies ahead.

The negative phenomena of the IT world would surely not be present at their current scale if the technologies that caused them had been better designed—if they had been more thoughtfully and prudently engineered and placed into the market more responsibly. Documents stored in company databases and private desktops as well as e-mail could have been encrypted by default for a long time (Zimmermann 1995). The collection of personal data through browsers, mobile phones, and video cameras could have been prohibited by default from the start. E-mail servers could have been configured in an attention-sensitive way to protect employees from information overload. And IT backend systems' data flows and connections could have been documented in a way that made it easier for company employees today to understand what their machines are doing. Modeling languages that would have enabled documentation and more robust system design, such as the Unified Modeling Language (UML), have been around since the 1990s (Hamilton 1999). However, these measures were not taken. Many systems have been deployed with too little testing and too little documentation, not to speak of a lack of stakeholder involvement. To use Jaron Lanier's words: "An aeronautical engineer would never put passengers in a place based on an untested, speculative theory, but computer scientists commit analogous sins all the time" (2011, p. 51).

Of course, the IT world has a valid excuse: Unlike aeronautical engineers, who can reference a century of experience, information technology in its current form has only just arrived. Our IT "mind children" (Moravec 1988) are only just in kindergarten. So we could view the first 20 years of the mass-market IT industry as an interesting start that has created a lot of exciting good and a lot of dreadful bad for our societies. The question now is how we want to progress.

This book gives a simple but extremely challenging answer: let's try to continue reaping the benefits of IT while systematically using its power to foster human growth. At the same time, let's do so mindfully and in a way that minimizes the downsides.

2.3 End of IT Innovation Based on Functionality

But how can we get better at creating the future if we do not know what lies ahead? "If a man does not know to what port he is steering, no wind is favorable to him," Seneca (1 BCE–CE 65) once said. A major challenge for companies, governments, and even industries in today's explosion of IT innovation is that they do not know what will come next. They may have some current cash cows, and they may excel in process innovation for those cash cows, but when it comes to true product innovation and anticipating competition, many have a hard time imagining the future. Sitting in our company boats, we often do not know what lies around the next bend in the river. We just sense one thing: the rapid flow of innovation charges the air with a feeling of constant change, which is often accompanied by a notion of haste and fear. Often we do not feel confident that our boat will be the one that makes it very far into the future.

This perception of change and uncertainty is well justified when looking at current industry dynamics: Digital innovation has caused once-heralded IT market leaders to tumble and even vanish at unknown speed. Motorola, Blackberry, and Nokia are well-known examples from the mobile industry that illustrate this dynamic. What hit these powerful company boats after so many years of market dominance? In the aftermath, we can see what happened: These companies flourished mainly on the basis of their engineers' talent to create functionality: telephony, for example, or texting. And they did not realize quickly enough that today functionality alone does not sell in the long term. People are so used to constant new functionality that its magic quickly wears off. And so the devices these companies offered were replaced by smartphones, which do not only offer more functionality but cater to one supreme human value: aesthetics. The desktop market is another famous example for this dynamic: The quasi-monopolist Microsoft lost important market share in the desktop market to Apple because Apple products cater to people's desire for perfected design. People simply value beauty. Yet this is not the only human value that Apple's success resides on. For a long time Apple users felt as part of a community of device avant-gardists, which created belongingness and the willingness to provide mutual help among Apple users. The beauty of the devices catered to users' self-esteem and the recognition from others (who also wanted to possess such beautiful technology). Apple products have indeed been catering to values at various levels of Maslow's pyramid: And finally the devices have less security issues such as viruses. Hence the success story of Apple is a good example to show that people buy into IT solutions that help them to move upward in Maslow's hierarchy.

Competing on functionality is not only a weakness for IT companies when they face direct competition. Another challenging dynamic in the digital market space is the dissolution of industry boundaries and the threat of what Internet economics scholars call "envelopment" of one's functionality by another industry (Eisenmann, Parker, and Van Alstyne 2006). Envelopment is a kind of absorption of functionality by another music player. Take Sony, which saw the music player industry go digital. Soon after digital players appeared on the market, they were enveloped by the mobile phone industry. And since Sony's players could only play music and do hardly anything else, it was easy for the mobile competitors to take over. The same happened to calendar functionality. Small PDAs like the 3Com Palm Pilot vanished and were enveloped by smartphones. However, there is one notable difference between calendars and music players: Some customers still enjoy stylish leather-bound filo-faxes, which they use to not only take notes at meetings but also to collect business cards and other small objects like paper notes, photographs, credit cards, and so forth. These day planners or diaries are still around because they offer more than just pure calendar functionality. They are about beauty, fashion, lifestyle, identity, hoarding, memories, and all those extremely human values that make people buy, enjoy, and keep products for reasons beyond functionality. Although it is easy to envelop functionality, it is difficult to envelop an authentic value identity.

Taken together, direct and indirect competition on the sole basis of functionality is a short-term strategy in digital markets that will not work in the long run. When people become saturated by functionality, they start to buy based on values. And if they do not get the values they want from the company they are with, they will turn to the competitor who offers it. This trend is an old story for the consumer goods industry. Any strategic management course teaches this. But for the IT industry this kind of thinking is only just starting to materialize.

2.4 Questioning "Function Hypes"

Going beyond the functionality is particularly hard for the IT industry because of a frequent rift between managers and engineers; between the "guys with ties" and their "geeks." Engineers are very respected and often trusted for their knowledge in IT companies and typically play an influential role in investment decisions. They are sometimes more influential than is good for a company, because coexecutives without a technical education or understanding tend to refrain from getting deep enough into understanding the technology they are offered to invest in. Technology is simply hard to understand for nontech people. As a result, executives in charge of the business often do not understand the true constraints of a technology deeply enough and let themselves be convinced by the technical enthusiasm of their engineering colleagues. But also the nature of engineers makes that they are thrilled by making something work, by getting things running, by building a platform, in short: by creating value through functionality. This enthusiasm and curiosity for the quickly expanding boundaries of the technically feasible drives the IT industry. Before a technology is hardly matured, industry buyers and end-customers are already driven by "hype cycles" (see Figure 2.2).

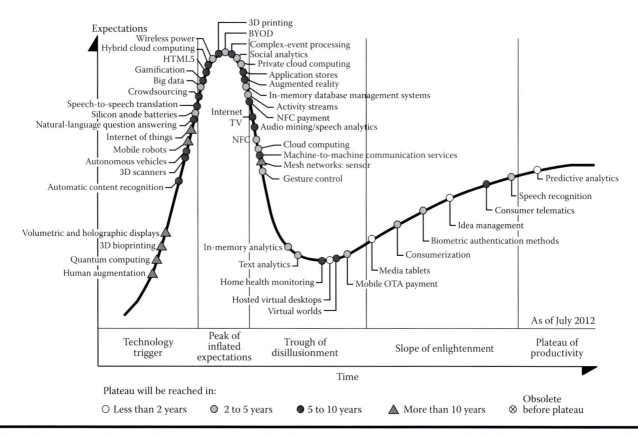

Figure 2.2 Gartner hype cycle. (From "Gartner's 2014 Hype Cycle for Emerging Technologies Maps the Journey to Digital Business," Gartner press release, August 11, 2014, http://www.gartner.com/newsroom/id/2819918. With permission.)

The concept of hype cycles captures the current technology buzz. Hype cycles are built around the latest marketing ideas of the IT market. Often (but not always) they promote minor technological advances. They are well-branded toys that excite engineers. Hype cycles create an almost irresistible force as everyone reads about them in the media and thinks that, therefore, they must be relevant (think for instance of the term "By Date"). This is not necessarily true. But engineers often feel obliged or are curious to try out the new functionalities. To do so, they convince managers that the new tech toys can generate tons of money and that a company can simply not go without it. But once they get their toys and use and test the functionality they often find that the technology is not necessarily working in the way originally expected; not delivering on its marketing promises. The result of this perennial phenomenon in the IT industry is well known: Hyped technologies on the top of the cycle soon sink in popularity as companies become disillusioned. Many people lose a lot of money and face, and then slowly, over many years or even decades, the once hyped technology diffuses into the market.

A nice example of the hype cycle trajectory is radio-frequency identification (RFID) technology, tiny chips that can be integrated into the fabric of products and read out from several meters, without a direct line of sight. RFID was hugely hyped when the U.S. military and retailers around the world announced its use in the early 2000s, asking that all retail products would need to be tagged. They wanted to revolutionize their logistics. They thought that being able to read out RFID tags from a distance automatically (instead of scanning barcodes manually) would quickly bring about real-time control over supply chains at low cost. Between 2003 and 2006, retailers and their suppliers invested millions in the technology, only to discover that the technology was not quite ready: RFID simply does not work in many distribution and retail environments, where metallic shelves reflect RFID signals. Moreover, in many European logistic centers, processes are so perfected that the returns on RFID investments could not justify the cost. Silently, the hype was buried, and many engagements were cut back. Today, RFID is slowly being introduced into the market in those areas where its real advantages play out well, such as the textile industry. Like the barcode technology it replaces, which took 30 to 40 years to reach its current level of deployment, RFID may take a few decades to be widely deployed and in fewer areas than initially anticipated. As this story shows, hype is hypothetical. Although hyped technologies do change the market, those players who rush to them first and without questioning do not win it.

My advice to deal with IT hypes is: Treat them with suspicion and think first. Envision the sustainable *value* that you want to create with the new technology. Don't invest unless you see that value. Only go to market with a fully tested, trustworthy and mature technical product. For sure that will be the less cost-intensive and more relaxed strategy.

2.5 True Cost of Hype Cycles

The storm of fantasy and curiosity that casts IT markets onto hype cycle waves has costs that extend beyond financial ones. One such cost is that they create artificial competitive pressure when a technology is first introduced. Companies fear the progress of competitors and do not want to miss out on a seemingly important trend. In turn, this line of thinking puts pressure on project teams and can lead to low-quality implementations of the technology. Too little time is devoted to reflective thinking and qualitative improvement of the standard solution. Much is rapidly developed and scrummed into the existing IT infrastructure or end-user service/device. The active involvement of external stakeholders who later need to use the technology is often perceived as more of a nuisance than a fruitful and important practice. The result of this self-inflicted pressure and haste is poor system documentation and too little consideration of the effects the technology will have on employees and customers. Even the most obvious and rudimentary values, such as the usability of systems, are still often neglected.

The second and even bigger cost is that the remnants of the hype remain in IT architectures, creating highly interwoven, complex, often undocumented, and error prone IT backend systems. Employees who use these IT systems are then given numbers and are confronted with workflows that they cannot fully comprehend any more (even if they want to!). A small private bank once revealed to me that their customer advisers get predetermined interest rates from the IT system for customers' credit, but they do not know how this interest rate is calculated or what justifies it. In fact, even the bank's CEO did not know. The CFO in another company received numbers for the financial statement, but he did not know how these were calculated even though he was liable for them. In fact, the numbers contained assumptions that did not truly reflect business reality.

Unfortunately, numbers pulled from corporate IT systems in many cases do not make sense to the employees who need to work with them either. Poorly documented IT systems do not even allow for understanding them if they wanted to. The result is human stress, burnout, and increased staff turnover as well as (in some cases) a loss of true control of operations (for more information, see Chapter 7 on health).

Against this background, it becomes clear that hype cycles are a phenomenon that wise company leaders should question. Ignoring IT hype cycles could be a first step toward taking a lot of stress out of organizations and whole industries. "Less is more" is an old saying. The "aesthetics of ommitance" is a signal from contemporary art that tells us that we should abstain from adding anything superfluous into our environment.[*] Instead we should rather think about where we can reduce our environment to bring it back to a comprehensible and primal status. Companies who stop basing their product development on hype cycles will have more time—time that they can then use to think more about the value they actually want to deliver to their customers beyond functionality.

This book presents methodology for this value-based approach. I will introduce a range of values in Chapters 5 to 12, and explain their meaning and application for IT. Chapters 13 to 18 explain how the values can be build into IT. However, the conceptualization of values relevant to an IT product and the successive tailoring of products to values take time. It is an exercise that requires deep thought.

2.6 Post Hype Cycles: Catering to Higher Human Values

Riding IT hype cycle waves is like being a ball in a pinball machine, which is why managers have such difficulty maneuvering through the explosive flow of innovation. But this does not need to be the case. Instead of accelerating their adoption of IT technology, managers have the power to tap the brakes. They can question the hype cycles and technical determinisms that seem to dictate many current decisions and instead apply caution to decisions about, if, when, and how to adopt technologies.

A good example for the questioning of hypes and numb digitalization is the German retailer Aldi, which holds about 36.5% market share in the country and is probably one of the biggest retail success stories of the past three decades.[†] For at least a decade Aldi resisted introducing scanner-based checkout systems because human employees were faster and more reliable. While other retailers rushed to invest in scanners, Aldi educated its staff to check out customers' goods extremely quickly memorizing price tags for all goods on shelf. The company even held weekly competitions among employees to see who was the fastest. Only when the number of goods sold went past the threshold where employees had a

---

[*] "Karin Sander—'Ästhetik des Weglassens' // GLOBArt Academy 2013," YouTube video, 17:29, October 29, 2013, http://www.youtube.com/watch?v=LF8eqi9RJuc.

[†] The market share includes the operations of Aldi North and Aldi South.

hard time remembering the full assortment did the company switch to scanners. As of 2015 Aldi still has no online shop while everyone else rushes to the new channel. The plain explanation of the company spokesperson is that the highest quality products at the lowest possible prices simply does not go in line with an online strategy for them.* The same conclusion is reached by Germany's drugstore market leader dm. Both companies remain unimpressed by hypes and even established technologies and stick to what counts for their customers: the highest quality products at reasonable prices. They do not pour money into technologies that do not cater to these values.

Applying this kind of value-centered approach to today's retail technologies, one could ask, for example, whether retailers should really introduce location-based services and mobile advertisements on their shop floors. Technologies such as Apple's iBeacon now allow for such a service, and the typical mainstream hype cycle approach would be for retailers to immediately test and deploy some of these services. It is at this very point that customer value needs to be questioned. Do location-based services in stores increase the customers' pleasure while shopping? Would not the service undermine their free decision-making and movement while shopping as well as their privacy? Does the additional knowledge gained from locally relevant ads outweigh the privacy drawbacks? Innovation managers and IT project teams can reflect on these questions. It may lead them to not invest. Or, it could lead them to discover that if customer knowledge is the value they want to cater to, then they need to concentrate on delivering it. This again could lead them to challenge whether iBeacon in particular is the right solution. What these questions show is that the concept of value can drive IT investments and can protect companies from overinvesting in things that are simply not enriching the customer.

To sum up: An alternative to conventional technology and innovation strategy is to start with the key values the company wants to create. Consequently, any technology investment must cater to these values instead of focusing mainly on new, technologically enabled functionality. I call this approach *value-based IT design*.

2.7 Values in IT Design

One could argue that product functionality is a value in itself. It is a practical value, a utility value. And when companies identify such functional values—for example, by innovating with lead users—don't they already pursue a value-based strategy? Lead users value product functionality to such an extent

that they are willing to personally modify an existing product (Lilien et al. 2002). For example, Gary Fisher was a professional bicycle racer who wanted to go downhill onto gravel roads for training purposes. He modified bicycles with parts from motorbikes to increase their stability and performance. By doing so Fisher pioneered mountain bikes, a great value-based innovation. Companies look out for such lead users today in order to innovate. They also involve larger customer groups, sometimes called "crowds," to get product suggestions (Bilgram, Brem, and Voigt 2008). However, I believe that we must question whether values identified around functionality (like speed, robustness, efficiency, etc.) will really create sufficient competitive advantage for IT companies in the long run. The term "feature fatigue" is spreading through the industry (Thompson, Hamilton, and Rust 2005). Economists like Tomas Sedlacek argue: "The whole new production seems to fill a void, which has created itself … We long for longing. The benefit from consumption could be exhausted. This well is already dried out" (2012, p. 302). George Joseph Stiglitz (the 1982 Nobel Prize winner for economics) said: "Individuals with common sense are desiring less for the satisfaction of hitherto needs, but for more and better needs" (cited in Sedlacek 2012, p. 274).

So what "more and better" values can be addressed by information technology? Steve Jobs began the *human* value revolution in IT by proving the power of aesthetics. By doing so, he buried a long-defended conviction of engineers that said "form follows function." His aesthetic perfectionism led Apple to introduce desktop computers, music players, and smartphones that took the markets by storm. His ambition was to only use IT artifacts that were perfectly designed from the outside as well as the inside. By reading his biography, it becomes clear how deeply he engaged with his engineers to achieve this; understanding their limits and objections and forcing them to do better than just creating functionality (Isaacson 2011). This mindset led Apple to offer completely new form factors, such as shuffling music or touching displays to dial. The displays and materials were perfected. Beauty entered IT (see Figure 2.3 for a historic comparison).

However, as Soren Kierkegaard (1813–1855) would have reckoned: aesthetics is only the first in a progression of various existential stages. The aesthetic, according to Kierkegaard, only gives way to the ethical (McDonald 2012). Whereas the prime motivation for the aesthete is to merely transform the boring into something interesting, the ethical recognizes a duty to social norms and hence a duty to comply with human values inherent in a community. From the point of view of ethics, aestheticism is "emptily self-serving and escapist" (McDonald 2012). This book will therefore not include aesthetics, even though I see it less critical and in constrast as an important value for IT. Throughout history beauty has been associated with

---

* Enrich Reichman, 10.02.2015. "Aldi und Primark: Die letzten Online—Verweigerer" URL: http://www.heise.de/newsticker/meldung /Aldi-und-Primark-Die-letzten-Online-Verweigerer 2544743.html.

Figure 2.3   Moving from function to form through IT aesthetics. (By Jojhnjoy via Wikimedia Commons.)

the "good" (Eco 2004). But as the title of this book suggests, I want to focus on ethical IT innovation. And my approach to do so is a value-based approach beyond functionality and beyond aesthetics.

Value-based IT design must be viewed from two perspectives: First, it can be used to consciously foster value creation through IT. And second it protects values that could be undermined or destroyed by thoughtless IT design. Take the privacy value, for instance. People who want to watch pornographic movies do not need to go into video rental stores anymore. Thanks to the Internet, they enjoy more privacy because they can download the films at home directly from porn sites. At the same time, the websites that offer this material or ad networks might track online viewers' porn consumption and porn preferences. So, although porn viewers think that they have more privacy, they might actually be exposed to spying entities they do not know or see. This example shows that online porn services can on one side create a greater sense of physical privacy while at the same time destroying it digitally.

Engineers know that IT system designs have such contradictive effects. Yet they also know that this does not need to be the case. To avoid the destructive influence, porn sites can encrypt the requests to their sites, choose to not log identifiers over time that would recognize returning customers, forgo advertisements on their sites, and so on. Many technical measures can be used to easily minimize the risks in the IT service so that customers reap the

benefits of private porn viewing while not encountering the negative risk of being exposed. What values are relevant for a value-based IT design approach? Figure 2.4 lists many potential values to start with. The figure is informed by earlier research on value-sensitive design (VSD) for machines (Friedman and Kahn 2003; Friedman, Kahn, and Borning 2006). Here scholars proposed many nonmoral values relevant for IT such as health, privacy, autonomy, trust, control, calmness, accountability, safety, security, and freedom also illustrates a shift from risk avoidance to value creation when we anticipate value effects and the impact of IT. In Chapters 17 and 18 I will show how risk analysis can be adopted to understand how an IT system undermines a value. And I will also show how technical as well as governance mitigation strategies can be used to avoid this loss and create value instead. My hypothesis is that if IT managers and engineers focused on creating IT values throughout the entire design process while rigorously controlling for value risks, they would support human beings in their flourishing much more than they do today. Machines would then be designed to strengthen people's values such as their health, increase their sense of privacy, freedom and autonomy, help them trust, and so forth. In the long run, we could even envision that machines support the development of cognitive skills such as learning, help them rediscover their senses, have more ethical integrity, be more just in their decisions, and so on. IT scenarios in Chapter 3 outline some potential developments in this direction.

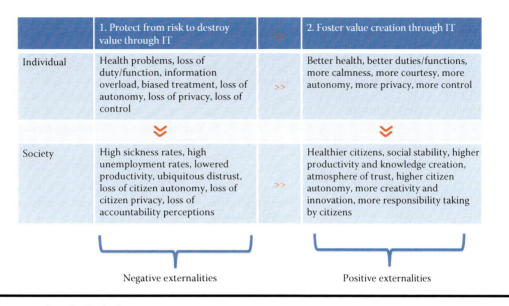

	1. Protect from risk to destroy value through IT		2. Foster value creation through IT
Individual	Health problems, loss of duty/function, information overload, biased treatment, loss of autonomy, loss of privacy, loss of control	>>	Better health, better duties/functions, more calmness, more courtesy, more autonomy, more privacy, more control
Society	High sickness rates, high unemployment rates, lowered productivity, ubiquitous distrust, loss of citizen autonomy, loss of citizen privacy, loss of accountability perceptions	>>	Healthier citizens, social stability, higher productivity and knowledge creation, atmosphere of trust, higher citizen autonomy, more creativity and innovation, more responsibility taking by citizens

Negative externalities — Positive externalities

Figure 2.4 Human values in IT design.

2.8 Necessity for Value-Based Design

Value-based IT design is not a philosophical exercise but an economic and political necessity. Value creation or destruction through IT regularly plays out not only on the individual level but also at the corporate and societal level. From the perspective of economics, one could argue that value creation through IT has positive externalities for society and companies, whereas value destruction through IT has negative externalities (compare Figure 2.4).

For example, consider the economic and social effects of free and unlimited attention consumption through our current IT systems. Attention is one of the most valuable resources we have as human beings. It is vital for survival, for raising healthy children and cultivating strong friendships, for learning, for creating knowledge, and for developing ourselves as well-rounded individuals. "What you focus on also affects who you are" (Gallagher 2009). However, thousands of advertising messages now compete every day for our attention via television, e-mail, the Internet, public space, mobile handsets, fixed-line phones, electronic games, and cyberspace. On average, employees send and receive about 120 e-mail messages a day (excluding spam; Radicati 2014). People interrupt themselves constantly, tempted by social networking sites such as Facebook and Twitter. Unfortunately, empirical evidence suggests that interruptions harm knowledge work performance. Interruptions lower intellectual capacity (Ahmed 2005), cause time loss and errors (Speier, Valacich, and Vessey 1999; Speier, Vessey, and Valacich 2003), and reduce employee satisfaction (Kirmeyer 1988; Zijlstra et al. 1999). About 40% of tasks that are interrupted are not resumed (O'Conaill and

Frohlich 1995; Czerwinski, Horvitz, and Wilhite 2004). As a result, one study estimated that the downtime associated with interruptions costs companies $588 billion per year in lost productivity in the United States alone (Goldes and Spira 2005).

The attention example makes plain that IT artifacts and services can cause negative feedback loops between individuals, companies, and society at large. Figure 2.5 illustrates this dynamic. It shows that feedback loops can be positive as well as negative. For instance, well-designed e-learning technologies (Chapter 5) or health applications (Chapter 7) as well as polite machines (Chapter 11) may create positive loop effects.

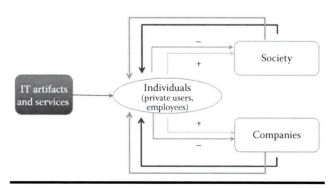

Figure 2.5 Positive and negative feedback loops caused by IT artifacts and services. IT artifacts and services can cause feedback loops between individuals, companies, and society at large.

2.9 Envisioning the Value of IT Services

Imagine we had a future in which we were no longer seduced by hype cycles. Instead we would be ready to embrace a value-based approach to IT design. In this future, we would want to build IT boats that were sturdy enough to handle the waters. However, how do we know that our particular boat will still excel farther down the river, where waters may be different, where some values may have changed? To really define the requirements for our boat we need to anticipate what lies ahead. We need to envision the future. And we will need to decide what values we want in this future. Taking a table like the one depicted in Figure 7.4 will not be enough.

IT companies need to put a lot of effort into understanding the future and building visions of what they want this future to be like, for themselves but also for society at large. It does not suffice to develop only scenarios for the transition of one's own company into the machine age. Since modern companies are so entangled with service networks, and since IT solutions can be so widely diffused, wise leaders need to reflect on the overall social and economic environment around them. They need to take responsibility and make judgments on what they consider to be good or bad for the future of entire industries and beyond. As Nonaka and Takeuchi wrote in the *Harvard Business Review*: "Managers must make judgments for the common good, not for profits or competitive advantage … Not only does a company have to live in harmony with society, but to be accepted, it must contribute to society. The majority of companies that have failed did not maintain that balance. Everyone is, first, a member of society before one of the company. Thinking only about the company will undoubtedly result in failure" (2011, p. 64).

Unfortunately, current management theory gives little guidance to envision the future. How can we wisely innovate for the future? What values should we cater to? How can we judge on value trade-offs? Books on innovation management recommend inviting experts to participate in visionary forecasting, to use Delphi panels, and focus groups to set horizons for new products (Ahmed and Shepherd 2012). In recent years companies have started to embrace the power of crowdsourcing to generate ideas. For example, the Lego company elicits ideas from customers, and then the innovation department decides what to produce. Asking the crowd is a fruitful exercise for envisioning and testing the next step into the future. But in none of the known cases, crowd intelligence is used to create the values, brand identity, or vision that successful IT innovation needs. In his online article "Digital Maoism," Jaron Lanier (2006) explains how the collective is smarter only when it is not defining its own questions. Value-based IT innovation cannot be outsourced.

To anticipate the future we must fully embrace our potential to form it. "What you think today, you will be tomorrow," says a Chinese proverb. For the German philosopher Martin Heidegger (1889–1976), known for his oeuvre *Being and Time*, the future presents the potentiality-for-being and is the most important temporal dimension. As we project our future into the present, both our present exercise and past experience guide us toward the fulfillment of that future we envision. So envisioning the future—a task only humans are well capable of—may be the right way to form it, to fully understand the setting in which our companies and products will be placed, and to determine the values that we want to create and protect. How does envisioning the future work?

While envisioning has a lot to do with intuition, we keep some indication in the literature that everyone can foster envisioning to some extent. Nonaka, Toyama, and Toru (2008) discuss two building blocks of envisioning in their theory of "managing flow." One is the concept of "having a purpose," which is related to personal values. The other is a practice they call "reflection in action." They write: "Deciding on specifications without thinking about essential purposes only leads to choices among existing options. To come up with a new solution … one has to answer existential questions that pursue essence based on one's own values and the values of the organization" (Nonaka, Toyama, and Toru 2008, p. 32). A value could be, for example, that we want to be healthy. We envision ourselves as healthy individuals and what this would mean for our lives and ourselves. Everybody can do that. We think about what *we* would look like and what *we* would do. Then, with this dream picture in our minds, we think about activities and technologies that could help us fulfill this dream. Chapter 14 contains some creativity techniques that help in practicing the envisioning of purposes.

Yet, once we have identified relevant values and potential future product purposes how do we decide on the right values? Which ones are important to pursue at all? And how can we ensure that by fostering one value we do not destroy another? Some people believe in the value of a health app for instance, like a nonsmoking app, but they build their business model around the sale of the health data they collect from users. They think that the health value is so important that it can be traded for the privacy value of their users. Others believe that privacy is one of the most fundamental values to defend and would therefore never use such a health app. So who is right? What values should the app be catered to?

Responsible and insightful leaders with high levels of respect, autonomy, and maturity are needed to make value decisions for our societies. Chapter 15 provides more insight into what such responsible and wise leadership implies. But no matter how good the leadership team of a company is, it is key for these people to understand what values resonate the most in society. They must engage themselves in person to understand where to avoid a value breach or clash of values. For their technological innovations companies must engage—together with their leaders—into a practice that Nonaka et al. call "reflection in action." This reflection requires people to actively and personally experience the innovation space. Nonaka, Toyama, and Toru (2008) describe such an exercise for a nontechnical product, the sports drink DAKARA produced by the company Suntory. Based on market research data, the innovation team of Suntory thought that the DAKARA drink should be promoted with the slogan "Gives the working man that one extra push." So a self-esteem or power value was suggested by market research, a relatively high place in Maslow's pyramid. However, participating in reflection in action, the innovation team at Suntory physically went out to actively observe how the DAKARA sports drink was consumed in the real world. They found that reality looks different than market research data had suggested. Its consumers were actually tired and had the DAKARA drink to make it through the stressful day. The product managers concluded that potential customers valued the idea of healing more than "an extra push." Based on this reflective action exercise, they marketed the drink with the slogan "To preserve your health when life gets tough" (Nonaka and Katsumi 2004, as cited in Nonaka, Toyama, and Toru 2008, p. 33). They chose a value much lower in the pyramid of needs.

Unfortunately, such a *physical* practice of reflection in action is not always possible when it comes to IT innovations. We simply cannot beam ourselves into the future. However, storytelling and scenario building can enable us to get close to the experience. For this reason, the next chapter uses short stories to envision future IT environments. The chapter features recurring characters using technologies that may be widely deployed between 2029 and 2035. Throughout the stories, different values are either fostered or destroyed. As you read the stories, put yourself into the shoes of these characters and consider the different sets of values on display.

EXERCISES

1. Envision a future IT product or service that you would like to use. Describe how it works. What benefits would it create for you? What harms might it cause? What underlying values does it touch upon?

2. Identify a current or past IT hype. Research how the hyped technology was presented and advertised, and how long the hype lasted. What value did it promise to create?

3. List current IT services that cater to each of Maslow's hierarchy of values. List IT services that undermine these values. Explain your reasoning.

4. Explain how a current IT service or artifact influences individuals directly or indirectly when feedback loops between individuals and society or companies are considered (see Figure 2.5). Give examples to illustrate your thinking.

References

Ahmed, T. 2005. "Abuse of Technology Can Reduce UK Workers' Intelligence: HP Calls for More Appropriate Use of 'Always-On' Technology to Improve Productivity." HP press release, April 22.

Ahmed, P. and C. Shepherd. 2012. *Innovation Management.* New York: Financial Times/Prentice Hall.

Asimov, I. 1991. *The Foundation Novels.* New York: Bantam Dell.

Bilgram, V., A. Brem, and K.-I. Voigt. 2008. "User-Centric Innovations in New Product Development: Systematic Identification of Lead Users Harnessing Interactive and Collaborative Online Tools." *International Journal of Innovation Management* 12(3):419–458.

Bruni, L. and P. L. Porta. 2007. *Handbook on the Economics of Happiness.* Cheltenham, UK: Edward Elgar Publishing.

Bynum, T. W. 2006. "Flourishing Ethics." *Ethics and Information Technology* 8(4):157–173.

Czerwinski, M., E. Horvitz, and S. Wilhite. 2004. "A Diary Study of Task Switching and Interruptions." Paper presented at ACM Conference on Human Factors in Computing Systems (CHI 2004), Vienna, Austria, April 24–29.

Diener, E. 2009. *The Science of Well-Being: The Collected Works of Ed Diener.* Social Indicators Research Series 37. Heidelberg: Springer Verlag.

Eco, U. 2004. *Die Geschichte der Schönheit.* München: Carl Hanser Verlag.

Eisenmann, T., G. Parker, and M. W. Van Alstyne. 2006. "Strategies for Two-Sided Markets." *Harvard Business Review* 84(10):92–101.

Frey, B. and A. Stutzer. 2001. "What Can Economists Learn from Happiness Research?" Center for Economic Studies and Ifo Institute for Economic Research, Munich.

Friedman, B. and P. Kahn. 2003. "Human Values, Ethics, and Design." In *The Human-Computer Interaction Handbook,* edited by J. Jacko and A. Sears. Mahwah, NJ: Lawrence Erlbaum Associates.

Friedman, B., P. Kahn, and A. Borning. 2006. "Value Sensitive Design and Information Systems." In *Human–Computer Interaction in Management Information Systems: Foundations,* edited by Ping Zang and Dennis F. Galletta. New York: M.E. Sharpe.

Gallagher, W. 2009. *RAPT: Attention and the Focused Life.* New York: Penguin Books.

Goldes, D. M. and G. Spira. 2005. *The Cost of Not Paying Attention: How Interruptions Impact Knowledge Worker Productivity.* New York: Basex Inc.

Hamilton, M. 1999. *Software Development: A Guide to Building Reliable Systems.* Upper Saddle River, NJ: Prentice Hall.

Isaacson, W. 2011. *Steve Jobs.* New York: Simon & Schuster.

ISO. 2008. ISO/IEC 27005: Information technology—Security Techniques—Information Security Risk Management. International Organization for Standardization.

ISO. 2014. ISO/IEC 29100: Information Technology—Security Techniques—Privacy Architecture Framework. DIN Deutsches Institut für Normung e.V.

Kirmeyer, S. L. 1988. "Coping with Competing Demands: Interruptions and the Type A Pattern." *Journal of Applied Psychology* 73(4):621–629.

Kurzweil, R. 2006. *The Singularity Is Near: When Humans Transcend Biology*. London: Penguin Group.

Lanier, J. 2006. DIGITAL MAOISM: The Hazards of the New Online Collectivism. Edge.org. http://edge.org/conversation /digital-maoism-the-hazards-of-the-new-online-collectivism.

Lanier, J. 2011. *You Are Not a Gadget*. London: Penguin Books.

Lilien, G. L., P. D. Morrison, K. Searls, M. Sonnack, and E. von Hippel. 2002. "Performance Assessment of the Lead User Idea-Generation Process for New Product Development." *Management Science* 48(8):1042–1059.

Maslow, A. 1970. *Motivation and Personality*. 2nd ed. New York: Harper & Row Publishers.

McDonald, W. 2012. "Kierkegaard, Soren." In *The Stanford Encyclopedia of Philosophy*. Stanford, CA: The Metaphysics Research Lab.

Moor, J. 1999. "Just Consequentialism and Computing." *Ethics and Information Technology* 1(1):65–69.

Moravec, H. 1988. *Mind Children: The Future of Robot and Human Intelligence*. Cambridge, MA: Harvard University Press.

National Institute for Standards and Technology (NIST). 2010. "Risk Management Guide for IT Systems." US Department of Defense.

Nonaka, I. and H. Takeuchi. 2011. "The Wise Leader." *Harvard Business Review*, May, 58–67.

Nonaka, I., R. Toyama, and H. Toru. 2008. *Managing Flow: A Process Theory of the Knowledge-Based Firm*. New York: Palgrave MacMillan.

O'Conaill, B. and D. Frohlich. 1995. "Timespace in the Workplace: Dealing with Interruptions." Paper presented at ACM Conference of Human Factors in Computing Systems (CHI 2005), Portland, Oregon, April 2–7.

Organisation for Economic Co-operation and Development (OECD). 2015. "Understanding Data and Analytics." In *Toward Data-Driven Economics: Unleashing the Potential of Data for Growth and Well-Being*, Chapter 3. Paris: Organisation for Economic Co-operation and Development (OECD).

Radicati, S. 2014. "E-Mail Statistics Report, 2014–2018." The Radicati Group Inc., Paolo Alto, CA.

Sedlacek, T. 2012. *Die Ökonomie von Gut und Böse*. München: Carl Hanser Verlag.

Smith, C. and G. Lenssen. 2009. *Mainstreaming Corporate Responsibility*. Chichester, England: John Wiley & Sons.

Speier, C., J. S. Valacich, and I. Vessey. 1999. "The Influence of Task Interruption on Individual Decision Making: An Information Overload Perspective." *Decision Sciences* 30(2):337–360.

Speier, C., I. Vessey, and J. S. Valacich. 2003. "The Effects of Interruptions, Task Complexity, and Information Presentation on Computer-Supported Decision-Making Performance." *Decision Sciences* 34(4):771–797.

Taleb, N. N. 2010. *The Black Swan: The Impact of the Highly Improbable*. 2nd ed. New York: Random House.

Thompson, D. V., R. W. Hamilton, and R. T. Rust. 2005. "Feature Fatigue: When Product Capabilities Become Too Much of a Good Thing." *Journal of Marketing Research* 42(4):431–442.

Wiener, N. 1954. *The Human Use of Human Beings: Cybernetics and Society*. Da Capo Series of Science. 2nd ed. Boston: Da Capo Press.

Zijlstra, F. R. H., R. A. Roe, A. B. Leonora, and I. Krediet. 1999. "Temporal Factors in Mental Work: Effects of Interrupted Activities." *Journal of Occupational and Organizational Psychology* 72:163–185.

Zimmermann, P. 1995. *The Official PGP User's Guide*. Boston: MIT Press.

Chapter 3

Future IT Environments: Five Scenarios

"Only those who know the goal, find the way."

Lao Zi (604 BC–531 BC)

In this chapter I will present five scenarios on how various technologies could be embedded in society by 2029 to 2035. This is of course only one view of how things could be. Having a shared vision for future information technology (IT) innovation with the help of scenarios is the best basis for practicing value analysis. Various value aspects of the scenarios from this chapter are coming back in later chapters of the book where their theoretical background is reflected.

3.1 Personas Used in the Scenarios

Table 3.1 introduces the personas used in the scenarios.

3.2 Scenario 1: Future of Gaming

It is the year 2029, and Stern walks across the square to enter the glass skyscraper that houses his employer United Games Corp. Stern is looking forward to his meeting today with a lead user, who shares his experience and proposals for innovation with United Games while also helping them watch out for the competition. Stern is confident that no competitive game can be a real threat to his own product.

Stern is the product manager for *Star Games*. Eighteen months ago, United Games launched this fully immersive virtual reality game that players enter through specially designed virtual reality (VR) glasses and partly control through sensor bracelets. Most households by now have their own VR tubes that look a bit like MRI scanners. Players lie in them with screens all around them in the tube and wear arm-covering sensor gloves; sliding into a 3D gaming experience of incredible power. The tube recognizes players' arm gestures and voice, which are used to manipulate the content

and flow of the multiplayer game. Far gone are the old days of living room sofas and manual controllers where other family members in the room would constantly disturb the gaming experience. Now players are completely sheltered, talk directly to their peers in VR, and hear their responses through the bone conduction element that holds their head in the tube. Bone conduction channels sound to the inner ear through electromagnet impulses to the bones of the skull. But this is not all: based on players' pupillary dilation, heart rate, and skin conductance, the tube determines a player's emotional state and automatically alters (morphs) the facial expression of the in-game avatar that is displayed to the other players.

Stern is recognized for his success throughout the company. About 800,000 players have enrolled for the game since it was launched a year ago. And more than 30,000 companies have signed up as well bringing in another 500,000 registered users. Thanks to this unusual success, United Games' holographic doorman in the headquarters' entry hall has welcomed and greeted Stern as a "chief innovator" now for three months each time he enters the building! "What a great job I have," he often thinks, as the company is the ideal place for him to combine his passion for game playing with his profession.

Star Games is a nostalgic reference to a science fiction classic of the 1970s. Players can assume any *Star Games* role they want to play and then have the freedom to do whatever they want: fighting in Jedi battles, flying spaceships, meeting for romance at Lake Naboo. The settings in *Star Games* are nearly endless. The game's major selling point is its rich spectrum of possible activities and the ability to change identities at any time. Still, a recent statistic showed that more than 70% of players stick with the identities they initially chose and then develop their own lives and relationships in the game based on that identity. On average, recreational players spend 4 hours in the game per day, with 10% at 7 hours a day. And—good for Stern—the hours are growing. The game is really addictive, or, as Stern would put it, "compelling."

Table 3.1   Scenario Personas

Persona	Description
Stern	Stern is a manager in his late 30s, unmarried, with no kids. He works for United Games Corp., where he is the product manager of *Star Games*, a fully immersive game that is a virtual reality version of the 1970s sci-fi classic *Star Wars*. Stern is highly career driven, makes decisions quickly, and is impulsive, cool, and pretty dominant. He believes in new technologies and doesn't like people he perceives to be Luddites, who oppose new technologies. He has the guts to leverage the benefits from new technologies for himself and his bosses.
Carly	Carly is Stern's colleague. She is in her mid-30s, also unmarried and without kids. She is extremely intelligent and knows her way around the business world. But she also has the air of a wise leader, which is why her colleagues perceive her as nonmainstream. She is typically calm but can be sharp in professional discussions. She questions how technologies are used and believes that it is more important for them to support the well-being of people and society than to generate profit.
Roger	Roger is the father of two kids, a son Jeremy (age 13) and a daughter Sophia (age 8). He is 43 and divorced. He lives with his two kids in a village 45 miles from a big city. He works as a software engineer for a major corporation in the city and commutes two days a week. The other day he works from home. He wants his children to be well acquainted with new technologies and is therefore willing to spend money on timely technology even though his budget is hardly sufficient. At the same time, he is very concerned about his kids growing up addicted to the tech stuff.
Jeremy	Jeremy is Roger's 13-year-old son. He gets along well with his father but also sometimes clashes with him. He is rather lazy and likes to stay at home playing online games. His peers recognize him as someone who possesses the latest technologies and excels at using them. He spends a lot of time online and does not like to be offline.

(Continued)

Table 3.1 (Continued)   Scenario Personas

Persona	Description
Sophia	Sophia is Roger's 8-year-old daughter and Jeremy's sister. She is really smart for her age and loves exploring and using technology. She thinks technology is a bit like magic and likes to spend time with it. She is very social and good at school. She often puts technology aside to enjoy physical play with others in the real world. However, she typically wears her 3D glasses, because she likes to be accompanied by her 3D virtual reality dragon called Arthur, who also acts as her personal agent.

Source: Stern (© Alex Navarro 2013, CC BY-SA 3.0); Carly (Lawrenceboucher 2007); Roger (© Tylerhwillis 2010, CC BY-SA 3.0); Jeremy (© Müller 2007, CC BY-SA 3.0); Sophia (© Maimidolphins 2012, CC0 1.0).

The hours are important for Stern because private players are not charged a subscription fee for joining the game. The game costs nothing, except for the purchase of the tube, and these are heavily subsidized by the government. The core of the business case for the consumer market is the purchase of virtual goods and premium-service use. For example, players pay extra to use the flight simulator at the VR's airports. The game becomes truly fun if players equip themselves with a Jedi sword and enter one of the spaceship cockpits, where they can fly across the gorgeous planets and dark skies of the universe. It is not very expensive to buy stuff in the game. Everything is based on micro-payments. But the longer a player stays, the better it is for United Games, because hours online and money spent are highly correlated. On average people end up paying around EUR 45 a month for the game, and the company has already earned more than EUR 300 million from the game just through virtual objects and services. But almost as much is coming in through the sale of players' emotion profiles and product placement.

Emotional profiling has become a core business for United Games because research revealed that people behave more authentically in VR than in the real world (they are more impulsive and less self-controlled). Hence, the data is really valuable for the placement of ads both inside and outside the game. The construction of emotional profiles is deeply embedded into *Star Games'* technical infrastructure and is used to understand the character and determine potential reactions of the people in front of the screen. How creative are they? What are their communication styles and educational levels? How violent or friendly are they in the game? Are they mentally stable? Stern cannot name all the variables offhand that are being measured and then clustered. The practice is not talked about a lot because Stern fears that people would behave less authentically in the VR if they knew. That said, he made sure that the practice is mentioned in the game's privacy policy, where players are informed that virtual characters in *Star Games* respond to their emotional state

and that this state must therefore be measured. In fact emotional profiling is outsourced to another company called The Circle, the best company in the market for mining emotions.

Stern himself is also profiled, not only when he plays the game but also while he works. He is profiled because all product managers at United Games host their meetings with colleagues and clients in their respective game's virtual reality facilities, which is really a lot of fun. Each time one visits a colleague, one goes to another virtual reality and enjoys the fantastic housing creations of that colleague. Stern has created a two-story penthouse for himself in Galactic City on the planet Coruscant. Physically, he actually works from home most of the time. He often spends 10 to 12 hours in the game per day. Today, however, the situation is different. Stern got a call to travel to United Games headquarters, and so here he is.

As Stern walks into the meeting room at United Games, however, he is a bit irritated. The faces of his colleagues and boss are rather nervous and do not quite fit Stern's good mood. At the back of the room, the wall-sized nano-based flex display shows a gigantic 3D image of a pink dragon in the woods. The picture stems from a new game called *Reality* that was just launched by the competitor company, Playing the World Corp., in the last quarter. Stern heard about the game already but did not know what exactly it would be about. When he looks more closely, he suddenly feels a slight shock. "Could it be that the dragon is actually displayed as a holograph in the real world?" he asks himself. Years ago, there was rumor of a cast augmented reality technology that would allow game players to wear a special glass and see holographic images placed right into the real world. But early attempts to deploy this technology failed. Mobile bandwidth simply was not sufficient at that time, and data volumes exceeding a gigabyte were too expensive. Also the glasses with the functionality were extremely ugly and cumbersome. "Could this have changed?" Stern wondered. He simply did not have the time and bandwidth to follow recent developments.

Stern sits down and listens to the presentation of the invited lead user, who has been playing the new game for the past 3 months and is obviously thrilled. *Reality's* game content is cast as an overlay onto the real world through players' AR (augmented reality) glasses. As players move through real urban and rural space, they can meet fantasy characters and discover 3D mysteries that are sheltered in different geolocations. Sometimes, they are also taken by surprise and a virtual character meets them or even attacks them. One cool thing about the game is the fostering of players' creativity. Anyone can design his or her own game characters with an easy-to-use programming tool and have these characters accompany them or engage with other people and their characters. The pink dragon displayed on the meeting room's screen is actually the toy of an 8-year-old girl named Sophia, who has chosen to be accompanied by this dragon in the game. She calls it Arthur.

Interesting things have happened due to the game lately. Children have started to meet outdoors again in the woods to fight virtual characters. This trend was covered in the press because most modern children had rarely left home lately, instead staying in VR tubes to play virtual games. Now children suddenly spent hours in fresh air, and the first medical studies show that the physical activity and emotional stability of players is significantly increased.

But playing outdoors is not the only asset of the competitors' game. It also integrates new forms of teaching. For example, many users have met with the game's holographic wise saint figure, who looks a bit like Dumbledore from the *Harry Potter* films. This figure, created by Playing the World, teaches players about nature and other real-life subjects. He provides wise and knowledgeable answers to most questions users ask about the world. Apparently, the character is hooked up to one of the MBI Nostaw machines, a supercomputer with a very high degree of artificial intelligence. The saint figure can also magically interact with the players' glasses and allow them to view thermodynamic and magnetic information as an overlay to the scenery. (This is really great fun to see, especially when other folks get angry or fall in love.) He can also grant players the ability to listen to the sounds plants use for communication and listen to the octave tones of planets. In any case, users of all ages are thrilled, and in the past 3 months 600,000 users have started to play even though the government would not sponsor the glasses needed for this game. Of course, the blogosphere is all over it, as are health care insurers and TV shows.

One new face in the room has actually just left the new competitor and joined United Games. Her name is Carly, a woman in her mid-30s with remarkable green eyes and a strong but calm aura. She tells the story of how *Reality* was developed. And as Stern listens to Carly speak, he gets half irritated and half impressed. "Since when does United hire this kind of lofty female breed?"

Carly informs the meeting participants that she used to be the product manager of the *Reality* game at Playing the World. *Reality* took many years to develop. It will be the flagship of a new initiative to bring kids (but also adults!) back to nature, and improve their social skills and their health. Other social issues have been considered in the game design: users must authenticate with an official ID and can play only as characters that match their age. Some tasks in the game can be completed only if several people meet up and cooperate. Payment is time-based. One hour per day is free, but users who want to play longer must pay EUR 1 per hour. In this way, the game helps to deter players from staying too long and getting addicted. When people play for the 3 hours that is normal for *Star Games*, the average revenue per user adds up to around EUR 50 per month. However, unlike United Games, Playing the World decided to not make any money on emotional profiles accumulated, because the company does not consider these sales as an ethical practice. Also product placement is not done in order to avoid information overload.

All profile information is logged in user-controlled personal data vaults that players can examine if they want to. The information is encrypted and not shared with anyone outside of the company, and personal 3D buddies created or chosen by players are stored in a database to interact and correspond to their owners. If players want to, they can download all of their transaction and profiling data in a structured form to their private clouds for EUR 3 a month and also rent out their 3D buddies for EUR 2 a piece per month. Whether they download their data or not, players are recognized by contract to own their personal data. The download of the data and the 3D buddy package then allows them to be accompanied by their 3D buddies outside the gaming sessions. A first set of 10,000 players have already used this service in addition to the game and seem to enjoy it a lot. They can speak to their buddies, who act as a kind of embodied personal software agent to them. Moreover, for another EUR 5 per month, the buddy is hooked up to the Wise Figure intelligence, which can adjust its answers to the knowledge level of the user. The whole service promises to be absolutely trustworthy. No personal data will be given to any third party.

Players can also upload and publicly display the virtual characters they create on their home computers. The player community can then rate others' creations. The best designs are added to the character catalog of the game. If that happens, players get unlimited free access to the game for 2 years. If a character is used by third parties for merchandising or other activities, creators get 50% of the revenue because they are recognized as the owner of the virtual artifact they created. One beneficiary of this policy was Sophia, the 8-year-old who created the pink dragon called Arthur that is now decorating the meeting room wall at United Games.

One challenge with the game is user awareness. Because users engage with virtual characters in real space, their attention is divided between the game and reality. In some cases, this requirement has caused players to bump into other people or fall over obstacles they did not see.

Listening to Carly, Stern does not know at first what to think. He alternates between astonishment, awe, and admiration for the competitors' proposal. "Yet, then again, their maximum revenue per user will be at EUR 60 per month while obviously incurring much more cost," he thinks to himself. Stern is relieved. "I wouldn't invest in such a philanthropic game," he continues thinking. "But could the absence of some of my players in *Star Games* recently be attributed to *Reality*?"

3.3 Scenario 2: Future of Work

It is the year 2030. Stern has a really hard time waking up. "My life is a big mess," he thinks to himself. As he turns around in his bed, he knows that his bracelet has now signaled to the coffee machine to prepare his morning café latte—a friendly nudge to get up and get going. But Stern does not feel like it at all today.

The last six months at his company United Games Corp. were pure hell for him, thanks mainly to an all-star devil named Carly, who seems to get all the honor and attention top management has to give. Shortly after Carly joined the company, she and Stern were staffed together on a new company project. As United Games recently bought a drone manufacturer, the company asked Stern and Carly to define an innovation strategy around the new technology. In fact, United Games invested in technology for small quadrocopters, flying vehicles of about half a meter in length that can be controlled by mobile phones, remote computers, and even voice commands.

The conflict started when Carly suggested that they use a crowdsourcing approach to get people's ideas on quadrocopter use cases. This idea was not in Stern's interest because he had anticipated the acquisition and had a long-standing vision of what United Games could do to leverage the drones' business potential. In his view, his company should ramp up the drones and sell them to police forces. In fact, Stern had a vision for how United Games could generally evolve to benefit from the highly lucrative e-government market. The powerful virtual reality worlds United Games created were already a virtual training terrain for military and police officers. So the contacts and sales channels would be there. The quadrocopter purchase was now an ideal opportunity for United Games to sell drones to municipalities, who would be looking to automate and replace some of their costly human interfaces anyways.

But Carly had a completely different idea. To her, United Games was a gaming company, one that brought people joy. That mission had brought her to the company in the first place. So her idea was to market drones to households. Drones could walk kids to school, look like fancy birds (colored in bright pink and blue colors!), and have built-in communication capabilities for people to control them directly through voice commands. Carly said that the timing would allow United Games to be a pioneer in the market and create a global voice-command standard for human–drone interaction. Drones could warn kids of dangerous situations and send real-time video to parents in case they wanted to see what their kids were doing. They could be sent home to fetch stuff in case the kid had forgotten something. The drones would be a bit like the owls from the *Harry Potter* movies but more beautiful and compliant. Stern thought the whole idea was completely girlish and ridiculous. What would his friends say if he told them over beer that he would now go into the business of turning drones into parrots? "And how do you want to resolve the incredible noise that drones are still making?" he said to Carly.

Stern's opposition to Carly's ideas caused their debate to turn vicious. Stern told his colleagues that Carly was a "childish bitch" who did not have the necessary mindset to work for United Games. She did not understand the corporate identity of United Games, which they had worked to define for years. "OK, I shouldn't have used the word bitch," Stern thought. But, in any case, he was mad after another frustrating meeting with her.

The next time he walked into the meeting room, he was greeted by her superior smile. "This lofty attitude, this untouchable arrogance," he told his buddies over lunch. "As if I was a child." The meeting went as bad as he had expected. But one bright spot from that meeting was that Carly suggested meeting online next time to cheer themselves up a bit. So she came to visit him in his Galactic City facilities in the *Star Games* VR. He spent a whole week preparing for the meeting, building a 3D simulation of what the military drones could look like and how they could be controlled, including a business case and roll-out strategy. "Anything to convince the lofty lady." But Carly was not impressed. Instead, she invited him to her virtual *Star Games* residence on planet Nanoo and had a VR simulation to show him a Heidi-like girl walking through the gardens with a fantasy drone that looked like an owl.

This was enough for Stern. He just felt that no professional means would ever make Carly see reason. And so, while out on joint space flights in *Star Games* or meeting physically for beers, he started to tell his longtime buddies in the company how he felt. "Enough is enough," he said, "somehow Carly should leave the company." His buddies agreed. They also felt that Carly was pretty arrogant and a bit over the top. They needed to find some way to cut her down to size. What about sending a military-style drone to her Nanoo virtual residence while she was there? As a product manager, Stern had access to all accounts in the game and could see where avatars were dwelling in real-time. It would be a great moment for Stern, and it would be completely harmless because it would happen in the virtual world. Second, if Carly knew the game well enough, she would be able to easily defend herself. Finally, third, they would not have the drone shoot, just hover around her house. She would not know their identities anyway. So Stern and his buddies put their plan into action, performing the attack on a Tuesday evening and having a lot of fun doing it. Afterward, they met for beers and football to cool down and celebrate their victory in the real world.

But then hell started for Stern. The next day, his boss called to ask about the progress on the drone project. His boss told him that he really liked Carly's idea of turning the drones into a nice household device. Pleasure and kids' safety were great messages for the markets, fitting United Games' image well.

Two days later, United Games' human resources (HR) officer dropped by Stern's Galactic City premises by surprise. First he chatted about Stern's recent attention scores. The company's attention management platform had found that Stern's attention span to his primary work tasks as a product manager was below average. "You seem to be interrupting yourself too often" the HR representative had said. "But what could I do?" thought Stern. There are simply too many messages, e-mails, social network requests, and so forth that would draw on his attention. So he obviously did not match the 4-minute minimum attention span that the company had set as a guideline for its employees. Employees' attention data was openly available to the HR department and management in order to deal with people's dwindling capability to concentrate. Then the HR manager asked for the rest of the activity logs, the encrypted part. Stern felt a bit awkward, but finally he decided that he had nothing to hide. So he handed over the secret key to his data and allowed the personnel officer and his staff to analyze his behavioral logs on *Star Games* as well as the logs taken in United Games' real office space.

Encrypted work activity logging was part of United Games' work terms and conditions for employment. In fact, the integration of activity logging into work contracts in many companies was celebrated years ago as a major achievement of the labor unions. The encrypted activity logging process came as a response to a steep rise in burnout and workplace bullying, which seriously impacted companies' productivity and damaged people's health, mental stability, and well-being. A compromise on the mode of surveillance was struck between unions and employers. Prior to these negotiations, employers had conducted video surveillance in a unidirectional way that undermined employees' privacy while providing no benefits to them. As part of the new process, employee activities and conversations would be logged in all rooms as well as VR facilities and stored in an encrypted

way under the full control of employees (in their personal data clouds). With this system no one, not even the CEO of the company, could view the original data. However, when a security incident happened, employees were informed and asked to share their data. In particular though when serious cases of burnout or bullying occurred, employees themselves could initiate a process of data analysis, handing over their secret key so that a designated representative could recover their data, text, and voice streams, and perform a conflict analysis. Data-mining technology would then look for patterns of behavior typical for mobbing or burnout as well as cognitive and emotional states. The streams could also be used to replay specific situations in which conflict had occurred. However, these replays would occur only in the presence of a trained coach or mediator. This practice had not only reduced bullying in recent years but also helped employees to better understand their own communication patterns and behavior. Finally, the encrypted data was also used to extract aggregated heat maps of the company's general emotional state. This practice helped upper management to better grasp the true emotional "state of their corporate nation."

Carly had handed in her secret key and initiated an inquiry into the attack on her Nanoo home. So this was what Stern was to confront today. "Perhaps I have gone a bit too far," he wondered. "But it is something totally normal for the *Star Games* anyway." After all, the idea of the drone attack had come from another colleague of his.

Stern slowly walks up to the kitchen. His café latte is not as hot he likes it, and the machine has not put as much caffeine as usual into his cup due to his raised emotional arousal. But never mind. He turns on his Roomba vacuum cleaner, which always cheers him up in the morning by roving around the apartment making some funny sounds, just like a maybug in the virtual world would do. "Perhaps a girlish drone owl isn't that bad after all," he thinks. "For sure, I need to offer something to end the war now, anyways."

3.4 Scenario 3: Future of Retail

It is the year 2030. Roger passes through the main entry gate of Halloville Mall. Sophia and Jeremy accompany him. They are excited to be allowed to join their father today. It is Jeremy's birthday, and he dearly desires a Japanese dragon sword to use in the *Reality* game. Since the family decided last year to leave the *Star Games* virtual world and instead play the *Reality* game, Jeremy has become a master dragon fighter in the offline world. The new Japanese sword would allow him to not only interact with holographic characters but also engage in a real sword fighting experience with peers without the risk of hurting the other person or being hurt. Sensors in the blade recognize when fighters cross their swords and harden the nano-based arms, turning the blade into a bright yellow light tube. In contrast, as blades get close to body parts, they instantly soften, stop shining, and feel like jelly. Kids just love to play with these devices, and Jeremy cannot wait to own one himself.

Jeremy and Sophia are now buzzing around the corridors. Jeremy has been joined by Martha, his favorite robot, who looks about his age. She is a humanoid resembling a beautiful blond teenage girl. She is so cool and well informed about all the latest products in Halloville that Jeremy has now hired her already 20 times over the past years and started to see her almost like a friend. Still, his dream would be to own a robot himself. Some people do and walk around with them proudly, like others walk their dogs. The more fancy robots are personalized in terms of voice, hair, eyes, size, and so on, and the more they have learned from their owners (including various software upgrades), the more people get attached to them.

Jeremy loves to come to Halloville. The mall is full of the latest technology: humanoid robots, holographic ads, beautiful waterfalls, and more. Sophia chats with her 3D software dragon Arthur, who gives her advice on what products and shops to avoid for bad quality and where to find stuff she likes and needs. Sophia almost cannot live without Arthur's judgment anymore. She really loves him like a friend even though he recently started to criticize her sometimes; for example, when she was lazy or unfair to a friend. He also helps her to avoid buying and eating too many sweets, because she really gained weight and deeply desires to lose a few pounds. Arthur is always extremely polite when advising her. His tone of voice is always soft and friendly. He relates his criticism to some history of her behavior and also garnishes his suggestions with some reference to philosophy, history, or statistics he is aware of. But most important, he is really selective of when he makes a remark.

Sophia has configured Arthur to run on top of her personal data vault. She could do so because of an agreement between the game company Playing the World and her personal data vault provider. The game company rents her the holographic figure for EUR 5 a month, and the personal data she generates through it remains in her private vault. All the data that is logged about her remains under her control. Arthur is a personal software agent that sits on top of this personal data, and it reveals data about her to the rest of the digital world only in an anonymized form and when absolutely needed, with privacy policies attached to the exchange. As a result, Sophia receives less information, but the information she receives is of higher quality and more tailored to her preferences. She can let Arthur know from what sources he should retrieve recommendations. She trusts Playing the World and believes that Arthur respects her orders, looks after her privacy, and recommends what is best for her.

Today, Sophia is looking for a nano-glove. Nano-gloves offer ultimate skin protection and give the person wearing them a lot of hand strength. Some people use them to repair stuff. But Sophia has plans to join her friends to hunt and crack some soft-material robot caterpillars that have recently turned up in the village. The robot caterpillars have been disseminated everywhere in the village to observe citizens' activities and listen to conversations. They are known to transmit audio and record meetings. They are just 1 cm in length and move and look like real caterpillars. Now the kids will take care of them.

Roger has decided to make his own choices today and will not use any of the recommendation devices. He has already consulted with his personal software agent at home to determine where to buy a Talos. The Talos suit tracks all body functions and analyzes his moves and progress. Unlike some of his neighbors, Roger does not think that he will get paranoid about the suit. Many of his friends have gone crazy. The textiles transmit everyone's activity data to a regional fitness database that displays everyone's performance. So, many of his peers became preoccupied about their physical condition when seeing how they perform in comparison to their peers. They feel like they have to meet at least the average performance standard in the region, which is pretty high. One of his friends was so thrilled by the Talos force that he exhausted himself in a 12-hour run in the woods. He later had to be hospitalized for his exhaustion.

Despite his own excitement, Roger has mixed feelings when visiting Halloville. Going through the mall's main gate gives the mall implied consent to read out his and his kids' data and send them tailored advertising and information. "Reading out" involves scanning clothes for radio-frequency identification (RFID) tags, recording movements and points of interest. Robots and on-shelf cameras analyze facial expressions and emotions. Video surveillance camera systems that embed security analysis screen their skin type and movement patterns. In return, Roger gets 3% off all his purchases in the mall plus free parking. The only exception is Sophia, who is able to use Arthur to reliably block her personal information exposure and provide her with neutrally tailored product information.

Roger does not like these ubiquitous data collection practices, even though the mall promises that they are done pseudonymously for him. As an engineer, he knows that if he possessed a Global loyalty card, all the data collected would be stored together with his and his kids' names in the GX2 global retail database from where it would be sold to data brokers. He did not opt to get one. Being on the pseudonymous scheme, instead, all advertisements Roger receives are based on locally generated transactional pseudonyms. These pseudonyms are derived from the International Mobile Station Equipment Identity (IMEI) numbers of his mobile phone, real-time responses to their physiognomies, and ads based on contextual factors such as season or temperature.

Roger sometime wonders whether he makes too much of a fuss about his personal data. Many customers have embraced the Global loyalty card, and their shopping experience is indeed different. All advertising screens in the mall and robot communication are customized to fit their (presumed) needs. Moreover, based on the purchase profiles of the Global loyalty card, an individual bundle price is calculated for the whole basket of goods purchased in the mall. Customers are told that customized bundle prices save them up to 5% on the list prices of individual goods and services (which Roger thinks are overpriced). When customers leave the mall, the cost of their bundle is automatically debited form their accounts.

Still, Roger dislikes this nontransparent form of pricing and the associated data collection practices. His dislike is based not only on privacy reasons. "If I am honest, I feel like a second-class citizen sometimes," he thinks. This is because rich people who are on a truly anonymous scheme can use a separate smaller mall entrance on the east side of Halloville that does not track any data. People's personal agents (a kind of app running on their mobile phones; function-wise similar to Arthur) block RFID read-outs and send their owners' data usage policies to the mall infrastructure, indicating that video and voice data must be deleted. However, when people go through that entrance, they do not receive the 3% discount and have to pay for parking, a luxury that Roger cannot afford. Personal robots are also available only for an extra charge and base their recommendations on the personal agents of those richer folks.

"But do I really care?" he wonders, pushing these bad thoughts to the side. Today is Jeremy's birthday, and the mall is the only place where they can buy the sword. Normally, Roger avoids these expensive shopping environments. Most of the products sold through these outlets are highly standardized. He typically buys only technical devices here. For everything else, he procures, barters, and borrows in the village where he lives with Jeremy and Sophia. "There, life is different," he considers.

"I just wonder how people who don't live in the villages feel." He knows that people who have very little time and live in the city center are more frequent customers at this kind of outlet. They buy everything here because they can shop efficiently and find everything in a short time. Often, they come in with preconfigured shopping lists on their mobiles. Robots then meet them at the entry and accompany them to their shops. The preconfigured shopping lists are aggregated from their fridges and other home appliances, and are routinely shared with the Halloville robots. Many people like to use this robot service because the robots are friendly looking humanoids and are extremely practical, even carrying the shopping bags. Some even carry the customers themselves around, which is nice for the elderly. This service, though, is available only to customers who use the Global loyalty card.

People can ask robots questions about all kinds of things, from the latest product offerings in the mall to purchase trends to politics and weather. The robots even entertain shoppers if the shoppers want entertainment. For this purpose, the robots are embedded with interactive emotion recognition systems. If they know, for example, that customers are interested in astrology, they can read them their daily horoscope and then combine an ideal purchase recommendation with their current "astro-mood." The emotional reaction of the customer to this kind of recommendation is stored in the customer's profile (along with data about subsequent purchases), and over time the customer relationship system—to which all the robots are connected—learns how to handle them. Football fans like Roger can have a short chat about an upcoming game combined with a beer promotion.

Recently, however, the robots' image has suffered a little. It became apparent that shoppers' financial situations and spending patterns were used by robots in such a way that they tried to steer customers into certain areas of the Halloville mall or, in contrast, tried to keep them away from others. For example, they would elegantly try to not accompany indebted customers to the high-end luxury parts of Halloville. They would pretend to enter funny failure modes or use cajoling practices to guide inappropriate customers away from purchase areas that they could not afford. This practice was done in part to provide rich folks with a less crowded shopping space filled with people of a similar socioeconomic status. Another issue is related to the humanoids' looks. They typically resembled people quite realistically and had all kinds of looks and sizes. Some of them looked like teenage girls and boys, and everybody thought that these younger looking robots were used in the mall as peers for kids. But then some men got teenage robots to be their shopping companions. And a whistleblower found that this robot choice, recommended by Halloville's IT system, was related to the system's knowledge of pedophilic tendencies for some male customers (because they had visited the child and teenage sex porn categories on porn websites).

Roger followed these revelations on blogs and was appalled. But he knows that most people just do not have the time and emotional bandwidth to be aware of this kind of company practice. If public attention fixes on a particular issue, it often quickly erodes. Most people just want to enjoy and get some positive energy out of the little time that is left to them outside of their jobs.

Roger watches his kids play at one of the waterfalls in the mall. A nice water sculpture of Niki de Saint Phalle accompanies a pool on the east side of Halloville, where they now take a walk. As his eyes glide across the area, he notices Carly. Carly was Roger's intern 5 years ago. Now she looks pretty mature and, given her expensive clothing, quite well off. "I always thought she would find her way," he thinks, smiling to himself. When Carly looks in his direction, she recognizes him as well and crosses the square to meet. Roger introduces her to Jeremy and Sophia. "So how is life?" he asks. Carly does not seem too happy about the question, telling Roger about an arduous bullying situation in her office at United Games Corp. The company's "cut-ties algorithm" detected her quarrels with a colleague over a drone investment, and personnel executives approached her to ask whether everything was OK. She denied feeling that way but revealed the major cyberbullying that she was enduring at work.

"Why don't you take a day off and come over to the Village? It is just 45 miles from the city center," Roger offers. "We really have a wonderful market on Saturday mornings, and the cafés and small shops are worth a trip." Roger loves to advertise his village. Even though many of the people living there do not have official jobs and can therefore only afford to live far out from the city center, the quality of life is pretty good. Villagers make money by selling their handmade goods to people from the city center that come on weekends. The villagers also offer therapies, massages, and all kinds of esoteric services. Around villages like his, farmers and small craft businesses have started to reappear to serve the local communities with food and products. Some of them also manufacture extremely high-end luxury products for the top 0.1% customers, but this market is rather small. Most villagers do not get far with the limited income they have. In former times when cash money still existed some people got around of some of their taxes, but now everything was paid electronically via mobiles. So bartering goods and services is a common practice, as is the sharing of higher-end devices such as drilling machines.

Governments have observed the rise of these communities, which are similar in some ways to Amish communities, with some wonder. Officials have debated about how technology could be used to tax bartering and sharing practices. However, they also know that the people living in the villages have little money to tax and, after all, the developments are a consequence of the high unemployment rates that automation caused. As long as crime rates are low, the rising rural communities are accepted. "My, how this world has drifted apart," Roger thinks. "Will you come to visit?" he asks Carly. "Sure, I'm looking forward to it," she replies.

3.5 Scenario 4: Future of University Education

It is the year 2035. Roger is standing in the kitchen cooking dinner when Jeremy storms into the kitchen. "What's the matter?" Roger asks. He seldom sees his son unnerved. "Can you imagine, Dad, I didn't get into Stanford!" Jeremy sits down at the kitchen table with an angry face. Roger steps away from the crackling pan on the stove, distracted.

When he was in his early 20s, not getting into Stanford was more of the norm. These days, hundreds of thousands of students enroll in the online versions of the old university brands. "I can't believe it, I just can't believe it!" Jeremy says. "What's the problem, Jeremy?" Roger asks. "I went to university around the corner from here and it hasn't done me any damage." "I don't have that choice, Dad. Hutchington University closed down 2 years ago, remember?" Roger nodded.

He remembered with grief the moment when the message reached him. The end of his academic alma mater. Twenty years ago he had studied economics there. Those were great days. The university had over 15,000 students enrolled. Of course, the students regarded some mechanisms of this old university system as a nuisance. They had to physically attend classes, even though many would have preferred to study the subjects online and avoid being cramped in with hundreds of other peers in the lecture halls. Some of the instructors were boring or only junior faculty, and some of the textbooks failed to keep up with the pace of innovation and the explosion of knowledge all around. As a result, a lot of the courses did not cover the material that Roger really needed when he joined the industry. But he did learn a lot, and many of the friends he had made during this time were still with him today. He would not want to miss this period in his life and the personal interaction with some of his university professors, whom he admired for their great knowledge and wisdom.

In the years since his university days, things had changed rapidly. A new university system suddenly appeared, and the old system imploded. It all started with Stanford Online. Stanford had a group of young and motivated Silicon Valley pioneers and a lot of funding. The concept was an elevator pitch, a huge global program of online courses featuring leading-edge thinkers as the instructors. Students used excellent video and written materials put together with great care by these "gurus" and their staff, who would respond quickly via chat to student questions. Students paid for the education, but it cost much less than in former times: tuition was a quarter of the traditional U.S. college fees, and no housing cost or moving expenses were required. The time it took to complete the remote courses was less than half that of the old systems. The biggest benefit was that all lectures were compelling—extremely so. A lot of money was put into them, and they had to undergo rigorous quality controls and peer reviews. The course material was jointly chosen by academics and industry. A course often consisted of material written by different thought leaders who looked at a subject from different angles.

Roger himself had enrolled in a Harvard Online course in innovation management 5 years ago. As part of this course, he received direct input from both Clayton Christensen and Nonaka Junior, major thought leaders in the field. In addition, chief innovation officers from 3M and Procter & Gamble presented as part of the online class. Because he wanted to apply to the IBM innovation department at the time, Roger also took a special add-on to the course that was sponsored by IBM and focused on how the company managed innovation inside the company. It was well known that people who did really well in these company add-ons enjoyed preferential treatment in recruiting by these companies.

Industry completely supported the idea of the online university and accepted the exams and degrees taken there as full equivalents of a physical university. This acceptance had been a key part of the rapid shift in the educational market. In 2019, industry had supported online education in a coordinated effort to fill the gap of 190,000 Big Data analysts and 250,000 data security engineers that were missing from the job market, simply because these skills were not foreseen in traditional universities' curricula. Also the standards of education were generally very different depending on the institution the individual came from, which led to an industry being more in need to conduct their own entry tests. By 2019 industry felt that the old educational system was simply not capable of providing them with people educated according to the latest knowledge and problem domains. The companies needed employees who understood material from about 40 to 50 basic courses, were trained according to the latest standards of innovation and could be on the job as young as possible. All the rest they would learn once being with the company.

Of course, a lot of discussion occurred at the time when this commercialization of education emerged. Some argued that research and education should be independent from industry, a norm that was even anchored in many countries' constitutions. But despite the opposing factors, the trend of students joining the online alternatives could not be halted. In parallel, unfortunately, governments and middle-class parents had less money to spend on education. Companies, in contrast, paid online universities a lot of money to be allowed to teach. More important, they ensured that learning materials were up to date. This was particularly important in subjects like biology, medicine, management, and informatics—actually, in all the natural sciences, where knowledge was exploding.

To be fair, many people loved the online universities. Students could finally learn from home and at their own pace, at the hours when they were most productive. They could interact online with university staff on an individual basis and not be exposed in front of others if they ever had a silly question. They were encouraged to work in online peer networks and often met in virtual reality for group work. Everyone could take the exams when they wanted to. They could easily earn money while studying. And at graduation, they were all invited to Stanford or to another attractive location for a weeklong celebration.

Seeing the developments at Stanford, a lot of old educational brands created similar schemes. A few new players appeared but had difficulty attracting students. Today, only 10 online universities throughout the world provided education to most people. A small group of no more than 5% of students still physically attend the old elite institutions, where they continue to be educated in small groups. But this level of education was reserved for the very rich or the very gifted. All other universities successively closed down simply because there were not any more students. Huge losses in free research and development (R&D) were the silent sacrifice of these developments. "What a pity," Roger thinks, "I wish I could afford to give my kids a true university experience."

"Did they give you any reasons for not letting you in?" Roger asks his son. "Well, yes Dad, and you won't believe it," Jeremy says. In fact, Jeremy's schools had registered their pupils' grades and behavior in online databases for the past 10 years. Everything had been recorded: the grades of every single test to various aspects of in-class behavior, such as playing tricks on teachers. "They sold my data to the universities," Jeremy explains. "And you remember how I fooled one of the teachers 6 years ago?"

Roger remembered. Jeremy had filmed one of his teachers with his new AR glasses when she made a mistake in front of the class, and he had then published this mistake on YouTube. All of his friends in class had found it really funny but not the teacher. Jeremy got in serious trouble with the school and almost had to leave. This incident was now obviously leading to Jeremy's exclusion from online enrollment in the university. "They also found some of my bad grades, and I guess they think I can't do any better because of my outdoor times." Jeremy had been a pretty bad student when he entered puberty. He also suffered from gaming addiction for at least 3 years that had left him physically in a bad shape until he fully recovered. In the last 3 years, though, he had improved a lot. "What do you mean by outdoor times?" Roger asks.

Jeremy's eyes drop to the floor. He does not know whether it is true, but a whistleblower software agent told him that Big Data analytics found that individuals' intelligence were highly correlated with their average time outdoors over the past 10 years. Since Jeremy had stayed indoors a lot when he played *Star Games*, his average outdoor 10-year rating was probably pretty low. And he now suspected that this data was being used to predict applicant performance. "There's this 360-degree personality screening that's now part of the admission system, and I guess they only let in people who pass it," he tells Roger.

"Why don't you go to one of the Goethe universities?" Sophia asks. Sophia, Jeremy's sister, had just entered the kitchen and listened in on their conversation. "I'd rather go there anyway," she says. Goethe universities were a reinvented form of presence-based universities. They had recently started to emerge as a response to the online universities. Goethe universities were explicitly independent from industry and cherished the positive aspects of old-school face-to-face academic education. The Goethe universities were organized as boarding schools where students had to spend at least 3 years full time for a bachelor's degree or 5 years for a master's. The universities therefore took about twice as long as online universities, which was not well regarded by industry. Goethe universities incorporated timely online courses as part of student's homework but not all of them were taught by famous instructors. The online-study materials, instructions, videos, and so forth were typically provided online before students came to class. They were given exercises and challenges to solve. In Goethe classes the emphasis was then that students were supervised in solving the exercises, asking questions and solving challenges around what they had learned by themselves online. The goal was to get them emotionally involved and interactive.

In addition to this reversed study model (in comparison to the traditional university system), Goethe universities had integrated two truly compelling innovations. One was their international teaching and peer program. Students often worked in teams of two to prepare for classes. Yet these peers typically lived on different continents and communicated through real-time translator interfaces. One student would speak into the computer or headset, and a personal agent then passed the speech to a translation engine that translated the sentence for the peer on the other side. Translations worked so well and instantaneously that even the most sophisticated formulations and words were completely understood. The use of this technology finally reintroduced high-quality conversation in classes and lectures between students from different countries. For a long time (and still in the English-speaking online universities), the level of the language was often miserable.

Another competitive advantage that Goethe schools developed was their lifelong learning scheme. Students' personal software agents recorded all of the material that the students studied during the time they were with Goethe. But the agents could also be linked to a learning-license scheme that tracked innovations in particular subject domains so that former students could stay current on the topics they had studied in school. Goethe schools had worked together with the big European libraries and Wikipedia to create a machine-readable knowledge pool from which personal software agents could pull information for free. Once students joined Goethe schools, they had access to a free, lifelong information service they could access through their personal agents.

Despite the use of these technological innovations, many hours each day were dedicated to personal interactions between teachers and students in the Goethe university facilities. One hour of physical exercise was encouraged every day as well as 30 minutes of meditation. Personal software agents were configured to support students and motivate them to participate in these sessions.

Goethe universities were financed by tuition, license fees from the learning updates, and donor money that people gave who were convinced of the importance of physical teacher–student interaction as well as a grief over the loss of the old university system. "I don't know what to think of these new institutions," Jeremy says. "What happens if nobody offers me a job afterward because I spent such a long time studying?" "Think about it," Roger says. "Studying in this way is a true gift." He was wondering how much the tuition for one of the Goethe schools would cost. He had heard that they screened all applications personally to decide who could join. This was for sure the more holistic approach. But could he afford to pay for such an education? And what would become of the holographic home-teacher that he had hoped to rent so that Sophia could take history classes? "We will have a hard time paying for both," he thought to himself.

3.6 Scenario 5: Future of Robotics

It is the year 2035. Stern walks through the glass doors of Future Lab's modern factory entrance. Future Lab is one of the world's leading robot manufacturers and is based on the outskirts of Zurich, Switzerland. The lab is an impressive building that allows visitors to see right into the robot manufacturing facilities that are situated behind a huge glass front behind the lobby desk. A small robot device that looks a bit like the R2D2 machine from the *Star Wars* movies immediately offers him a coffee while one of the last human receptionists in the country winks at him. A warm memory passes through Stern's body. For many years he had been the product manager of the *Star Games* virtual world at United Games Corp. and R2D2 was one of the most beloved figures in that game.

R2D2 is also one of Future Lab's bestselling devices. It was the device the company launched with the slogan "The human use of machine beings" 3 years ago. Future Lab's philosophy is that robots should be devoted human servants; completely owned and controlled by their owners and never replacing humans. This product and sales philosophy has gained the company wide respect and recognition from the public; a public that has increasingly become wary of remote-controlled robot devices that replaced many industry jobs.

The idea of the human–robot hierarchy (with humans always on top) is deeply embedded in Future Lab's design process. For example, robots manufactured by Future Lab never look like human beings. The company's perspective is that the organic nature of humans is much too beautiful to ever attempt to re-create it with a machine. However, Future Lab's robots are embedded with powerful artificial intelligence (AI) technologies including voice, face, and emotional recognition. But these AI functions run independently in dedicated sandboxes contained in the device that does not need to be networked to function. The robots learn locally after their initial setup and so become pretty unique creatures depending on their owners. The decentralized architecture also prevents Future Lab's robots from ever being taken over by a central computer, which increases their security. Finally, Future Lab's robots are designed with a view to total user control and excellent feedback functionality. Users cannot only command Future Lab's robots through easy and direct voice control but also switch them off completely through a designated "off command." Users can easily repair and replace most of the fully recyclable plug-and-play hardware components by using 3D plotters. This possibility to deconstruct robots like LEGO parts has also led to very fancy personalization efforts of the community. The related software is completely open and constantly improved by the opensource community.

With these user-centric control principles, the decentralized architecture as well as the recycling strategy, Future Lab gained tremendous ground in selling its robots. In particular, elderly individuals have started to buy the R2D2 model. They use it in the household (lifting heavy items), as gardening support (mowing the lawn and weeding), as a means to travel over short distances (sitting on it), to play music, and to serve as a security device (video monitoring of the apartment at night or in their absence). Stern was impressed by the model's recent sale figures.

"Future Lab is clearly not affected by the general public's ambiguous and rather negative attitude toward robots," he thinks. The robot industry has had difficulty in recent years gaining user acceptance. The initial trend had been very promising. Humanoid robots started to be deployed everywhere without people openly rebelling against them. Stern himself had joined Robo Systems 4 years ago, shortly after his quarrels with a colleague (Carly) at United Games Corp. Robo Systems was the leading robot manufacturer back then; by now, it has probably sold more than 100,000 robots. Stern's responsibility was to manage the Alpha1 series, a humanoid that Robo Systems sold to support the police force. Alpha1 systems look like tall steel men. They can act autonomously and in response to their immediate environment. If needed they can also be taken over remotely by an operator who then embodies the machine. Initially, the business went really well. Crime rates and social unrest, which had shot up so rapidly due to rising unemployment, dropped dramatically in areas where the robots were deployed. However, people hated them and were afraid of them. Many people did not want to go out at night in fear of meeting such a device. Retailers and restaurants complained. Gangs attacked the devices and broke them. The public indirectly supported these violent actions, even when human police officers accompanied the robots. They were

seen as enablers of state control. As a result, Robo System's sales went down.

Stern's strategy to protect the Alpha1 series and policemen (but also Robo System's business) was to bundle Alpha1s with mini drones called Bees, Robo System's cash cow. The 2.4 cm long flying Bee drones could transmit video and audio perfectly, and they used pattern recognition algorithms to detect any unusual activity. Remotely authorized personnel, including police stations, could access the video streams and control the Bees. By accessing the criminal faces database Faceshock, the mini drones could notify local police stations if a suspect was seen moving around town or hiding in the forest. To detect suspects, the Bees regularly patrolled the streets, urban outskirts and national parks, or just sat unrecognized on old electricity poles.

Less well known than the ordinary Bee drone model was the BeeXL. The BeeXL drone was a bit bigger than the regular device but carried small doses of extremely powerful teargas combined with a hypnotic. When gangs attacked Alpha1s, these Bees came to the robot's support and sprayed gas onto the attackers, who quickly fell and could be picked up and arrested by the police. To optimize reaction times, BeeXL drones were set to autonomously intervene and spray gas as soon as they detected violence. Recently, though, a debate started in the press on this kind of autonomous action by Bee robots. When an old lady with dementia danced violently in a public square, a Bee drone had mistakenly identified her as a criminal, intoxicating her in front of a stupefied crowd of witnesses. Even though the lady woke up safely in the police station, the case was filmed and uploaded to the net. Since this incident happened, the public, press, and some politicians had called for nonautonomous devices in public services. In particular, people in the villages came up with more ideas on how to catch and destroy Bee drones.

Last but not least, the whole industry got into huge trouble last year when a quarrel broke out between the government and its most important supplier of military robots, Boston Flexible. Boston Flexible is part of The Circle, a powerful IT conglomerate. After many years of reticence, the U.S. government decided to split The Circle to break its monopoly. This step was widely heralded by the press and other governments throughout the world. But people had dramatically underestimated The Circle's power because, as *The Guardian* termed it, "the first government–company war" soon broke out. The whole issue did not last long, but The Circle's ultimate response to the government's orders was to recall the entire Boston Flexible robot fleet, which formed the heart of a military operation abroad. As a result of this step, thousands of soldiers who relied on the robots' protection died. The U.S. military stepped in and took control of Boston Flexible's structures. But then, The Circle went on strike and with it all its robot cars stopped moving throughout the world. Since 45% of all active self-driving cars were operated by The Circle, this meant that the traffic broke down across the world. After 5 days of strike and $800 billion of financial loss, governments gave in and The Circle remained untouched.

"But this is not my business now," Stern thinks. Stern walks past the glass windows that line Future Lab's entry hall. He sips at his coffee and watches the smoothly moving robots and manufacturing belts. No human being is in sight. "How are you?" a voice calls from behind him. He turns around and can hardly believe his eyes. Carly is standing in front him. The colleague that had caused so much trouble for him at United Games Crop. "Could Carly be the head of R&D I was supposed to meet today?" he thought, feeling flashes of envy. "Wondering why I'm here?" she asked, a radiant smile on her face. "I was recently appointed as Future Lab's chief operating officer and the head of R&D is reporting to me. But I thought that since we are old colleagues we could meet directly to discuss." Stern was stalled. "How come she could make such a career?" he wondered to himself.

Carly walks Stern to the top floor of Future Lab, where a staircase gives way to another splendid view of the manufacturing hall. At the back of the factory, Stern sees an R2D2 speeding up to one of the belts where a defective component threatens to halt the belt. "We designed Future Lab as a single digital entity," Carly explains, following him down to manufacturing. "This means that we tried to take all human intervention out of the company's processes. Supply, logistics, manufacturing, and even sales are done by machines," she continues. "Systems can detect simple process failures and repair themselves. Only when problems get too complex, a human employee delves into the problem space. But even this activity is highly supported by R2D2 robots or Hal, our AI-based knowledge management system. We named him after the computer from Arthur C. Clarke's *2001 Space Odyssey*. I'll introduce you to him. Anyway, 9 of our 17 employees are devoted to performance monitoring. They typically communicate with Hal, who has the overview of all areas of expertise."

Carly demonstrates Hal to Stern. "Hello Hal," she says, "How are you today?" "Good Carly," the system answers. "We had an energy–supply problem last night because our local energy feed was too low for manufacturing. We are currently running on 93% capacity." "Hal always starts out by giving me some general important information that I need to know," Carly explains. Then she asks, "Hal, how are R2D2 sales doing?" "The conversion rate of automated sales calls is now at 13%," Hal explains, "plus 10% in the last month since introduction of the last emotion-enhanced persuasion algorithm."

A graphic showing call-center performance appears on the glass screen in front of them. One chart that shows only recent dates seems to be about the AI-based customer calling system. A second one refers to the conversion rate of human sales agents. Carly's face darkens. "Hal recommended to us to experiment with the automation of the sales force," she

said. "This is a bit against our product philosophy of not replacing human beings. But what can I do? The owners of the company often believe Hal more than they believe the human management team. A trend that goes unfortunately against our initial company philosophy."

As they walk on, Carly tells Stern more about the new customer calling system, which appears to be extremely sophisticated in understanding people's emotions based on the way they speak. "The Circle system we license for sales interprets callers' words and intonation so that it can respond with empathy. The system also calls people for R2D2 acquisitions only if it knows that the person is in a good mood that day and has received his or her paycheck within the last 3 days. We can calculate a daily mood for any potential customer who owns a Circle phone. As you probably know, phones from The Circle log their owners' bodily conditions and emotions, and this data is then shared with The Circle call center system." Stern has heard about the new Circle system but has never seen it live. He now considers suggesting it to Robo Systems as well. Still, he is surprised to see Carly reporting about it in such a neutral voice. He remembers very well how opposed she was to automation and privacy abuse back when they both worked at United Games. "Obviously she has matured a bit," Stern thinks to himself. He finds her more attractive now than he did years ago.

Stern sits down with Carly and tells her about his reason for visiting. The Alpha1 system's street fight fiascos have left Robo Systems with a huge drop in sales and a series of lawsuits. "The sales channel is pretty burned out," he explains. So Robo Systems is wondering whether Future Lab might be an ideal partner for building a new robot model in a joint venture, bringing together Future Lab's know-how in ethical robot design and Robo System's powerful hardware. Robo System's Alpha1s are known for their extremely reliable construction. They can walk across any type of terrain, they are fast and lightweight, and they are extremely strong. They have the advantage of acting autonomously while still being taken over remotely when needed. Stern's idea is to keep Alpha1 body parts, but integrate a picture-frame size display to the head structure through which the remote controller can then directly communicate with the people facing the robot. The robots could hence become a nurse, for example, in elderly homes and relatives could visit their elderlies, communicating with them through the robot's display. Alternatively, it would be possible to put some fur on them, add emotional capabilities, and integrate Future Labs layered, mostly decentralized control and intelligence architecture. Such animal robots, he explains, could serve as sales clerks in the malls, pets, companions, nannies, or teachers. They might look a bit like Chewbacca from *Star Wars*.

Unfortunately, Carly's face does not light up as Stern outlines his ideas. He is painfully reminded of how they clashed at United Games 5 years ago. "Why does she not want robot pets?" Stern wonders. But Carly seems more relaxed now than she was back then, not wholly opposed or emotional. She explains that Future Lab's philosophy of the human-robot hierarchy implies that robots should not be of the same size and stature as human beings. "Our experiments have shown that humanoid robots inspire fear," she explains. Of course, putting fur on top of the steel structure and adding friendly faces to the Alpha1s could mitigate this consumer reaction. Another concern could be that Future Lab is opposed to using robots as human companions. "This might decrease people's motivation to make real human friends," she says. Despite these concerns, Carly promises to present Stern's request to Future Lab's philosophical advisory board.

Future Lab's philosophical advisory board, Carly says, is a group that has absolutely no stake in the business, and many years of experience and deep knowledge about how to handle ethical questions. "They are a bit like the governors in Plato's philosopher state," Carly reckons, smiling as she accompanies Stern back into the entry hall.

The sun goes down slowly as Stern leaves Future Lab. It has been a rich day, highly informative. But he is also a bit frustrated. Frustrated, because he sees his own fortunes doomed with Robo Systems. Frustrated because he sees a former colleague succeed so much in her career. And frustrated by all these stubborn ethical principles that seem so outdated to him. "But then, at the same time, they do seem to pay off," he thinks to himself.

3.7 IT Innovation Lying Ahead

The scenarios include a long list of relevant information technologies we can expect to permeate First and Second World societies by 2029 to 2035. Embracing Marc Weiser's vision of "ubiquitous computing," data may be collected pervasively (Weiser 1991). As the scenarios describe, most human and device activities will be recorded. Stern's sensor bracelet recognizes his movements and emotions (based on skin conductance); VR and AR glasses enable gaming, shopping, and work activities; Jeremy's school behavior, grades, and outdoor times are logged; smart textiles, such as Roger's Talos suit, register people's fitness levels; and Bee drones record public activity for security reasons. This data is then processed to enable new services that are supposed to facilitate everyday life, such as preparing morning coffee. But the data is not only collected and processed, but also stored in "Big Data" pools. These data pools can then be used to better understand real-world social and psychological phenomena.

One critical example of enhanced machine-comprehension of real-world phenomena is the correlation between human intelligence and outdoor time that Jeremy suspects is the reason that he does not get into Stanford Online. Here the Big Data collected by AR glasses or sensor bracelets (outdoor

time) and prep schools (pupil grades and behavior) is proprietary and not controlled by the data subjects. Jeremy does not know whether the two data pools were combined; he can only suspect that this happened. Of course, Big Data pools could also be user controlled. Many of the scenarios feature Sophia's personal agent Arthur, a software system that sits on top of Sophia's personally controlled data vault. The data vault and Arthur agent's software participate in all data collection activities involving Sophia and accumulate "her personal Big Data" on the basis of which analysis is allowed for her. All of this analysis, such as the product recommendations at the mall in scenario 3, is under her control.

Agent Arthur and the Wise Figure in the gaming scenario incorporate three technologies that, when combined, may have a disruptive market potential. Arthur is a figure (a dragon) that Sophia can see in 3D through her AR glasses, which cast the figure in front of her current geolocation. To achieve this effect, cast AR technology ("Castar Kickstarter"*) is combined with AR glasses (Tang and Fadell 2012). For Sophia to fully benefit from Arthur, the software agent needs to be intelligent, embedding powerful AI functionality that allows it to recognize her voice, emotions, and cognitive load (i.e., with the help of pupillary dilation [Marshall 2002] or skin conductance [van Dooren, de Vries, and Janssen 2012]). It calculates an appropriate user model of Sophia, which it then uses to adjust its communication with the girl and support her decision making. The agent engages in knowledge retrieval to answer Sophia's questions, using both structured and unstructured information sources.

Sophia's personal agent software is described in the scenarios as a privacy-sensitive system because its knowledge about Sophia is based on the girl's personal data vault, which she rents from a trusted provider (see, for example, solutions promoted by the Personal Data Ecosystem†). It learns locally. It exchanges Sophia's data based on her privacy policy preferences (Casassa Mont, Pearson, and Bramhall 2003; Cranor 2003; Cranor et al. 2006). And it retrieves (pulls) product recommendations and information for her from the net while safeguarding her privacy (Canny 2002; Searls 2012). It uses a strong authentication mechanism and it uses identity information to supply information in an age-appropriate way.

In contrast to this personal agent architecture, the retail scenario includes a personal data and recommendation system architecture that is centrally managed by the global loyalty program GX2. Similar to today's advertisement networks and loyalty card schemes, GX2 centrally collects

---

* "CastAR Kickstarter," film produced by Technical Illusions, published on YouTube on October 14, 2013. 4:54 minutes video length, available at https://www.youtube.com/watch?v=AOI5UW9khoQ (accessed August 11, 2015).
† Personal Data Ecosystem Consortium, accessed April 14, 2014, http://pde.cc/.

and controls all personal sensor, RFID and purchase data from mall visitors and combines this data with external data sources (for example, information about porn consumption). It does so with limited user control and transparency, and then broadcasts advertisements and adjusts robots' behavior for each mall visitor. Foucault would refer to this kind of technology as a "technology of power" because it determines the conduct of individuals and forces them to submit to its actions (Foucault 1988).

Both the mall scenario and the robot scenario contain a multitude of potential robot systems that could soon be deployed. Humanoid robots are presented as potential replacements for police and military forces (see, for example, Atlas* and PetMan† robots developed by Boston Dynamics). Soft material drones may be distributed everywhere, taking the shape of small unnoticeable animals (see, for example, MIT earthworm project‡). Using the same design philosophy as soft material drones, are the bee drones, which are now developed for surveillance purposes (i.e., DARPA micro drones§). In contrast to robots that are used for surveillance, the scenarios also present robots that could provide a new level of practical value to households, such as the R2D2 system that supports elderly people at home. The home-servant drone suggested by Carly in the work scenario could help people fetch things and improve household security by walking children to school or hovering around neighborhoods (e.g., quadrocopter).

Throughout the scenarios, people communicate with their IT artifacts by using voice recognition. This technology involves a new interaction paradigm. Voice recognition as well as VR and AR glasses will replace significant elements of traditional web interaction. Classical web interaction comes back only in the educational scenario and when Sophia configures her agent Arthur. End-user programming of devices and managing personal agents may very well continue to be PC-based. But overall, classical PC-based interaction will probably be less dominant than it is today.

Voice-based communication with software agents or in virtual worlds is also an important interaction paradigm in the corporate world. Two scenarios describe how work environments will change: The total recall scenario outlines how people could meet more often for professional purposes in virtual worlds. People would enter these virtual worlds from home through their VR glasses. Real-time translation

engines and easy-to-use, end-user-programmed VR artifacts could help to enrich and facilitate professional communication. The same is true for education. Yet, almost all scenarios mention that there will be less work as well due to the automation of many company processes. The scenario 5 case outlines an enterprise that fully automates processes, including self-healing processes (for early developments in this direction; see, for example, Mendling 2008). Strong AI systems, such as the agent Hal or The Circle sales system, accumulate the information needed to run the company and perform traditional management activities autonomously. Robots manage the facilities and production. Humans may not even dominate the sales force or call centers any more. The few remaining employees supervise the machines.

Finally, a number of promising new artifacts are presented, enabled through IT and material innovation. Among these are new nano-based gaming devices, textiles, 3D plotters, and real-time translation services. Innovation in the field of genetics is explicitly not covered in the scenarios to avoid introducing too much complexity. This is despite Ray Kurzweil's vision that IT-based innovation will be centered on three core areas: genetics, nanotechnology, and robotics (the so called GNR revolution) (Kurzweil 2005).

3.8 Positive Potential Outcomes of IT Innovation

The scenarios make plain that IT has a lot of positive potential for the development of people and society at large. One of the core positives is the very high probability that we will be able to continuously learn and improve ourselves. If we have ubiquitous access to information through voice-based personal agents, we can engage in self-paced, lifelong learning, and become much better educated as a society.

But this access to knowledge is just one form of self-empowerment. Technology could also help us to better understand social phenomena. In Chapter 2, I questioned whether Big Data will be the ultimate source of innovation. But I do believe that it is very well suited to helping us use technology to better understand who we are and how social phenomena work. It may get us closer to finding the "truth about ourselves." I flesh out this concept in the second scenario on total recall, where I describe how a company's data recordings could be used in bullying cases in the office. Couples, colleagues, and all kinds of social groups that agree on sharing their data may be able to understand their conflicts and use this understanding to improve their behavior. Today's sporadic video feedback on body movements in sports or video-based educational sessions is a very rudimentary version of what could lie ahead.

Self-empowerment could also be enabled through enhanced decision making. Personal agents could serve as sparring

---

* "Atlas Revealed: Boston Dynamics' DARPA Robot," YouTube, accessed April 14, 2014, http://www.youtube.com/watch?v=w40e1u0T1yg.

† "Darpa PetMan Robot," YouTube, accessed April 14, 2014, http://www.youtube.com/watch?v=5S4ZPvr6ry4.

‡ "Soft autonomous earthworm robot at MIT," YouTube, accessed April 14, 2014, http://www.youtube.com/watch?v=EXkf62qGFII.

§ Micro-Drones 2.0, accessed April 14, 2014, http://microdrones.tumblr.com/.

partners and help us improve decision making, providing information that we can use to weigh pros and cons, especially in complex situations. Of course, such cognitive decision support must be viewed (and configured) with caution. Agent behavior should be supportive, but it should not dominate or nudge us too much in one direction (Thaler and Sunstein 2009). Technology paternalism must be avoided (Spiekermann and Pallas 2005), as must be "IT-configured citizens" (Cohen 2012, p. 1913) who move only within "filter bubbles" that make them conform to political and ideological ideas that are not their own (Cohen 2012, p. 1917). Rather, agents should be built to help people think autonomously. Agents can provide people with information and structure rather than make decisions for them. Agents should not be "manufacturing consent" (Cohen 2012, p. 1917) but rather manufacturing well-informed individuals.

IT could also empower us by extending our senses. The gaming scenario describes how AR glasses can show the environment in thermodynamic or electromagnetic terms. It describes how players can listen to the sound of plants' and planets' octave tones. Accessing this layer of information, which is embedded in our natural world but not perceivable to our senses normally, would allow us to potentially develop a higher sensitivity to our environment and natural surroundings. In turn, we might be motivated to learn more about nature and experiment with causes and effects that were hitherto left to speculation.

Finally, smart textiles such as Roger's Talos suit or Sophia's nano-glove may endow people with a physical force that has never been available to humans before (see, for example, Talos suit*). The question, of course, is how such increased bodily strength impacts our perception of self. Will our sense of self become distorted?

Whatever our reactions, taken together the technologies have the potential to bring a lot of joy to people: joy at play through new nano-based gaming devices and textiles; joy from ubiquitous holographic representations of our fantasies; joy in nearly endless virtual worlds we can design ourselves. The joy may even extend to the work environment. People will need to commute less often to work, as they will be able to conduct some of their duties and meetings in virtual space. In many respects, the new technical world increases convenience. Robots could carry our bags through the mall or bring merchandise directly to our houses. They could carry elderly people around and help them at home. Personal drones could fetch things we forgot. Our household devices can already brew coffee while we are still in bed.

This increased convenience and the time savings that automation and robotics provide give us as societies have the potential to finally leave us with more free time for ourselves. We may be freed of the "realm of necessity," the necessity to labor as much as we do today. If the financial benefits of the productivity gains from technology are shared fairly in this future society, and if people can afford these technologies and services, we may soon reach the point envisioned by many thinkers as the "realm of freedom" or "paradise of leisure" (German: *Muße*; Jonas 1979/2003). This freedom could be used to develop our human potential.

3.9 Negative Potential Outcomes of IT Innovation

Although the positive potential outcomes sound almost like a utopia we cannot wait to embrace, it will be a challenge for our societies to ensure that the flow of innovation moves in this direction. All of the scenarios contain hints that the wealth created by gains in technological productivity may not be evenly distributed in society. This development is something that we have observed in reality for many decades. Erik Brynjolffson and Andrew McAfee (2012) have outlined the economic perspective in their recent book *Race Against the Machine*. To date, our societies have tried to address the trend toward economic inequality, but the question is whether the next levels of automation (as described in the scenarios on the future of robotics and university education) can be absorbed by our current social systems. If companies move to the level of automation that will become technically feasible, a lot of white-collar work will likely be taken over by machines. Companies may evolve to be digitally integrated entities. Against this background, wise leaders (see Chapter 15) will have to decide whether to automate processes. And if they do decide to automate, they must determine how to make this choice economically and socially acceptable to employees and societies at large.

In the scenarios, I mention "villages" that spring up as a result of unemployment and where the cost of living is lower. If we really had such a dramatic and even geographic fragmentation of societies (into those with work and those without work, the rich and the poor), the villages certainly might not be as peaceful and cozy as I describe them in the retail scenario. The reader will recognize that I describe indirect opposition to governments, for example, the attacks on Alpha1 robot systems or destruction by kids of the soft-material robots used for surveillance. So, my scenarios do describe an erosion of citizens' trust in government.

This vision of citizen behavior, of course, is based on the premise that citizens will actively shape their environments. In fact, the tremendous learning potential and prospect of ubiquitous information access can give us hope that people

---

* Denise Chow, "Military's 'Iron Man Suit' May Be Ready to Test This Summer," livescience, accessed April 14, 2014, http://www .livescience.com/43406-iron-man-suit-prototypes.html.

will have higher levels of political maturity than today. At the same time, the information overload we would face in a world of ubiquitous connectivity could also lead to the opposite: people could become passive and lose their ability to follow the events that are important to them. I hint at this potential in the fifth scenario, where people do not even recognize that the government is at war with a company. (This scenario might not be so far-fetched given that some companies are already larger, richer, and more powerful now than many nations.)

To avoid a passive society, attention-sensitive system design (Horvitz et al. 2003) will be key. Attention-sensitive systems filter the incoming stream of information, and they protect us from constant external and self-induced interruption. In doing so, they protect the most valuable resource we have: our mental capability, our ability to think, to be creative, and hence—at least according to Descartes—our ability "to be." Attention-sensitive systems should ensure that the load of information and communication is adjusted to the right level for each individual. So far, few attention-sensitive systems have been introduced in the market, though spam filtering is an early current example. And our current business models, which trade our attention for free IT services, do not support the introduction of such systems. As a result, we might soon see a digital divide between people who are "media competent" (those who can abstain from digital media and forgo addiction) and people who seek to plunge themselves into a constant digital information stream. In Chapter 6 on freedom I will expand on what attention-sensitive systems could look like and how they can be designed.

Another key way to protect *democratic* societies will be to protect and even strengthen open access to information and knowledge. Information access has two dimensions: one is affordable access to competitive views and perspectives through an unbiased blogosphere or the remains of a free press. Also affordable access to current scientific knowledge is important. The other dimension of information we need to leverage is increased transparency of the digital world. The first dimension is well recognized today and is already provided for in many ways. Never before in human history has the public been so aware of their governments' and leaders' practices. And this awareness supports people's political engagement and consciousness (see, for example, the revolutions in the Middle East and public debates of the Snowden revelations). On the other hand, the digital world is inherently invisible and opaque. If transparency is not consciously created and designed into systems, data flows will remain opaque. Almost all of the scenarios include some hints to this lack of transparency. Jeremy can only suspect that his outdoor time is being used to assess his intelligence. United Games Corp. engages in emotional profiling of their VR players and sells the profiles to external parties without

players knowing about it. Bee drones hover around suburbs without people seeing them. These practices, which are in many ways direct extensions of what is being done today, would obviously lead to a tremendous distrust in companies and governments deploying such systems.

Because information flows are highly complex, it is not easy to control them or make them more transparent. They have a level of complexity that can be difficult or impossible for human beings to fully grasp. In a future world where data collection becomes ubiquitous, people would likely need a powerful AI, such as the agent Arthur, to maintain a perception of control. However, such agents would need to be fully controlled by the individual, and they would need to have access to unfiltered information sources. People will need to be able to trust such agents without reservation. Current web technologies do not allow for such a level of trust, because they were not built to adequately address privacy and anonymity.

One could argue, of course, that people still adopted these current web technologies. A lack of privacy and security did not halt their adoption. But laptops can be closed, and mobile phones can be turned off. Such actions make us believe that, to some extent, we are still in control. However, the current web's infancy is nothing compared to the level of surveillance that will be feasible by 2029. As you may have observed in the scenarios, mall environments may embed holistic behavioral and emotional tracking. The same is true for workspaces. Robots will be able to recognize our emotions and behaviors and respond to us accordingly. We may be confronted with robots not only in malls, but also on the streets, in the woods, and at home. Textiles and bracelets will log our everyday existence. We might have AR buddies like Sophia's dragon, Arthur. And virtual reality surroundings record every single step we do in them in order to function.

In short, the virtual world and the real world may blend together. And if this blending is not done carefully, in a way that addresses our need for privacy and anonymity, we could be rushing toward an environment of total surveillance that would be so powerful that we cannot even imagine its social and cultural effects. This book's later chapters will extensively cover how we can build privacy into machines to do everything possible—from the bottom up—to avoid such a dystopia.

EXERCISES

SCENARIO 1: GAMING

1. Name and describe the benefits and harms created by the two online games.
2. Map the benefits and harms you identify to human values.

3. Describe whether governments should subsidize gaming. In this scenario, why do you think the government subsidized VR glasses but not AR glasses?
4. In-class role play: Prepare for a debate between Roger, Jeremy, and Sophia on whether the family should purchase *Star Games* or *Reality*. What are the arguments for and against the two games? Conduct the debate in front of the class, with the roles of Roger, Jeremy, and Sophia played by students. Vote in class for one of the games after the debate.

SCENARIO 2: WORK

1. Name and describe the benefits and harms of the systems used in the future work environment described.
2. Map the benefits and harms you identify to human values.
3. Write an essay about whether you would invest in an activity logging system if you were in charge of United Games Corp. Explain your decision. Also discuss whether you perceive the prediction algorithms that work on encrypted employee data as a privacy intrusion.
4. In-class role play: Prepare for a debate between Carly and Stern on the potential drone use. Conduct the debate in front of the class, with the roles of Carly and Stern played by students. Vote in class on how United Games Corp. should use the drones.

SCENARIO 3: RETAIL

1. Name and describe the benefits and harms of the robot system deployed in the Halloville Mall.
2. Map the benefits and harms you identify to human values.
3. Write an essay on the social, economic, and technical benefits and drawbacks of the various privacy-management schemes offered by the Halloville Mall. What scheme do you prefer? Which one should Halloville Mall use as a default?
4. In-class role play: Prepare for a debate between Jeremy and Roger. Jeremy wants to use the robot services and apply for a Global loyalty card, which is available to kids starting at the age of 14. Roger is deeply opposed to Jeremy getting the card. Conduct the debate in front of the class, with the roles of Jeremy and Roger played by students. After the debate, vote in class on whether Jeremy should be allowed to obtain the Global loyalty card.

SCENARIO 4: EDUCATION

1. Discuss the benefits and drawbacks of research and education driven and financed by industry.
2. In-class role play 1: Prepare for a debate between Roger and the dean of Stanford Online. Roger is against 360-degree personality screening. The dean of the online university is defending the practice. Conduct the debate in front of the class, with Roger and the dean played by students. After the debate, vote in class for or against the screening practices.
3. In-class role play 2: Prepare for a debate between the dean of Stanford Online and the dean of the first Goethe university about which educational scheme is better. Have the debate in front of the class, with the two deans played by students. After the debate, vote in class for or against the screening practices.

SCENARIO 5: ROBOTS

1. Name and describe the benefits and drawbacks of Robo System's Alpha1 systems as well as Future Lab's R2D2 approach.
2. In-class role play: Prepare for a debate between Stern and a member of the philosophy board of Future Lab. Stern's role is to convince the philosopher to create a pet robot that serves people as nannies, pets, or friends. The philosopher does not want to replace humans with such pets. Have the debate in front of the class, with the roles of Stern and the philosopher played by students. After the debate, vote in class on whether Future Lab should change its philosophy.

References

Brynjolfsson, E. and A. McAfee. 2012. *Race Against the Machine: How the Digital Revolution Is Accelerating Innovation, Driving Productivity, and Irreversibly Transforming Employment and the Economy.* Lexington, MA: Digital Frontier Press.

Canny, J. 2002. "Collaborative Filtering with Privacy." Paper presented at 2002 IEEE Symposium on Security and Privacy, Oakland, California, May 12–15.

Casassa Mont, M., S. Pearson, and P. Bramhall. 2003. "Towards Accountable Management of Identity and Privacy: Sticky Policies and Enforceable Tracing Services." HP Laboratories Bristol.

Cohen, J. E. 2012. "What Privacy Is For." *Harvard Law Review* 126(7):1904–1933.

Cranor, L. F. 2003. "P3P: Making Privacy Policies More Useful." *IEEE Security & Privacy* 1(6):50–55.

Cranor, L. F., B. Dobbs, S. Egelman, G. Hogben, J. Humphrey, and M. Schunter. 2006. "The Platform for Privacy Preferences 1.1 (P3P1.1) Specification: W3C Working Group Note 13 November 2006." World Wide Web Consortium (W3C)—P3P Working Group.

Foucault, M. 1988. "Technologies of the Self." In *Technologies of the Self: A Seminar with Michel Foucault*, edited by L. H. Martin, H. Gutman, and P. H. Hutton. Amherst, MA: University of Massachusetts Press.

Horvitz, E., C. M. Kadie, T. Paek, and D. Hovel. 2003. "Models of Attention in Computing and Communications: From Principles to Applications." *Communications of the ACM* 46(3):52–59.

Jonas, H. 1979/2003. *Das Prinzip Verantwortung: Versuch einer Ethik für die technologische Zivilisation*. Frankfurt am Main: Suhrkamp Taschenbuch Verlag.

Kurzweil, R. 2005. *The Singularity Is Near*. New York: Viking Press.

Marshall, S. P. 2002. "The Index of Cognitive Activity: Measuring Cognitive Workload." In *IEEE 7th Human Factors Meeting*, 7-5–7-9.

Mendling, J. 2008. *Metrics for Process Models: Empirical Foundations of Verification, Error Prediction, and Guidelines for Correctness*. Heidelberg: Springer Verlag.

Searls, D. 2012. *The Intention Economy: When Customers Take Charge*. Boston: Harvard Business Review Press.

Spiekermann, S. and F. Pallas. 2005. "Technology Paternalism: Wider Implications of RFID and Sensor Networks." *Poiesis & Praxis–International Journal of Ethics of Science and Technology Assessment* 4(1):6–18.

Tang, J. G. and A. M. Fadell. 2012. Peripheral treatment for head-mounted displays. U.S. Patent 8,212,859, filed October 13, 2006, and issued July 3, 2012.

Thaler, R. and C. R. Sunstein. 2009. *Nudge: Improving Decisions About Health, Wealth, and Happiness*. New York: Penguin Books.

van Dooren, M., G.-J. de Vries, and J. H. Janssen. 2012. "Emotional Sweating Across the Body: Comparing 16 Different Skin Conductance Measurement Locations." *Physiology & Behavior* 106(2):298–304.

Weiser, M. 1991. "The Computer for the 21st Century." *Scientific American* 265(3):94–104.

Chapter 4

Value Ethics and IT

"Imaginary evil is romantic and varied; real evil is gloomy, monotonous, barren, boring. Imaginary good is boring; real good is always new, marvelous, intoxicating."

Simone Weil (1909–1943)

In his famous book *The Imperative of Responsibility*, Hans Jonas (1979/2003) wrote that we must always consider the potential risks of technological innovation: "The recognition of the *malum* is so much easier than that of the *bonum* ... Primitively speaking, the prophecy of doom receives more attention than the prophecy of salvation" (p. 70). With this perspective and emphasis of the negative potentials of technology, Jonas influenced a long line of technology assessment work that focused on the negative consequences or potential risks of technical advancements. Less thought has been invested by philosophers and political scientists into building positive visions for the future, visions in which we constructively focus on our positive values while avoiding the malum.

Negative thinking rarely motivates people to do any better. In contrast, engineers are typically driven by their desire to build things they enjoy for themselves or find useful. They want to create value, not destroy it. So a better way to frame ethical system design is to embrace a desire to create positive value through technology. Jonas (1979/2003) wrote, "it is not the moral law that motivates moral behavior, but the appeal of the good-per-se in the world" (p. 162). We need to concentrate on how technology can benefit society while addressing its risks along the way.

4.1 An Introduction to Values

So what would be "good" or "ethical" IT? Where does value come from? Reading the five future IT scenarios in Chapter 3, you will have noticed that many values can be promoted by new IT devices and services. I showed how IT can promote

learning and health (the Wise Figure), coach us good ethical conduct (Arthur Agent) or support our convenience (robots that carry people). On the negative side, future IT has the potential to brutally undermine most of the values we currently cherish. Think of the Alpha1 humanoid robots used by the police force against civilians, the soft robots spying on villagers, Big Data analysis of outdoor times preventing university access, and so on.

But what is a value actually? Might we not argue that some of the noted issues and effects of IT, such as convenience, seem more like functionality than a value? Is health not a state of being rather than a value? Is control or transparency logically the same as liberty or knowledge? And are some values more important than others?

These questions show that we have to clarify what a value is. In this chapter, I outline the term *value* as it is understood in philosophy. I will discern intrinsic and extrinsic values. I will briefly discuss the role of values in moral philosophy and how values relate to and differ from virtues. Finally, I will choose those values that seem to be the most vital for human eudemonia (flourishing) and discuss those in detail.

4.2 What Is a Value?

The term *value* comes from "treasuring" something. It implies a degree of worthiness. It is derived from the Latin word *valere*, which means "to be strong" or "to be worth." A value hence denotes something that is perceived as good. Clyde Kluckhohn defines a value as follows: "A value is a conception, explicit or implicit, distinctive of an individual or characteristic of a group, of the desirable which influences the selection from available modes, means and ends of action ... A value is not just a preference but is a preference which is felt and/or considered to be justified—'morally' or by reasoning or by aesthetic judgments, usually by two or all three of these" (Kluckhohn 1962, p. 395).

A value is not equivalent with "the absolute good." In fact, a value implies a threshold level, whereas the absolute good is beyond any thresholds (Shilton 2013). The threshold level of how strongly something is valued depends on the culture of a group or a society at a specific time. Nietzsche (1844–1900), for instance, discussed the *ascetic* ideal that reigned in nineteenth century Germany (Nietzsche 1887). As a result of this ascetic ideal, he observed that charity, humility, and obedience were important values in his society. In contrast, we can observe today's capitalist societies, where economic success informs the dominant ideology and where almost opposite values like competition, pride, and autonomy are favored. Charity, humility, and obedience do still exist as values, which shows that values persist. But their importance fluctuates over the course of history and depends on the ideals of a society. As we move into the machine age, we must consider our current ideals. Our ideals will influence how we regard values such as privacy or freedom and how much importance we grant to them. Value ethics involves asking the question of "what is desirable, good or worthwhile in life? What is the good life as distinct from a morally good life? What values should we pursue for ourselves and others?" (Frankena 1973, p. 79).

Some values are considered so important over time by some societies that they become *rights* and enter countries' legal systems. This is the case, for example, for human freedom and dignity. Other values transcend individual countries' legal systems and become international conventions, encouraging societies to cooperate on the basis of common values. Such conventions are particularly important in times of significant globalization, like today. For example, the right to a private life has entered the European Convention of Human Rights (Article 8; Council of Europe 1950). Further examples are freedom of thought (Article 18), freedom of peaceful association (Article 20), and the right to be protected from unemployment (Article 23), all of which have entered the Universal Declaration of Human Rights (United Nations General Assembly 1948). In this book, I will not speak of rights, because I assume that anything that has been recognized as a legal right today is also considered to be a current value (at least by all those countries and cultures that signed the agreements).

4.3 Intrinsic versus Extrinsic Values

Scholars make a fundamental distinction between intrinsic (final) values and extrinsic (instrumental) values. An *intrinsic* value is something that is valuable "in itself" or "in its own right" (Zimmerman 2010, p. 3). When someone asks "What is (the value x) good for?" the answer goes beyond the mundane for an intrinsic value. Happiness is such an intrinsic value. "What is happiness good for?" The answer is

that happiness is simply there as an ultimate goal of human kind. Scheler (1874–1928) argued that some values are simply given a priori and are anchored in each person's *ordo amoris*, an "order, or logic, of the heart" that does not need a of reason (Frings 1966).

Complementary to intrinsic values, scholars recognize *extrinsic* or *instrumental* values. Instrumental values lead causally to intrinsic values. They are not good for their own sake, but they relate to and enable something else that is good. Philosophy scholars therefore say that an extrinsic value is *derivatively* good. It derives its value from the fact that it leads to a higher (intrinsic) good (Zimmerman 2010, p. 4). For example, in the stories in Chapter 3, many of the IT applications create convenience. This convenience is an extrinsic practical value because it can increase happiness. An old lady that cannot walk through a mall by herself will find shopping to be much more pleasurable if she is accompanied by a robot that carries her bags and even carries her around when she is physically unable.

Value theory often questions how many values there are. Scholars agree that there are many extrinsic instrumental values, but how many intrinsic values are there? Monists believe that there is only one final value or "super value" to which all other values relate or are instrumental. Epicurus (341 BC–270 BC) and Jeremy Bentham (1748–1823) are famous proponents of this view. They held the view that only the value of human happiness finally counts. This view on human nature has also been called "psychological hedonism" (Frankena 1973, p. 83). However, most philosophers who have written about intrinsic values have not been monists or even monistic hedonists. Instead, they outline other values besides happiness that have intrinsic value. Frankena (1973) identifies many intrinsic values such as knowledge, beauty, health, truth, power, and harmony, which have all been considered as intrinsic values. He outlines that many philosophers regard the "presence of some kind of degree of excellence" as a characteristic for an intrinsic value.

4.4 Intrinsic Values in Philosophy and Psychology

Philosophy is not the only discipline to study values. Psychologists study human values to understand human behavior and motivation. Milton Rokeach developed an extensive value catalogue that has been tested throughout the world (see Table 4.1). In his work on values, Rokeach (1973, p. 1) held five assumptions: "(1) the total number of values that a person possesses is relatively small; (2) all men everywhere possess the same values to different degrees; (3) values are organized into value systems; (4) the antecedents of human values can be traced to culture, society and its institutions, and

Table 4.1   Nonhierarchical Collection of Intrinsic Values as Summarized in Philosophy and Psychology

Human Values Discussed in Philosophy (Frankena, 1973)	Human Values Identified and Measured in Psychology
Life, consciousness, and activity	Comfortable life (prosperous life)
Health and strength	N/A
Pleasures and satisfactions	Pleasure (an enjoyable, leisurely life)
Happiness, beatitude, contentment	Happiness (contentedness)
Truth	N/A
Knowledge and true opinion of various kinds, understanding, wisdom	Wisdom (mature understanding of life)
Mutual affection, love, friendship, cooperation	True friendship (close companionship); mature love (sexual and spiritual intimacy)
Harmony and proportion in one's own life	Inner harmony (freedom from inner conflict)
Power and experiences of achievement	Self-respect (self-esteem)
Self-expression	A sense of accomplishment (lasting contribution)
Freedom	Freedom (independence, free choice)
Peace, security	National security (protection from attack); family security (taking care of loved ones); a world at peace (free of war)
Adventure and novelty	Exciting life (a stimulating active life)
Good reputation, honor, esteem	Social recognition (respect, admiration)
Beauty, harmony, proportion of objects contemplated; aesthetic experience	A world of beauty (beauty of nature and the arts)
Morally good dispositions or virtues	N/A
Just distribution of goods and evils	Equality (brotherhood, equal opportunity for all)
N/A	Salvation (belief in God, eternal life)

Source: Krobath, H. T. 2009. *Werte: Ein Streifzug durch Philosophie und Wissenschaft.* Würzburg, Germany: Könighausen & Neumann.

personality; (5) the consequences of human values will be manifested in virtually all phenomena that social scientists might consider worth investigating and understanding."

Rokeach's stance that all men everywhere possess the same values has been challenged by proponents of "ethical relativism." Ethical relativists believe that there are no universally valid norms and values. Instead, they argue that different cultures, beliefs, and practices lead to different values and that all of them should be tolerated. In today's global and postmodern world, this respect for other cultures and their doings is a very important and timely perspective. Yet, as Charles Ess argues, ethical relativists establish their own global value and that is *tolerance* for other cultures and individuals. Thereby, they indirectly admit that some values may be universal (Ess 2013). Some scholars have also warned

that ethical relativism could lead to "moral isolationism" (Midgley 1981). If everyone is allowed to do as they please in the name of personal values, the willingness and necessity to cooperate falters. Ess warns that such developments can lead to a "paralysis of moral judgment" (2013, p. 217). Relativism would require us to accept many cultural and individual practices and preferences in the name of tolerance that run counter to our intuitive and emotional judgment. As a result, we would be unable to develop true common ground for joint decision making.

The extreme opposite of ethical relativism is ethical absolutism. Ethical absolutists believe that there are universally valid values that define what is right and good for everyone, everywhere, and at all times. Extremely religious communities sometimes tend to argue along these lines.

The middle ground between these two extreme positions is ethical pluralism. Ethical pluralists agree with ethical absolutists that some values are universal. But, embracing ethical relativism, they argue that the degree to which such values are important in a society differ between cultures. They may also differ between individuals depending on where and how those individuals live in a society. For example, the value to belong to a family is probably universally felt in all societies and in most individuals. Yet the degree to which this belonging to a family is important for a person and determines his or her lifestyle differs between cultures and between social subgroups.

In this book, I embrace ethical pluralism. There are universal values that all cultures and individuals can agree on and strive for, and it is this set of values that should be respected by our globally distributed IT systems. These universal values may still be of different importance from one country and subculture to another. All users should have the choice though to tweak and set their machines in a way they need it to have their particular value emphasis respected. Take the example of the privacy value: We know that there is a perception of privacy around the world. Machines can be set to respect this value. Yet, every user should be allowed to change the machine settings to be more or less open according to his or her individual preferences.

In order to identify the universal values that count for us globally I draw from knowledge about intrinsic values accumulated over the past 2500 years of scholarship in philosophy. I then combine this philosophical knowledge with insights gained in psychology. As mentioned earlier, psychologists have studied values and come up with their own proposals of what is important for people. Even though psychologists pursue a different scientific method than philosophers, there is considerable overlap between the two disciplines when it comes to values (see Table 4.1).

Philosopher William Frankena argued that everything that has value could somehow be related to his list of intrinsic values. Recently, some scholars have argued that new intrinsic values have emerged and should be added, for example, the ecological value of "natural environment" or "untouched wilderness" (Zimmerman 2010). In the face of current IT innovations, and considering that privacy reappears as a dominant value throughout our scenarios, we might consider adding privacy to the list as well. However, as I will show, privacy is not an intrinsic value. Instead it is an extrinsic value that is highly instrumental to the intrinsic values of knowledge, freedom, and security (see Chapter 12).

Producing finite lists of values can be problematic because it risks excluding relevant concepts. "We should give up the attempt once and for all to make atomic lists of drivers and needs," wrote Abraham Maslow (1970, p. 25), who is known for having established one of the most popular lists of human values himself. Maslow's main criticism of value lists is that values in themselves can be broken down into subcomponents: We can have multiple extrinsic values that cater to multiple intrinsic values. "If we wished, we could have such a list of drivers contain anywhere from one to one million drivers, depending entirely on the specificity of analysis" (Maslow 1970, p. 26). That said, I still believe that lists structure our thinking. IT innovation teams can use the list of intrinsic values in Table 4.1 as a starting point to creatively reflect on what goals they can cater their system designs.

4.4.1 Nonmoral Values versus Moral Values

The values I have described so far are *nonmoral* values. Nonmoral values are properties, states of affairs, or facts that we consider good or desirable in our society. Nonmoral values are important because they give a frame and identity to our lives, telling us what is good and worthwhile to strive for. At the same time, nonmoral values are not morally obligatory. They do not force us to act in a certain way (unless they have entered the legal system). In contrast, *moral values* imply an expectation of how people should behave *relative* to others. They exist as a response to the presence and needs of others (Krobath 2009, p. 178). They tell us what *ought* to be. Examples of moral values are honesty or fidelity, respect, and responsibility. The most well-known moral rules of behavior are those embedded in our religious systems, such as the Ten Commandments guiding Christian and Jewish tradition or the rules embedded in the Quran. For example, "you shall not lie" is a rule that corresponds to the moral value of honesty.

4.4.2 Values versus Virtues

We have seen that ethics recognizes the normative character of nonmoral and moral values. However, some scholars criticize "value ethics" in general. Karl R. Popper (1902–1994) said, "much of what is written about values is just hot air" (Popper 1976/2002, p. 226). Some philosophers do not accept values as having a normative status. Kurt Baier (1969), for example, wrote "The assessment of the value of a thing does not, by itself imply that one should do anything" (p. 53). The German philosopher Martin Heidegger was radically against value ethics, saying that "Here as elsewhere thinking in values is the greatest blasphemy imaginable against being" (1919/1998, p. 265).

This criticism is not unfounded. Values are empty shells if people do not act on them. I would even add that a major threat to value-based ethics is its potential of abuse by nonvirtuous actors. Nonvirtuous but powerful actors often claim values they really do not pursue. They also establish values in a society that are unethical in the end. For example, the Nazi regime in Germany established the value of being of Aryan descent, and murdered and persecuted those parts of the population that were not. Another example is Darwin's principle of the "survival of the fittest" that is propagated by some members of the elite today as a social value. The arrogance inherent in this value can lead to bitter discrimination against the handicapped, less intelligent or less wealthy. Finally, "transhumanists" in the IT world argue that humans generally are suboptimal biological systems as compared to digital machines. Transhumanists go on to establish an ideal of superior machines and inferior humans, a philosophy that could lead to a shift in how machines or the owners of a machine infrastructure would treat the rest of human society.

The success of a value-based approach therefore depends on a broad social, historical, and philosophical consensus on what constitutes a value or what ideals form an epoch. At any time, the relative importance of values depends on the virtue and wisdom of top decision makers: high courts, politicians, journalists, entrepreneurs, managers, bloggers, artists, scientists, and IT engineers, to name a few. These figures are in a position to decide whether our IT systems live up to the relevant values of the time. Their courage, generosity, high-mindedness, healthy ambition, truthfulness, and perception of justice determine whether machines will promote or undermine our values.

What does it mean to be virtuous and wise leader who fosters the right values to the right extent? A leader who is interested in human growth and advancement beyond his personal bottom line? Leading management scholars like Ikujiro Nonaka have called for reembracing Aristotle's thinking to understand what is expected from wise leadership (Nonaka and Takeuchi 2011). Aristotle (384 BC–322 BC) extensively discussed how we reach eudemonia, how we make human flourishing possible by developing an ethical habitus. The virtues Aristotle identified in his *Nicomachean Ethics* are summarized in Table 4.2. Note that Aristotle believed in holding to the principle of the Golden Mean. This means that all virtues derive parts of their rightness by being in the middle between two extreme forms of behavior. Table 4.2 therefore includes not only the virtue a leader might possess but also the extreme forms of behavior he or she should avoid.

Throughout this book I will be coming back to some of these virtues. I will describe the importance of virtue ethics for wise leadership in Chapter 15. I will also mention some of them as they become important throughout the IT system design process.

4.4.3 Necessary Values for Human Growth

Computer ethics scholar Katie Shilton accompanied IT project teams over many years and tried to discern the reigning values. She was constantly confronted with the "paralysis-of-moral-judgment" phenomenon of whose values should actually govern the project. Based on this experience, she

Table 4.2 Aristotelian Values: Aristotle's List of Virtues as Outlined in *Nicomachean Ethics*

Generally	*Courage* (andreia)—Acting neither foolhardily nor cowardly
	Temperance (sophrosyne)—Serenity and calmness vis-à-vis life; find the right balance between desire and indiscipline
In relation to money and property	*Generosity* (eleutheriotes)—Sharing goods appropriately in relation to what one possesses; find the right balance between wastefulness and stinginess
	High-mindedness (megaloprépeia)—Find the right balance between being too grand and too narrow minded/petty
Concerning reputation and honor	*Inflatedness* (megalopsychia)—Appropriate self-confidence; find the right balance between faintheartedness and inflatedness
	Healthy ambition (philotimia)—Find the right balance between too little and too much ambition; controlling the urge to be superior
	Gentleness (praotes)—Not too much, not too little
In communication with others	*Veracity/truthfulness* (aletheia)—Find the right balance between showing off and irony; that what is can be recognized as such
	Humor (eutrapelia)—Pleasantness of conversation/ready wit
	Kindness (philia)—Brotherly love; neither too smarmy nor too hardheaded
In political life	*Justice* (dikaiosyne)—Righteousness

Source: Krobath, H. T., 2009, *Werte: Ein Streifzug durch Philosophie und Wissenschaft*, Würzburg, Germany: Könighausen & Neumann, p. 22.

developed a framework to describe the various forms in which values can be discussed in technology design (Shilton, Koepfler, and Fleischmann 2012; Shilton 2013). The first dimension is *agency*: Who holds a value: a subject or a machine? Is the value stable or likely to change? Where does the value originate: from a cultural background or from an engineer's preference? The second dimension along which values can be discussed is the *unit*. Is the value held by an individual or by a collective? By a user or by an engineer? The third dimension is the *assemblage*, which asks for the final set of values catered to. Often people do not agree on the importance of a respective value, or they prefer to emphasize different kinds of values in the design of a system; the resulting system is an assemblage of these views. Shilton's work reflects that project teams suffer from disorientation as to what values to embed in technology. Agency, unit and assemblage question the source and justification of values held by different members of a project team. Any hierarchy of values seems to be subjective for each team member.

To prevent the hierarchy of values from being arbitrarily chosen by engineers or project leaders, it is important to create common ground on values. The starting point of such a common ground could be the intrinsic value categories that have been recognized as vital by both philosophy and psychology, as summarized in Table 4.1. In fact, the engineering community has already approached some of these value categories. The value of *beauty*, for example, entered computer science after a long battle with those who believed that "form follows function." Human–computer interaction scholars have found ways to build systems in a usable way and create a positive user experience that is based on our knowledge of aesthetics (Norman 1988; Nielsen 1993). Emotional and affective computing approaches also work with our perceptions of beauty (Zhang 2005). The triumphal march of the beauty value in machine design serves as an encouraging example of how values can be embraced and embedded into machines.

An open question is whether some of the 18 intrinsic value categories listed in Table 4.1 are more important than others. Can we identify any priority or hierarchy among them? One way to do so is to combine this list of cross-disciplinary intrinsic value categories with the needs that humanistic psychology has identified as particularly important for human eudemonia. As I mentioned earlier, Maslow found some values to be triggered by humans' *basic* needs (Maslow 1970). These basic needs are fundamental to human life, and the values they correspond to should therefore be prioritized in the design of IT systems.

Note that needs and values are not the same, but they are directly related. A need is a necessity or a strong want for something, and as long as this object of desire is unfulfilled, our valuation of it is particularly high. For example, when humans are hungry, they do not value their safety, sense of belonging, or esteem as much as they normally would;

they just want food. Once the hunger is satisfied, it becomes unimportant (is less valued) in the current dynamics of the individual. Then, people turn to the next higher need they value. In many instances, I may value things, but I do not need them; for instance, I may value property or fame but not need it. The needs described in Maslow's pyramid are more fundamental. All of these values are needed for human flourishing, which is why he described all of them as "basic."

Maslow's hierarchy allows us to prioritize some values. We know that if the lower values are not fulfilled, the higher ones cannot materialize either. For example, we know that people need some self-esteem. Yet this desire is less valued as long as people's basic needs for health, food, and safety are not (at least partially) fulfilled. So we need to ensure that the lower values such as health are ensured, to not cripple human flourishing at higher levels.

Finally, Maslow identified two further values in addition to the five layers of needs in the pyramid. These values are knowledge and freedom, which he considered preconditions for the satisfaction of basic needs. We need knowledge and freedom, he argued, "because without them the basic satisfactions are quite impossible, or at least, severely endangered" (1970, p. 47).

Concentrating on Maslow's work, I reduce the list of 18 intrinsic human value categories (Table 4.1) to 7, which are summarized in Figure 4.1: knowledge and freedom as preconditions for human growth, health, safety, friendship, self-esteem, and self-actualization. In the following I will outline how these intrinsic value categories are affected, created, fostered or potentially undermined by IT systems. I will not write about self-actualization, because the way in which this concept is understood by Maslow, I hardly believe that we can build it into machines (Maslow 1970). The remaining concentration on 6 final intrinsic values is useful because it allows us to focus IT projects on what are indubitably important values across cultures. It also reduces the complexity of this book.

Looking at Figure 4.1, some computer ethics scholars will wonder why many of the values that are most commonly discussed in the discipline are not mentioned in the pyramid. For example, privacy and autonomy are not depicted here, even though they are frequently debated values in the computer ethics literature. However, note again that the values shown in this synthesis pyramid are all values of *intrinsic* character. This means that many extrinsic values help people to achieve these higher-order constructs. For example, autonomy is instrumental for creating a perception of freedom. Therefore the section below on freedom and liberty integrates a subsection on autonomy. So, in the following, I will not only define and discuss the intrinsic value categories noted in the value pyramid in Figure 4.1, I will also present in full detail core extrinsic values that cater to these. The summary of the concepts finally covered is shown in Figure 12.2 in Chapter 12.

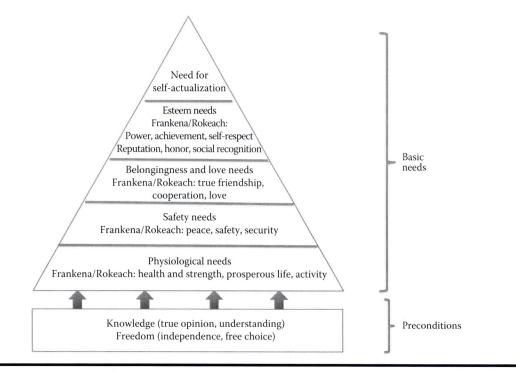

Figure 4.1   Frankena's and Rokeach's list of values combined with Maslow's hierarchy of human motivation and flourishing.

EXERCISES

1. Identify values from the robot scenario (scenario 5) in Chapter 3, and then discern which values are extrinsic and which are intrinsic. Align the extrinsic values that you identify with the parts of Maslow's pyramid depicted in Figure 4.1.

2. Using Table 4.1, compare the values of American philosopher Frankena to the values of psychologist Rokeach. Discuss the differences between philosophy and psychology in how they frame values.

3. Explain the difference between values and virtues. Reflect on why value ethics can be abused if it is lived without virtue ethics. Give an example where this happened.

4. Describe a value that used to be important in your society but has lost significance, even if it is still around.

5. Take the value of tolerance and critically discuss its importance for modern society. How might tolerance apply to the scenarios in Chapter 3? Can you relate tolerance to Aristotle's view on virtuousness?

References

Baier, K. (1969). What is value? An analysis of the concept, in K. Baier and N. Rescher (Eds.) *Values and the Future*, Toronto, 33–67.

Council of Europe. 1950. European Convention on Human Rights.

Ess, C. 2013. *Digital Media Ethics*. 2nd ed. Hoboken, NJ: Wiley.

Frankena, W. 1973. *Ethics*. 2nd ed. Upper Saddle River, NJ: Prentice-Hall.

Frings, M. S. 1966. "Der Ordo Amoris bei Max Scheler. Seine Beziehungen zur materialen Wertethik und zum Ressentimentbegriff." *Zeitschrift für philosophische Forschung* 20(1):57–76.

Heidegger, M. 1919/1998. *Pathmarks*. Cambridge, UK: Cambridge University Press.

Jonas, H. 1979/2003. *Das Prinzip Verantwortung: Versuch einer Ethik für die technologische Zivilisation*. Vol. 3492. Frankfurt am Main: Suhrkamp Taschenbuch Verlag.

Kluckhohn, C. 1962. "Values and Value-Orientations in the Theory of Action: An Exploration in Definition and Classification." In *Toward a General Theory of Action*, edited by Talcott Parsons, Edward Albert Shils, and Neil J. Smelser, 388–433. Cambridge, MA: Transaction Publishers.

Krobath, H. T. 2009. *Werte: Ein Streifzug durch Philosophie und Wissenschaft*. Würzburg, Germany: Könighausen & Neumann.

Maslow, A. 1970. *Motivation and Personality*. 2nd ed. New York: Harper & Row Publishers.

Midgley, M. 1981. "Trying Out One's New Sword." In *Morality and Moral Controversies*, edited by Johan Arthur, 116–119. Upper Saddle River, NJ: Simon & Schuster.

Nielsen, J. 1993. *Usability Engineering*. Mountain View, CA: Morgan Kaufman.

Nietzsche, F. 1887. *Zur Genealogie der Moral*. Leipzig: C. G. Naumann.

Nonaka, I. and H. Takeuchi. 2011. "The Wise Leader." *Harvard Business Review*, May, 58–67.

Norman, D. A. 1988. *The Psychology of Everyday Things*. New York: Basic Books.

Popper, K. 1976/2002. *Unended Quest: An Intellectual Autobiography.* London; New York: Routledge.

Rokeach, M. 1973. *The Nature of Human Values.* New York: Free Press.

Shilton, K. 2013. "Values Levers: Building Ethics into Design." *Science, Technology & Human Values* 38(3):374–397.

Shilton, K., J. Koepfler, and K. Fleischmann. 2012. "Chartering Sociotechnical Dimensions of Values for Design Research." *The Information Society* 29(5):1–37.

United Nations General Assembly. 1948. Universal Declaration of Human Rights.

Zhang, P. 2005. "The Importance of Affective Quality." *Communications of the ACM* 48(9):105–108.

Zimmerman, M. J. 2010. "Intrinsic vs. Extrinsic Value." In *The Stanford Encyclopedia of Philosophy*, edited by Edward N. Zalta. Stanford, CA: The Metaphysics Research Lab.

Chapter 5

Ethical Knowledge for Ethical IT Innovation

"He that knows nothing, doubts nothing."

Randle Cotgrave (1611)

Men's relationship with knowledge (episteme) has been an ambiguous one for millennia. There seems to be a deep fear that wanting to know too much can be dangerous, or at least knowing too much about what Ginzburg (1976) has coined "high knowledge"; that is, insight into the secrets of nature (*arcana naturae*), secrets of God (*arcana Dei*) and power (*arcana imperii*) (Ginzburg 1976). The Bible tells us the story of Adam who followed his curiosity and ate from the tree of knowledge of good and evil (Figure 5.1). And as he did so, humanity was tossed from paradise. In his Epistle to the Romans, St. Paul warned the Romans *noli altum sapere, sed time*, which has been translated and interpreted as an appeal to not know too much. But, at the same time, Aristotle famously begins his *Metaphysics* declaring that "All men by nature desire to know" (Aristotle 1984, Book 1). Maslow (1970) talks about "the reality of the cognitive needs" (p. 49).

Figure 5.1 *Adam and Eve* by Lucas Cranach the elder.

And Kant called out provocatively, *Sapere aude*, "Dare to know!" (Kant 1784/2009, p. 1). History shows that the value to accumulate knowledge is a contested one.

5.1 What Is Knowledge?

Philosophers ascribe the established definition of knowledge to Plato who saw it as a justified, true belief (Plato 2007). In this definition three core components of knowledge become apparent. First, knowledge needs substantiation and justification so that the one knowing it can be sure that what he knows is in fact knowledge and not just an attitude or a fake. Aristotle said: "Men do not think they know a thing unless they have grasped the 'why' of it (which is to grasp its primary cause)" (Aristotle 1984, 194b16–194b23). Second, knowledge needs to be present in a knowing subject; a person who "believes" in the knowledge artifact and for whom it is relevant. In other words, knowledge needs a beholder. And third, knowledge needs truth: "We can say that truth is a *condition* of knowledge; that is, if a belief is not true, it cannot constitute knowledge. Accordingly, if there is no such thing as truth, then there can be no knowledge" (Truncellito 2015).

With the explosion of IT capabilities and an unprecedented capacity to collect data and information, analyze it, store it and combine it, the term *knowledge* has gained tremendously in importance. Scholars write about "the knowledge creating company" (Nonaka and Takeuchi 1995), politicians propagate "the knowledge society" (Stehr 1994) and the creation of "knowledge commons" (Hess and Ostrom 2006). IT folks market databases that promise to be "knowledge management systems." "Big Data" sets lead scholars to talk about an "industrialization of knowledge," which is supposed to result from a confluence of (1) Big Data generation and collection, (2) data processing and analysis, and (3) data-driven decision making and control (OECD 2015).

The promise of knowledge created by IT systems is compelling. The traditional understanding and connotation of the term *knowledge* carries weight in people's mind, because it stands for believable, justified and true phenomena. We have to be very careful though to not overstrain the term *knowledge* when we use it in the IT world as a synonym for all sorts of data processing. In many cases, IT investments have gone astray in past years when IT managers believed that they could really procure knowledge when buying a knowledge management system or a means for knowledge discovery, when really they only got a database or a visualization tool. So one first ethical question when it comes to a discussion of knowledge in an IT context is to ask about the conditions under which we are actually allowed to use the term in such a way to not mislead investors and users. Deceptive wording (i.e., in advertising) is a well-known problem in marketing. Based on an analysis of "deception by implication" in marketing communication (Hastak and Mazis 2011), I would argue that there is a risk that the term *knowledge* has in most cases been misleading in an IT context. A "semantic confusion" is often created among recipients who associate knowledge with a true and justified belief, a property that IT systems cannot deliver.

The scientific community makes a very clear distinction between data, information, and knowledge, which is particularly important when analyzing machine capability. Meyer (2007) summarizes the established view: data is observed symbols, for example, raw sensor data, sensor metadata, a birth date, a name. Information is interpreted symbols and symbol structures, that is, aggregated and "cleansed" sensor data or structured data sets. Floridi (2005) goes even further and defines information as "well-formed meaningful

data that is truthful." Knowledge is then interpreted symbol structures or patterns that are used within a decision process (Meyer 2007). Knowledge is created, for example, when data scientists take aggregated and cleansed sensor data and put it to statistical analysis to extract (if possible) causal models or when they develop higher-level indices that can support decision making. Knowledge can reenter the information base for further knowledge elaboration (for example, statistical factors reentering a database for further analysis). Figure 5.2 illustrates this vital distinction of terms.

5.2 Ethical Challenges in IT-Driven Knowledge Creation

The scenario descriptions in Chapter 3 show that future societies bear tremendous potential for us to become more knowledgeable through our machines. We will probably have ubiquitous access to knowledge and information when we need it and where we need it. Potentially, we have agents like Sophia's Arthur or the Wise Figure that search and filter information for us and make us see phenomena we cannot perceive naturally (such as thermodynamic and magnetic information). They may aggregate and interpret information to some extent and then coach us and help us to learn at our individual level of knowledge capability. Ideally, machines have access to everything humanity knows at a low cost.

The amount of information to know about is exploding though. I mention in the scenarios that people have gotten passive toward new information or also elapse some of it. They may stop to know due to overload. Or they may make false judgment as a result of too much competing information. Such a loss must again be overcome by machines, such as Hal (in the robot scenario), who confronts Carly actively with the information she needs for her job.

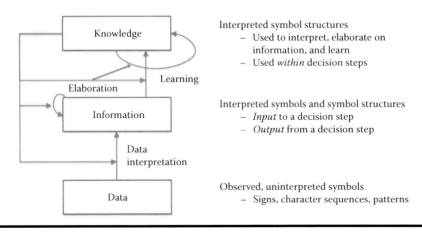

Figure 5.2   The distinction of data, information, and knowledge. (From Meyer, Bertolt, 2007, "The Effects of Computer-Elicited Structural and Group Knowledge on Complex Problem Solving Performance," Mathematics and Natural Science Faculty, Humboldt University Berlin, p. 6. With permission.)

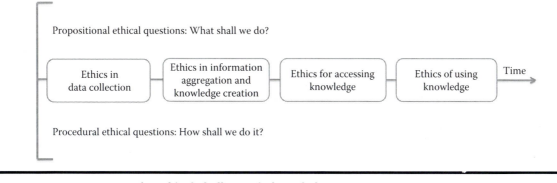

Figure 5.3   A process structure for ethical challenges in knowledge management.

Seen the tremendous responsibility inherent in knowledge creation through IT the question is what is an ethically correct way to select information that is relevant for a person and how can we ensure that such an information selection process by a machine does not integrate any bias? Who should be allowed to select what we should know? How transparent does this process need to be? And what data and information should be used at all? The next sections in this chapter give some answers to these questions.

The collection, aggregation, interpretation, access, and use of information and knowledge can be depicted as a process structure (Figure 5.3). At each stage of this process we can observe distinct ethical challenges. More precisely, we have ethical challenges on two levels: One is looking into whether we should do something (propositional ethical questions). Is it ethically legitimate that we collect certain data, aggregate, interpret, access, and use it? The other level of ethical challenge is of procedural nature. This puts the spotlight on how we actually create knowledge. How should we collect data? How should we aggregate it and make it accessible? How should we use what we know?

5.3 Ethical Challenges in Data Collection

When discussing data collection it is vital to first distinguish between data that is ethically sensitive and data that is not. Data collected for knowledge creation can be of personal and impersonal nature. Personal data, according to European data protection law, either directly identifies an individual (i.e., through a social security number) or it is indirectly indicative of an individual.*

Typically, ethical questions arise only around the collection of personal data, and only so when that personal data is put to uses that go beyond the original purposes and reasons for which it was collected and is needed. For example, let's

think back of the work scenario in Chapter 3. The initial reason for collecting the virtual-reality data is to make the virtual world work and to make it technically interactive and responsive. There is not ethical issue in this. However, if that same data is logged and analyzed for a secondary purpose, which is to monitor employees' moods and behaviors and calculate "cut-ties probabilities," then we get into an ethically problematic space. In this case a user's consent for data use is required (see Section 5.3.1). Figure 5.4 gives an overview of problematic and unproblematic data collection practices.

When personally identified or identifiable data is collected about us for secondary uses, then the European member states as well as the United States and all OECD member states have acknowledged that there could be a potential for it to be used against us (harm us) and they have therefore set up protective regulations or guidelines (OECD 1980; European Parliament and the Council of Europe 1995; Federal Trade Commission [FTC] 2000; Greenleaf 2011). In Europe this restrictive legal approach to data collection has historical roots. The extent of the Holocaust in Europe was driven in part by the availability of personal data records about Jews, which fostered their systematic persecution by Nazi officials (see Figure 5.5). For this historical reason as well as a legal case history of privacy

Data collected and used only for primary service delivery:

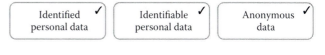

Data collected and used for *secondary* purposes beyond service delivery:

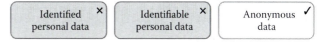

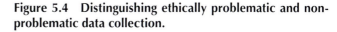

✕ Ethically and legally problematic; requires user consent

✓ Ethically and legally not problematic

---

* "Identifiable" personal data allows for uniquely reidentifying a person from a larger pool of data. For example, there is probably just one woman that lives at my address and is born on the same date as me.

Figure 5.4   Distinguishing ethically problematic and non-problematic data collection.

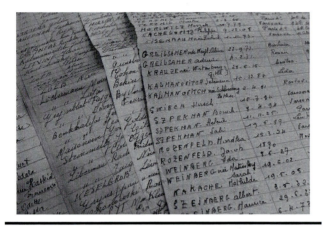

Figure 5.5 Records of Jews' birthdates, names, and towns registered by the Parisian police. (© APA PictureDesk.)

breaches in the United States (Solove 2001, 2002, 2006), personal data collection is regulated to some extent in most countries (Greenleaf 2011). Typically, data collectors are required to minimize the personal data they collect about us and they are only allowed to collect it for a specified and legitimate purpose (see, for example, Article 6 of EU Data Protection Directive). Some data categories are not allowed to be collected at all, except for exemptions or if people give their explicit informed consent to do so. These include personal data revealing racial or ethnical origin, political opinions, religious or philosophical beliefs, trade-union membership, and data concerning health or sex life (Article 8, European Parliament and the Council of Europe 1995).

Today's data-rich service world makes this data scarcity approaches more difficult. In many cases companies and people find it beneficial to collect personal data for more than just service delivery. Think of the employee mood barometer that was described in the work scenario in Chapter 3 where top management is made aware of the emotional state of the company. Or agent Arthur who gives Sophie advice on the quality of products received from other buyers. Think also about the fitness feedback that Roger can receive from other wearers of the Talos suit. All these service examples have in common that they use (at least initially) personal data and aggregate and analyze this data to build valuable secondary information services on top of it.

Does this mean that innovative future services rely on personally identified or identifiable data? No. Most of the data-rich services I have described in Chapter 3 can be built with the help of people's anonymized data sets. It is not necessary to maintain the "personal" nature of data sets in order to foster innovation around data. Let's take the Talos suit example. Roger's personal fitness data could be collected by his own personal agent and then passed on to the Talos service platform in an anonymized form. On that platform all anonymous Talos customers would then pool their fitness

data for comparison and benchmarks without revealing their identities to Talos Corp. They would hence technically exclude the risk that their Talos data could ever be sold to or shared with third parties. Companies like Talos Corp. on the other side do not run into any ethical or legal problems while being able to deliver intelligent data-driven services.

Summing up, the propositional ethical question around data collection is whether we should collect personally identified or identifiable data. Ethical system design, as I will introduce it in Chapter 18, would start with this question by challenging whether the collection of personal data is really necessary for the provision of a digital service or whether it is not possible to only use anonymous user data in the first place so that no ethical conflict or legislative issue can arise. Box 5.1 informs about the technical details of anonymization of personal data.

5.3.1 Informed Consent

Let's assume that personal data needs to be collected and that anonymization or pseudonymization is not feasible. Then an ethical question arises about how that data is collected. There is widespread legal agreement that personal data should only be collected with the *informed consent* of data subjects.*

Consent is historically rooted in the Nuremberg Code.[†] Conceptually the obtaining of consent can be grouped into two distinct activities: One is to inform data subjects about data use intentions. The second is to obtain the consent from the data subjects. Informing about data uses means that consent seekers have to give accurate information about the specific purposes and reasons of personal data use as well as potential benefits and harms resulting from that use. Friedman, Felten, and Millett (2000) and the European Article 29 Working Party on Data Protection outline that companies need to meaningfully disclose about their data usage practices before people consent (or decline to consent). "Meaningful disclosure" requires a company to state: (a) what data will be collected, (b) who will have access to the data, (c) how long the data will be archived, (d) what the data will be used for, and (e) how the identity of the individual will be protected (Friedman, Felten, and Millett 2000, p. 2). This

---

* See Article 19 of Directive 95/46/EC (European Parliament and the Council of Europe 1995), U.S. Fair Information Principles including principles of "notice" and "choice" (FTC 2000), and OECD Privacy Guidelines (Organisation for Economic Co-operation and Development 1980).

† The Nuremberg Code was formulated in response to Nazi doctors' experimentation with human subjects. The Nuremberg Code outlines how informed consent must be obtained and constituted for medical and health research purposes. The code has been adopted by the U.S. National Institutes of Health. This means that for health research it is required that human subjects consent to the collection and use of their data.

BOX 5.1 TECHNIQUES FOR PSEUDONYMIZATION AND ANONYMIZATION OF PERSONAL DATA[1]

Data can have different degrees of identifiability (Table 5.1). Pseudonymous or anonymous use of data for service delivery protects individuals' privacy and makes data collection ethically or legally unproblematic.

Pseudonymous data means any personal data that has been collected, altered, or otherwise processed so that of itself it cannot be attributed to a data subject without the use of additional data. This additional data should be subject to separate and distinct technical and organizational controls. Any reattribution should require a disproportionate amount of time, expense, and effort according to timely technical standards.

Technically the creation of pseudonyms could imply that separate databases for profile and contact information are created in such a way that common attributes are avoided. Steps should also be taken to prevent future databases from introducing common identifiers. Identifiers should therefore be generated at random. Any information that is highly specific to an individual (e.g., birth dates or contact data) should be avoided whenever possible. The general guideline for pseudonymous data is to minimize the granularity of long-term personal characteristics collected about an individual.

Even so, it may still be possible to individually identify a person based on transaction patterns. Pattern matching exploits the notion that users can be reidentified based on highly similar behavior or on specific items they carry over time and across settings. For example, mobile operators may be able to reidentify a customer by extracting the pattern of location movements over a certain time span and extracting the endpoints of the highly probable home and work locations. Typically, only one individual will share one home and work location.

Pattern matching does not always result in the identification of a unique individual. Often, a pattern may match multiple individuals. k-Anonymity is a concept that describes the level of difficulty associated with uniquely identifying an individual.[2] The value k refers to the number of individuals to whom a pattern of data, referred to as quasi-identifiers, may be attributed. If a pattern is so unique that k equals one person ($k = 1$), then the system is able to uniquely identify an individual. Detailed data tends to lower the value of k (for example, a precise birth date including

Table 5.1 Different Degrees of Data Identifiability

Data Types	Definition in Terms of Linkability	Protective System Characteristics
Personally identified data	Linked	• Unique identifiers across databases • Contact information stored with profile information
Pseudonymous data	Linkable with reasonable and automatable effort	• No unique identifiers across databases • Common attributes across databases • Contact information stored separately from profile or transaction information
	Only linkable with disproportionate amount of time, expense, and effort according to timely technical standards	• No unique identifiers across databases • No common attributes across databases • Random identifiers • Contact information stored separately from profile or transaction information • Collection of long-term person characteristics on a low level of granularity • Technically enforced deletion of profile details at regular intervals
Anonymous data	Altered or otherwise processed in such a way that it can no longer be linked to a data subject	• No collection of contact information • No collection of long-term person characteristics • k-anonymity with large value of k • l-diversity with large values for l • Differential privacy

day, month, and year, will match fewer people than a birthday recorded without year of birth). Long-term storage of profiles involving frequent transactions or observations also tends to lower the value of k because unique patterns will emerge based on activities that may reoccur at various intervals. The values of k associated with a system can be increased by storing less detailed data and by purging stored data frequently.

In some cases, large values of k may be insufficient to protect privacy because records with the same quasi-identifiers do not have a diverse set of values for their sensitive elements. For example, a table of medical records may use truncated zip code and age range as quasi-identifiers, and may be k-anonymized such that there are at least k records for every combination of quasi-identifiers. However, if for some sets of quasi-identifiers, all patients have the same diagnosis or a small number of diagnoses, privacy may still be compromised. The l-diversity principle can be used to improve privacy protections by adding the requirement that there be at least l values for sensitive elements that share the same quasi-identifiers.[3]

Anonymous data refers to any personal data that has been collected, altered, or otherwise processed in such a way that it can no longer be attributed to a data subject. Anonymity can be provided if no collection of contact information or long-term personal characteristics occurs. Moreover, profiles collected need to be regularly deleted and anonymized to achieve k-anonymity with large values for k or l-diversity with large values for l.

NOTES

1. Sarah Spiekermann and Cranor Lorrie, "Engineering Privacy," *IEEE Transactions on Software Engineering* 35 (2009): 67–82.
2. Latanya Sweeney, "*k*-Anonymity: A Model for Protecting Privacy," *International Journal of Uncertainty, Fuzziness and Knowledge-Based Systems* 10, no. 5 (2002): 557–570.
3. Aswin Machanavajihala, Daniel Kifer, Johannes Gehrke, and Muthuramakrishnan Venkitasubramaniam, "*l*-Diversity: Privacy beyond *k*-Anonymity," *ACM Transactions on Knowledge Discovery* 1, no. 1 (2007).

information should be provided to users in such a way that it can actually be understood by them, according to the principle of comprehension.

The second activity required for consent is to actually obtain it. People need to give their consent freely and voluntarily. They need to be able to exercise a real choice and there should be no risk of deception, intimidation, coercion, or any other negative consequences if he or she does not consent. Friedman outlines that opportunities to accept or decline one's personal data usage should be visible and readily accessible. The European Article 29 Working Party on Data Protection comments that "consent must leave no doubt as to the data subject's intention" (2013, p. 3). Hence, the menus to accept or decline personal data uses should not be buried under myriad website layers or hidden in obscure locations such that data subjects cannot find them.

Figure 5.6 summarizes the requirements for obtaining consent as outlined by different U.S. and European scholars and regulators.

To meet the requirement of meaningful disclosure, engineers can embrace protocols such as P3P 1.0 (P3P stands for Platform for Privacy Preference Project) as specified by the W3C in 2002 (Cranor et al. 2006). P3P describes web browsers that can read standardized machine readable privacy policies published by companies or governments. The openly accessible privacy policies are accumulated according to an XML format encoding a privacy taxonomy that embraces 17 possible data categories collected for 12 possible purposes, 6 possible types of recipients, and 5 possible types of retention policies. Company information is pulled off the privacy policy by the user's web browser and can then be translated into a "privacy nutrition label" of the kind displayed in Figure 5.7. An alternative for how to transmit privacy preferences and find the sources of how data is handled has been proposed by Time Berners Lee and his group in the form of HTTPA (where the A stands for accountability; see also Section 8.3.2.2 on audit trails; Seneviratne and Kagal 2014).

5.3.2 User Control over Data Collection

Informed consent is a challenge as IT becomes more ubiquitous. "Weaving computing into the fabric of everyday life" is a high vision of the IT world called "ubiquitous computing" (Weiser 1991). This vision implies outspokenly that our natural environments should collect data about us human beings, and the nature and infrastructure around us at all times. I describe this vision in the mall scenario (Chapter 3):

> Going through the mall's main gate gives the mall implied consent to read out his and his kids' data and send them tailored advertising and information. "Reading out" involves scanning clothes for RFID tags, recording movements, and points of interest. Robots and on-shelf cameras analyze facial expressions and emotions.

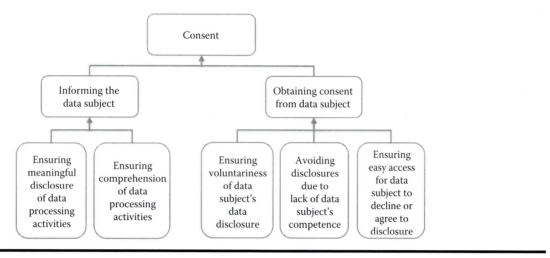

Figure 5.6 Conceptual analysis of informed consent.

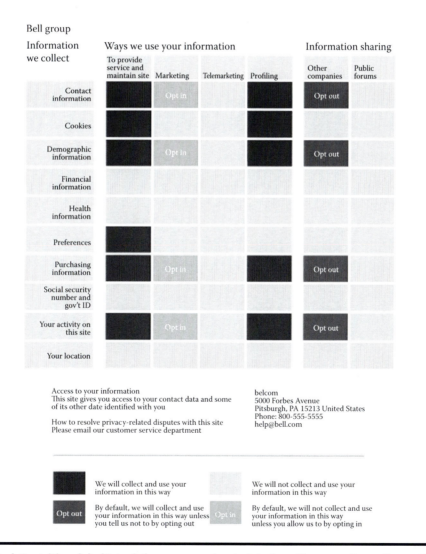

Figure 5.7 An adapted "nutrition label" to inform users about data handling practices. (From Cranor, Lorrie, 2012, "Necessary But Not Sufficient: Standardized Mechanisms for Privacy Notice and Choice," *Journal of Telecommunications and High Technology Law* **10(2):273–308. With permission.)**

Video surveillance camera systems that embed security analysis screen their skin type and movement patterns.

How can informed consent be organized in such an environment and in such a way as to not overwhelm the user? Current scientific proposals foresee that personal software agents serve as mediators between the intelligent infrastructure and us (Langheinrich 2003, 2005; Spiekermann 2007). Agent Arthur is an example for this kind of mediating software entity. Personal agents can learn and store our privacy preferences and then permit or block requests to collect data about us. Requests for our data as well as data sharing can be logged on the client side (Danezis et al. 2012) and with the requesting data collecting entities. The latter may receive a kind of "sticky policy" with our data (Casassa Mont, Pearson, and Bramhall 2003). These policies travel as metadata tags with the information that is collected from us indicating to data controllers and processors whether, to what extent, and under what conditions we allow for our data use (Nguyen et al. 2013). Figure 5.8 broadly illustrates the kind of privacy mediation process that could be implemented.

What is crucial in this invisible and ubiquitous data collection process in the long run is that people continue to exercise and perceive control over what is happening. How can this be done? To answer this question it is helpful to first conceptualize the construct of perceived control generally and then apply it to data collection.

Perceived control is the conviction that "one can determine the sequence and consequences of a specific event or experience" (Zimbardo and Gerrig 1996, p. 385). According to Averill (1973) three types of control can be distinguished: cognitive control, decisional control, and behavioral control. *Cognitive control*, which has also been coined "information control," implies that a person has the possibility to understand and interpret an (potentially threatening) event. This type of control can also be described as "to know what's going on." It is a function of comprehensive and complete information on one side but also depends on people's ability to absorb and understand that information on the other side. *Decisional control* is the opportunity to choose an action among several true choice options. True choice means that the options available must be affordable by the individual. Finally, *behavioral control* is gained when one is able to take an action and thereby directly affect, modify, or regulate an event. For example, when one's Wi-Fi signal is weak and one can just walk to another room where the signal is strong again, then behavioral control is experienced.

Figure 5.9 summarizes the three dimensions of perceived control. Depending on the IT service, either all or just some of the controls need to be provided in order for people to feel comfortable. The influences of IT designing companies are in the bolded boxes. It is in their hands to decide what information they provide to customers, what choices they offer, and how they implement user feedback.

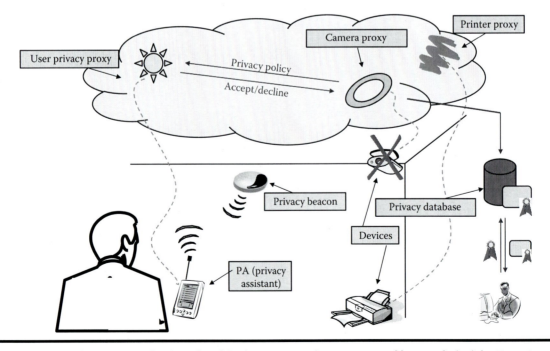

Figure 5.8 Overview of privacy mechanisms for ubiquitous computing as proposed by Langheinrich. (From Langheinrich, Marc, 2003, "A Privacy Awareness System for Ubiquitous Computing Environments," paper presented at 4th International Conference on Ubiquitous Computing, UbiComp2002, Göteborg, Sweden, September 29–October 1. With permission.)

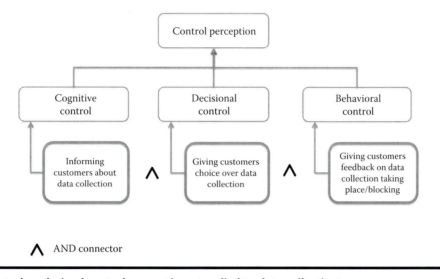

Figure 5.9   Conceptual analysis of control perceptions (applied to data collection).

Let's now transfer this conceptualization of control over data collection to an example: I relate the control dimensions to the case where RFID is used in mall environments; a scenario I described in Section 3.4. RFID chips are as tiny as grains of sand (see Figure 5.10). Industry plans to introduce them as replacements for barcodes in the long run in many areas. The chips can be embedded into products where they take much less space than traditional barcode labels while containing potentially much more information on the product. Depending on the frequency the regular read range is 6 to 8 meters. Readers do not need a line of sight for reading the tags information, which potentially allows reading a tag's information through cloths or even walls. All of these technical traits promise a lot of benefits for logistics, on retailers' shop floors, and beyond the point of sale (i.e., the intelligent fridge and other smart home services are enabled through this technology). The deployment of RFID technology will be one running example throughout this book.

Studies on RFID technology have shown that customers are concerned that their personal belongings could be read out by RFID readers without their knowledge and consent (Guenther and Spiekermann 2005). The challenge of future retail environments therefore is to give people some perception of control over such invisible data collection practices. Information control would imply that a retailer informs customers of read processes taking place. This information should

be provided in a form that is easily accessible and easy to read. For example, a notification could be sent to visitors' mobile phone. Alternatively, there could be signs at the mall entry and so forth. Such first layer notifications could be linked to more detailed descriptions for all those customers who are really interested to understand the details of data collection and processing. In the scenarios, Roger is such a customer who wants to know what is going on and acts accordingly. The mall in my scenario is transparent as to its data collection and use practices. Besides informing customers about read processes, the mall gives its visitors a true choice over being read out or not. Customers who want to stay anonymous use a special entrance to the Halloville mall. Alternatively, they can use an agent, such as Arthur, to mediate data collection. The retail story I tell hints to the problem of financial affordability though. Roger sees himself forced into the data collection process because he cannot afford to forgo the discounts he receives in return for the data collected. If the mall gave price discounts to anyone, regardless of whether data is read, then visitors would have a true choice to participate in data collection or not, and they would probably feel positively more in control when entering the mall. Fostering behavioral control perceptions over RFID is more difficult than giving entry choices and information. For example, it is hard for a mall operator to prove that RFID read processes do not take place once a customer has opted out. As a result, RFID technology was shown to produce feelings of helplessness in people, which could turn out to be an RFID implementation challenge for retailers in the long term (Guenther and Spiekermann 2005). Recording and visualizing data requests that have been blocked may be of relief to untrusting customers. Chapters 16 to 18 contain a methodology for how we can systematically think about building privacy and perceived control into an RFID-enabled mall like Halloville.

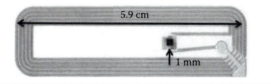

Figure 5.10   Example of an RFID chip. (By Kalinko, CC-BY-SA-3.0.)

5.4 Ethical Challenges in Information Aggregation and Knowledge Creation

Once data has been collected it needs to be aggregated to become information and knowledge. It is in this step of the knowledge management process that the main added value is created for companies and society at large. However, to truly create this value and speak of "knowledge" creation, several challenges need to be overcome. One is that the collected data entering the information processing phase as well as the resulting information product need to be of high quality. The other is that information aggregation and knowledge creation should be as transparent as possible to ensure that what we create is actually true and does not distort reality.

5.4.1 Data Quality

Thomas Jefferson (1743–1826) once wrote: "He who knows nothing is closer to the truth than he whose mind is filled with falsehoods and errors." He reminded us that an important prerequisite for knowledge creation is that the data used is of high quality. Data quality can be characterized as data's *fitness for use* in a respective application context (Wang 1998). This fitness is not always given. "Data quality problems plague every department, in every industry, at every level, and for every type of information … Studies show that knowledge workers waste up to 50% of time hunting for data, identifying and correcting errors, and seeking confirmatory sources for data they do not trust" (Redman 2013, p. 2). This observation must be complemented by the fact that even commercial computer programs rarely come without bugs. Statistical surveys suggest that on average two to three mistakes can be observed for every 1000 lines of code, even in professionally commercialized software products.* Against this background, using digitally produced knowledge must always be regarded with a critical distance, respecting the potential for software mistakes, distorted data sources, or misinterpretations of data sources.

Incidents of misinterpretation of data sources can be reduced if data is well described with the help of metadata (see Box 5.2). Metadata create what Wang (1998) calls "representational information quality": interpretability, ease of understanding, concise and consistent representation (p. 60). In addition to metadata it is advisable to flag information with quality indicators that signal the degree of reliability to its users and provide the contact details of those units that actually produce the data (Redman 2013).

Figure 5.11 shows what data quality criteria there are and what common sources of error exist (Scannapieco, Missier,

and Batini 2005). The accuracy of data can be compromised due to simple mistakes in the syntax and semantics of data entries as well as duplicates. Often, data is not as fresh as it should be. As soon as data is not stable (like a birth date) but time variable (such as an address or age) it needs regular updates. Time-related quality dimensions of data (which could be signaled to users) typically include the currency and timeliness of data. Currency measures how promptly data is updated. Timeliness measures how current the data

---

BOX 5.2 IMPORTANCE OF METADATA AND DATA CREDIBILITY

For data to be meaningful and comprehensive and also in order to challenge its truthfulness, metadata is highly important. Metadata is "data about data"; for instance, units of measure. Metadata allows data analysts to recap the conditions under which the data has been originally collected and to understand the true meaning of the data collected. Only against the background of this knowledge it is possible for data analysts to further combine and analyze the data they use and draw meaningful conclusions from them.

Harvard Business Review[1] reports what can happen if metadata is not specified: NASA lost $125 million with the Mars Climate Orbiter because one group of engineers used English units for distance (feet and miles) while another group used metric units for key operations. Another corporate example illustrates how important metadata is to assess a company's success: Let's take a typical firm in which several words may be used to describe a "customer." One employee talks about a customer as soon as repeated business has been achieved with a client. Another employee at the same firm may talk about a customer once a first contract has been signed. A third colleague may be very bullish and will speak about a customer when there is only a single sales lead. So what is a customer for a company? Metadata rules specify what a company will mean by a customer and IT systems then reflect this uniform understanding upon which the market and sales operations can be monitored and stock market reporting can be done in a meaningful way.

The examples illustrate that it is important that metadata rules are defined and it is equally important that they are documented and communicated within a firm.

NOTE

1. Thomas C. Redman, "Data's Credibility Problem," *Harvard Business Review*, 2013, pp. 2–6.

---

* "Software Bug," *Wikipedia*, accessed January 25, 2015, http://en.wikipedia.org/wiki/Software_bug.

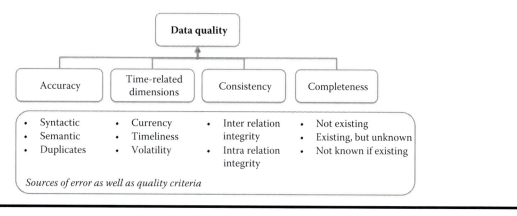

Figure 5.11 Conceptual analysis of data quality.

is relative to a specific task. The latter quality criterion is also relevant for web content (i.e., blogs or news articles), which, unfortunately, often lacks the date of publication.

A common problem in knowledge creation is that the data collected about a phenomenon or about a person is not complete. Let's take the example of an advertising network that has been able to collect the gender of most ad viewers, but for some ad viewers this attribute could not be observed. In this case the knowledge about these respective viewers is not complete. To follow Figure 5.11, the attribute "gender" exists and it is known that it exists, but it is not known to the advertising network. Completeness can be measured by the ratio of known attributes about a phenomenon divided by the total number of attributes.

Finally, data must be consistent. Relational databases often allow for automatically checking the consistency of data sets by looking at their integrity. A distinction is made between intra- and interrelation integrity depending on whether data is part of the same relation or domain (Scannapieco, Missier, and Batini 2005). For example, the original year of a film publication must be before the remake year of a film.

To ensure high levels of data quality, scholars have proposed total data quality management procedures that should be implemented in companies. They advise to equate today's "information manufacturing" to traditional product manufacturing (Wang 1998). This proposal is important, because it allows companies to realize that data assets are highly valuable and deserve bigger care and attention than they sometimes receive today. Figure 5.12 visualizes this thinking.

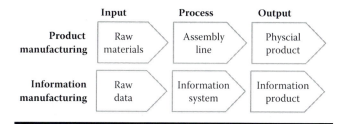

Figure 5.12 Data is considered as a raw material in information manufacturing.

5.4.2 Truth

I have outlined in the introduction to this knowledge chapter that knowledge requires truth. Spence even argues, "information without truth is not strictly speaking information, but either misinformation or disinformation" (2011, p. 264). Yet, as the data quality section above has shown, errors, and misinterpretation of data create risk that what we believe to be knowledge is really misleading. Hence, when working with machines we always need to be cautious as to the extent we can fully trust their output. We also need to be aware that a lot of machine output that appears as perfect knowledge at first sight is really just a spotlight on probabilities. Take the Google search engine's result of a person query: the search results page is not a perfect overview of who the person is that is queried. It is a just a selection of data on a person with a high probability of relevance in a specific life-domain of that person. Unlike the common belief that machines are right and hence more reliable than people, data scientists and experts will agree: Machine output is rarely perfect.

If we seek truth on the basis of the kind of ubiquitous data collection described in the aforementioned scenarios, a number of ethical challenges can arise. Let's go back to the retail scenario where I describe how the mall's robot choice algorithm determined that female child-robots would be best suited to accompany some adults with pedophilic tendency in their shopping trips.

> Another issue is related to the humanoids' looks. They typically resembled people quite realistically and had all kinds of looks and sizes. Some of them looked like teenage girls and boys, and everybody thought that these younger looking robots were used in the mall as peers for kids. But then some men got teenage female robots to be their shopping companions. And a whistleblower found that this robot choice, recommended by Halloville's IT system, was related to the system's knowledge of pedophilic tendencies for some

male customers (because they had visited the teen sex porn categories on porn websites).

Several questions arise out of this scenario: Do we want this kind of knowledge to be created at all? Are we sure that we want to produce knowledge on the basis of all the data we collect? Who should be allowed to establish such truths about us? And who is liable if truth damages our reputation?

As Michael Lynch (2008) said: "Some may want the truth all the time, and many may want it some of the time, but not all will want it all the time."

Some scholars have taken a rather critical view on creating truth arguing that it may paralyze people and impact their productivity: "The question whether truth has value or whether knowledge thereof has a destructive effect is old. Who increases knowledge, increases pain; knowledge paralyzes action; consciousness entangles in fear and disturbs the natural course of lively processes; the reach out for knowledge is the fall from grace" (Jaspers 1973, p. 142). How would an adult feel if he read in the press that the mall assigns child robots to pedophiles and he had a young robot girl walking the mall with him the day before? Or another example: How would you feel if a genetic screen test showed that you may have cancer in the next year or so with 80% probability? What does this knowledge do to you? What does it do to us? Will it change our lives to the better or to the worse? Has knowledge of our own truth the potential to lead to self-fulfilling prophecies?

As a society we have not found final answers to these questions that will be relevant for us in the years to come. Popper took a very positive perspective in this regard. He argued: "It is only through knowledge that we are mentally set free—free from enslavement by false ideas, prejudice and idols. Even though this endeavor of self-education does not exhaust our meaning of life, self-education can decisively contribute to make our life meaningful" (Popper 1974, p. 201). If an algorithm found a pedophile tendency in a person, would it not be good for that person to find out about herself or himself more clearly? In the future work scenario I gave an example for enhanced self-awareness: What if algorithms identified bullying behavior in companies and allowed employees to learn about and delve into their behavior retrospectively? Machines are dispassionate about truth. So they are able to hint toward a version of the truth that may be different from the one that we will sometimes want to create about ourselves or remember in a certain way. Machines will force us into a different perspective on ourselves. I presume that one of the biggest challenges of the future will be how we humans will be able to handle this perspective that some would claim to be our objective truth.

A very important question in this context is how and by whom we are to learn about our own presumed truth. Should everyone be allowed to establish (a presumable) truth about us and tell or not tell us about it? Shouldn't we have a say in who is allowed to know something about us? In the work scenario I outline how knowledge creation about us could be organized: Companies may collect a lot of data about us to provide services or increase security, but in order to be allowed to and be able to analyze the data further or use it for secondary knowledge creation, they could be required to ask for our permission. Such an obligation to request permission could be organized as outlined earlier, with the help of sticky policies (Casassa Mont, Pearson, and Bramhall 2003), dynamic consent mechanisms (Kaye et al. 2015), personal agents (Langheinrich 2003), and metadata architectures (Nguyen et al. 2013). In the work scenario I go even so far as to foresee full data encryption policies that allow personal data only to be used when an individual provides her private key to decrypt her data. Some analysis, such as the company's emotional mood analysis, could be done on encrypted data (Gentry 2009).

Another angle to look at truth and the ethics of knowledge creation is to question whether all data sources should be equally used in machine calculations? Especially in future times where it may be that almost all of our real-world activities will be recorded by some computer system, it is questionable whether we want to process and use all of this data. Some thinkers have proposed an "ethics of ignorance" along the lines of George Pettie (1548–1589) "So long as I know it not, it hurteth me not."* This approach would imply that we simply decide that some data will not be used. We could abstain from collecting this data, we could delete some of it right after collection, give it very little algorithmic weight, or forget (delete) it over time. European data protection law at the moment effectively integrates a similar approach called "data minimization" (that applies however only to personal data). Also according to U.S. legal case history, not using all data is justified on two grounds: one is that some data may simply be too confidential by nature to be collected. The other one is that some data may be too sensitive to be transferred as such a transfer would be equal to blackmailing someone.

Let's transfer these arguments to the pedophile scenario: Collecting, storing, transferring and analyzing porn category choices made on sex-video hubs is technically easy to do. However, we could consider it as simply too confidential to be done and so we—as a society—could decide that knowledge about sexual orientation will simply not be created. We may outlaw it even. A search engine company would then simply and automatically delete all search queries that relate to sex or sex categories right after the query has been made. The problem is though that if we go down this

---

* Nicklas Lundblad, Google "The Ethics of Ignorance," presentation at the Oxford Internet Institute, May 2014.

road, then—as a global society!—we need to agree on what categories of data we do not want to know about and that we really do not want to know about them. Our personal and corporate curiosity will be our own biggest enemy in making this decision wisely. The emblem depicted in Figure 5.13 stemming from Ovid's *Metamorphoses* depicts the case where three women's curiosity led them to open a box of knowledge that they later regret to have opened, after seeing its horrific content.*

A way out of this dilemma (to decide what to ignore) could be to not ignore but instead create more transparency and awareness around what data is collected and used and by whom. Do sex-video hubs actually store and share category choices? If we had such a primary transparency of corporate practices, we would be better prepared to take responsibility for the truth. Potentially we would be able to adjust our behavior according to our counterpart, just as we do it in the offline world. When we learn that some people leak our secrets, we avoid them and don't share information with them anymore.

The search for truth is not only a matter of what data is collected, but also a matter of the kinds of analysis we put the data to. Should algorithms be allowed to put our data to any analysis possible no matter the ends? Often we may be surprised about the results of data analysis. Even having seen the raw data, data scientists often do not expect a specific outcome. Take the example of a German bank that created seven psychological profiles about its customers to better sell to them. Customers were characterized as preservationists, hedonists, adventurers, wallowers, performers, tolerants, or disciplined.† Do we want our identities to be classified like this? And more important, do we want to be systematically treated according to them?

The choice over data analysis remains very much with the personal ethics and virtues of the data analysts today as well as the companies they work with. Company policies can provide guidance as to what kind of data analysis should be allowed and what not. In order for social norms and pressures to play out, it would be good to log and openly publish analysis practices that are being done regularly on personal data. Julie Cohen writes that we should have "protocols for information collection, storage, processing and exchange" (2012, p. 1932). Such transparency would be an important lever to control for personal and company curiosity.

Figure 5.13 Picture from Ovid's *Metamorphoses* where Herse, Pandrosos, and Aglauros are too curious and open a box whose content horrifies them. (With permission from University of Glasgow Library, Special Collections.)

5.4.3 Transparency

In the section on truth in knowledge creation, I outlined why transparency is vital. True knowledge can be created only with reason, scrutiny, and high-quality data. Once knowledge is created, it needs regular monitoring, challenge and incremental improvement, especially as technology advances. For these reasons, transparency has been embraced as an important political target. The EU Commission views transparency as essential for achieving corporate social responsibility, social justice, environmental security, true democracy and well-being (European Commission 2001). In his first memorandum after assuming the U.S. presidency, Barack Obama embraced transparency as a key tool for promoting accountability (The White House 2009). Most multinational companies have embraced the principle of transparency in their codes of conduct (Kaptein 2004). Figure 5.14 gives an overview of the information quality criteria necessary for transparency.

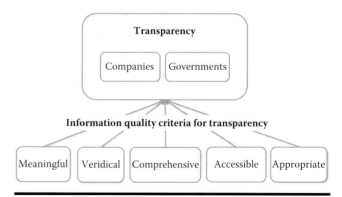

Figure 5.14 Conceptual analysis of the transparency value.

---

* Herse, Pandrosos, and Aglauros, daughters of Cecrops, were given the task of guarding a box carrying the infant Erichthonius (which was the result of an attempted rape of Athena by Hephaestus). The girls opened the box and were horrified to see a half-man, half-snake. See Ovid's *Metamorphoses*, 2.252ff. http://www.emblems.arts.gla.ac.uk/french/emblem.php?id=FANa075, accessed June 4, 2014.

† Verkaufshilfe: Sparkasse sortiert Kunden in Psycho-Kategorien. Der Spiegel. 4.11.2010. URL: http://www.spiegel.de/wirtschaft/unternehmen/verkaufshilfe-sparkasse-sortiert-kunden-in-psycho-kategorien-a-727133.html.

Note that the word *transparent* has an almost diabolical ambiguity. In modern times we mostly associate transparency with *visibility*. However, transparency originally meant invisibility. We look through something without recognizing it, like a window glass. This ambiguity in the word reflects the reality of many transparency initiatives: Companies and governments can provide us with a lot of information, but that information might not give insight into what is really happening within corporate walls. Take the case of Enron, a widely heralded U.S. energy group that went bankrupt when false accounting practices were uncovered. The company was regularly audited, so it did provide apparent transparency into its business practices. But in truth, it used opaque instruments to obscure the real (and fraudulent) basis of its profits.

To create meaningful transparency, "the information disclosed … is supposed to consist of meaningful, veridical, comprehensive, accessible and useful data. This is not a mere litany of properties" (Turilli and Floridi 2009, p. 108). *Meaningful data* conveys a message that has significance for a recipient in a particular context. This transparency through meaningfulness is complemented by the quality criterion of *usefulness*. Unfortunately, we find a lot of "Open Data" today that is simply not useful because it lacks the context or explanation needed to understand it. In contrast, a positive example for providing meaningful and useful information around data processing is Google's information about how it calculates the price for its advertisements.* Even complex formulas around the pricing mechanism are explained.

Comprehensiveness means that the information is easy to read and understand. For human readers, clarity increases when information has been properly edited. But a lot of information that is made available today is not edited nor structured for a nonexpert reader. It is possible that much of the data published now is not even created for humans anymore but rather for machines. But even for machines, the data is incomprehensible unless sufficient metadata is provided for the machine to interpret the data in a reliable way (see Box 5.2). Even though machines are advancing in their ability to extract meaning from unstructured data, it is unclear whether this ability gets us far enough. To what extent will machines be able to comprehend unstructured data that is not annotated? What kind of meaning will they extract? And what are the limits of this processing effort? For ethical knowledge creation, transparency requires us to understand how knowledge was created. What data and what assumptions were used to build a version of the truth we can reasonably believe in?

To reach transparency on knowledge creation, another major precondition is *accessibility*. In fact, accessibility has been a prime focus of transparency advocates. So far, access to company information is often granted only with special permission. Theoretically, people in Europe have the legal right to access the personal data that a company holds about them and should be informed about the logic of automated decisions made about them (see "Right to Access" Article 12 of European Parliament and the Council of Europe 1995). Some similar rights have been legally recognized in the United States.[†] But in truth, the right to access is currently hard to exercise. Most of the information is difficult to obtain and unstructured. Companies do not like to reveal how they process and aggregate data and what knowledge they create. They consider knowledge creation of any kind as a kind of trade secret. It took Max Schrems, an Austrian law student, 2 months and 22 e-mails to get at least a partial insight into his Facebook data. When he finally got access, he received 1222 pages in print. Theoretically speaking, Facebook did not provide for transparency here, because the piles of paper it sent were hardly comprehensible. Since then, the company has changed its operations and worked toward more transparency, providing much more information to customers on its website about what data the company holds and processes.

In fact, accessibility is gradually improving, at least for the types of data where companies have a legal obligation to publish data handling practices. For example, the U.S. Gramm-Leach-Bliley Act requires financial institutions to publish information about their privacy practices and gives guidance on the structure this published information should have. Academics were hence able to analyze the data and make meaningful bank comparisons (Cranor et al. 2013). The EU and some U.S. states legally require companies to publish notifications about data breaches. Some companies in the EU might soon be required to publish the results of their privacy impact assessments.

Of course, only standardized and *machine-readable access* to such company information will create the kind of transparency we need. Standardized and machine-readable information about corporate data-handling practices would allow for remote analysis and comparison of what companies do with (at least the personal) data they collect. Today, early precursors of this practice are quality seals, which signal the trustworthiness of processing after an audit of company practices is made by the seal provider (for example, TRUSTe or EuroPriSe). But as the machine age advances, machine-readable information on more granular practices might become more common. The pressure to publish information on internal practices may incentivize companies to abstain from unethical data analysis that would otherwise remain obscured.

---

* "What Is Google AdWords? How the AdWords Auction Works," http://www.wordstream.com/articles/what-is-google-adwords.

[†] For example, the U.S. Privacy Act provides people the right to access their records. So do the Cable Communications Policy Act, the Fair Credit Reporting Act, and the Children's Online Privacy Protection Act.

Economically, transparency in the form of standardized and machine-readable accessibility should benefit the economy. It helps to reduce information asymmetries in the market between companies and consumers. Customers get a clearer view of whom they are dealing with and can make informed choices on whom to entrust with their personal data. Competition between companies is fostered, as a structured analysis of competitive practices becomes feasible. IT service companies can compete on the basis of ethical conduct.

Finally, accessibility, meaningfulness, and usefulness are not enough to create full transparency. As Enron's fall illustrated, information must also be *veridical*, meaning truthful. But truth is not a simple concept. As discussed in the section on truth, it is easy to avoid telling lies or half-truths. The more sophisticated dimension of truth is *appropriateness*, which involves selecting those essential pieces of available information that best reflect reality. "Credibility does not arise from details, but from appropriateness," writes Armando Menéndez-Viso (2009). Menéndez-Viso refers to Descartes (1596–1650), who once described how a good portrait does not necessarily reflect every detail of a person, but the main lines: "We must observe that in no case does an image have to resemble the object it represents in all respects, for otherwise there would be no distinction between the object and its image. It is enough that the image resemble its objects in a few respects. Indeed the perfection of an image depends on it not resembling its objects as much as it might" (Descartes, as cited in Menéndez-Viso 2009, p. 158). A painter must therefore always judge what is appropriate to draw. Depending on his sympathy for his subject, he might be tempted to embellish him or her. Or, in contrast, he might be tempted to create a caricature. A judgment of the appropriate information to display is a tightrope walk.

If machines are put in charge of transparency, they will not be able to evade this balancing act either. A machine must judge what is appropriate to show or to output so that users can build a truthful perception of that output. This matter is illustrated in a recent judgment made by the European Court of Justice, which ordered the search engine Google to delete a link to outdated information about a Spanish man that portrayed him in a bad light. Google helped to display material that was outdated and hence inappropriate for display. In its ruling, the European Court of Justice has recognized that some personal information embeds a "right to be forgotten." Information that is outdated can be inappropriate to display.

5.5 Ethical Challenges for Accessing Knowledge

One of the most important levers for the flourishing of people and society is our access to the information and knowledge exploding around us. Philosophers consider the access to information as a *primary good* in modern societies: "Access to information is relevant to every conceivable plan of life" (van den Hoven and Rooksby 2008, p. 383). Through the Internet we already feel the benefits of efficient access. Many of us can download scientific articles at the click of a button where in former times we had to go to physical libraries that often did not even have what we were looking for. We have free encyclopedias such as Wikipedia at our fingertips. We can instantly and easily search for what we do not know and want to learn about. Most important: A search has become so easy that we mostly find what we are looking for (which has not necessarily been the case in earlier days). In short: Many of us have access to a lot of the world's information and knowledge and the way we are accessing it is hugely more efficient than it used to be in analog times.

As the stories in Chapter 3 showed, this current situation is just the beginning. In the future we may have technologies in the form of virtual bodies, such as agent Arthur or the Wise Figure, that can explain to us what we want to learn about. They may be so smart that they can adapt the way they communicate with us to our level of knowledge (just as online games bring us to the next higher level of performance depending on how we succeed at lower level tasks). Accessing knowledge in such a playful and instantaneous way bears huge potential for the development of cognitive skills in our future societies. Yet, I also mention in the educational story that Roger wonders whether his financial means suffice to afford the holographic private home teacher he wanted to hire for Sophia and himself. I hint to the fact that access to knowledge may not come for free and be affordable for everyone.

In fact, access to services that transmit valuable information and knowledge is rarely for free. Even if the "free" culture of the Internet suggests this on the surface, access has been and is conditional in most cases. *Conditional* access means that—even if we have broadband Internet connectivity—we somehow pay directly or indirectly for service use. We do so either by paying money for information. For example, many news portals now charge for articles. Alternatively, many Internet services condition their service use on the right to monetize our personal data traces. Personal data is a new currency with which we indirectly pay for access. Meglena Kuneva, the EU's former Consumer Commissioner, expressed this economic reality when she said: "Personal data is the new oil of the Internet and the new currency of the digital world."

In contrast, *unconditional* access to services would imply that services are truly free of cost. Besides the monetary aspect, this would indicate that we can access the information and knowledge services without paying for them with our personal data and that we can access them publicly (from home or from a public library) without needing to be a member of a specific group, such as a university (educational institution), a company, or an association (e.g., a standardization organization). One of the most valuable sources of knowledge we have today is our scientific knowledge that is published in academic journals, conference proceedings, or standardization documents. These journals, proceedings, and standards are, for the most part, not publicly accessible. Independent innovators, small companies, consultants, or interested individuals often fail to meet the condition of access to this knowledge, because they either cannot afford the amounts requested or, they are not part of the privileged groups with access. Figure 5.15 summarizes how access to knowledge can be classified today as conditional or unconditional.

Distinguishing conditional versus unconditional access is important for service design, marketing, and public policy development, because it untangles the true complexity of access to information and knowledge. Conditional knowledge access has consequences for the digital divide in societies as well as for innovation. Hereafter I first want to concentrate on the effects of conditional access on the digital divide. In the following, I reflect on the effects of conditional access for personal and social development. In Section 11.6, I describe how conditional access to software, patents, and content can hamper creativity and innovation.

5.5.1 Conditional Access to Knowledge and the Digital Divide

The "Digital Divide" is defined as the "stratification in the access and use of the Internet" (Ragnedda and Muschert 2013, p. 1). It is of concern to most governments that want to give citizens equal opportunities to participate in the digital service world. Heavy investments have been made to continuously improve the infrastructure coverage of countries in terms of Internet connectivity. As a result, the Digital Divide has been significantly reduced in many developed countries, at least in terms of connectivity to the Internet. Yet, there still is an important Digital Divide in the way people use their access to the Internet. Many people officially own an Internet connection, but do not use it. Many use it, but only for gaming and entertainment and not for accessing knowledge. The reasons for this "use divide" are often attributed to individual factors of users, such as their age, sex, ethnicity, educational level, and employment. Van Dijk (2013) critically points out that we tend to "(simply blame) inequality of access on attributes of individuals" (p. 29). One way to explain the Digital Divide in how people use the Internet is to say that those who are unemployed or uneducated are simply lazy. Due to a lack of motivation, it is not surprising that they do not use the Internet or only use it for chatting and gaming (not effectively participating in the knowledge part of the web). But in truth, the divide is also a matter of people's position in society and whether this position allows them to meet the conditions of access to knowledge. A lot of valuable knowledge on the web is really conditioned on being a member of a privileged group, such as a university or a company that holds access rights to the richest parts of digital content. People who are unemployed often simply do not have access to knowledge through their employer. Even if they wanted to access the Internet, they could not. For example, a 2014 study showed that only 27 million (24%) of the 114 million English-language scholarly

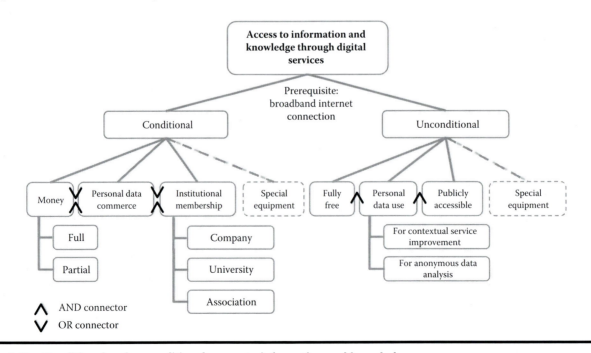

Figure 5.15 Conditional and unconditional access to information and knowledge.

documents on the web are available free of charge (Khabsa and Giles 2014).

The scenarios in Chapter 3 describe how gamers of Playing The World exclusively have access to the saint figure that can answer all questions. In addition, only those players who can afford an extra €5 per month can port their personal agent to the real world where it continues to give them access to information (Sophia's agent Arthur). This sounds like a small amount. Note though that the price tags could be much higher.

Another form of conditional access to information is described in Chapter 3 in how the mall functions in the retail scenario. Here, the conditon of access is to be willing to share personal data:

> Going through the mall's main gate gives the mall implied consent to read out his and his kids' data and send them tailored advertising and information. "Reading out" involves scanning clothes for RFID tags, recording movements, and points of interest. Robots and on-shelf cameras analyze facial expressions and emotions. Video surveillance camera systems that embed security analysis screen their skin type and movement patterns. In return, Roger gets 3% off all his purchases in the mall plus free parking. The only exception is Sophia, who is able to use Arthur to reliably block her personal information exposure and provide her with neutrally tailored product information. …
>
> Rich people who are on a truly anonymous scheme can use a separate smaller mall entrance on the east side of Halloville that does not track any data. People's personal agents (a kind of app running on their mobile phones; function-wise similar to Arthur) block RFID read-outs and send their owners' data usage policies to the mall infrastructure, indicating that video and voice data must be deleted. However, when people go through that entrance, they do not receive the 3% discount and have to pay for parking, a luxury that Roger cannot afford. Personal robots are also available only for an extra charge and base their recommendations on the personal agents of those richer folks.

The description of the future mall outlines how people's access to unbiased information and a protection of their privacy may cost them both money and effort (to use a separate mall entrance). This is problematic from an ethical perspective. As Edward Spence outlines, "The epistemology of information … requires the equal distribution of the informational goods to all citizens" (2011, p. 264). Currently such an ethical standard is not the reality though. Instead, a Digital Divide has started to widen between those who can pay for information access and consciously chose to protect their personal data versus those who need to trade personal data and live in "filter bubbles."

5.5.2 Objectivity versus Filter Bubbles

The term *filter bubble* was coined by Eli Pariser (2012). It refers to the fact that many information services, such as the social network service Facebook or the search engine Google proactively filter the information that is provided to us. More precisely, they personalize the information that is displayed to us. Google has been reported to use 57 different variables to decide what search results are shown (Halpern 2011). This means that every user gets a different and personalized answer to the same search query term. Large data brokers and advertising networks have accumulated databases that hold and track our daily behavior online and represent each of us with at least 2000 personal attributes. This information is then used to select individually relevant ads for us as we surf the web.

Although it may be OK to receive personalized ads on presumed purchase preferences, the question is how far such filtering should go. Is it OK to also receive information online that is prefiltered according to political or religious opinions for instance? According to Pariser, Facebook characterizes its users on multiple dimensions, including political attitudes. When a Facebook user is observed by the platform to hold a conservative political attitude, she or he will from thereon receive mainly the conservative news from her network (Pariser 2011).

In her article "Mind Control & the Internet" Sue Halpern (2011) analyzes why this way of serving filtered information can be problematic for political and democratic stability. She describes the example how Republicans and Democrats in the United States have gained very different perceptions about climate change between 2001 and 2010. While in 2001 49% of Republicans and 60% of Democrats agreed that the planet was warming, this relatively similar perception of reality fell apart in the following years. Exposed to probably different news feeds only 29% of Republicans believed in global warming by 2010. In contrast, Democrats' awareness of climate change increased in the same time period. By 2010 70% of Democrats believed in global warming. "When ideology drives the dissemination of information, knowledge is compromised," Halpern writes. In other words, when people are enveloped in their personal ideologies by personalized digital media, then their perception of reality is narrowed to what they have always tended to believe anyways. They are not challenged any more. They are captured in a narcissistic spiral of "information cocoons" that continuously reinforce themselves.

To avoid such developments, scholars have pointed to the importance of objectivity in the way information and knowledge is provided online (Tavani 2012). The goal of objectivity is to capture reality truthfully and free from bias and emotions. In its definition of "objective evidence" Anglo-American law gives a hint to two characteristics of objectivity: Information can be used in court as objective evidence only when it is relevant and genuine. Furthermore we can distinguish between (1) objectivity in the way information is selected and (2) objectivity in the way information is distributed. Figure 5.16 summarizes this conceptualization of objectivity. That said, there are various ways to define objectivity. Box 5.3 gives a short historical introduction into how perspectives have varied over time on what objectivity actually is.

We can apply the conceptualization of objectivity as depicted in Figure 5.16 to online information and knowledge access. Let's take search engines again as an example. Tavani (2012) points out that two traditional criteria to rank search results are relatively objective in the way they select information. According to these criteria a site is more *relevant* the more other pages link to it and the more visits or click-through-rates it has. In addition, *genuineness* of the selection is provided when search engine providers make an effort to include (index) a maximum number of available sites regardless of their content.

The distribution of search results is objective in terms of genuineness if we as users can all access the same sites scrolling down the list of search results and see them unfiltered and in the same order of relevance as anyone else. To be genuine, the search results would hence not be personalized to each user. They would just be ordered according to objective (and ideally transparent) relevancy criteria. The decision of relevance of the distributed content would then not be predetermined by the machine but taken by the users themselves. We as users would need to train our judgment in the way described in Box 5.3. As in the old public library days we would be forced to scroll through long alleys full of search results to find what is relevant for us. In doing so, we

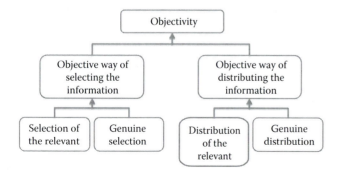

Figure 5.16 Conceptual analysis of objectivity value.

BOX 5.3   ON OBJECTIVITY AND ITS RELATION TO MODERN DATA-DRIVEN BUSINESS

In their book *Objectivity*, Lorraine Daston and Peter Galison distinguish between three kinds of representation of objectivity in science that have dominated over the course of history: truth-to-nature, mechanical objectivity, and trained judgment.[1] These different ways to "objectively" describe reality can be best understood when looking at the three different pictures shown in Figures 5.17 to 5.19.

Originally scientists used illustrations to describe nature. They aimed to capture the underlying ideal truth in nature (Figure 5.17). This meant that they did not work out all the specificities that each specimen could have, but they were more interested in the perfect, fundamental recurring, and essential characteristics of nature's reality. This scientific period was followed by a time of thinking in terms of "mechanical objectivity" (Figure 5.18). Here scientists were trying to not let their own imagination influence their perception of reality but to stay as close as possible to what they could see. At the same time, of course, both their selection and perspective did influence objectivity. Mechanical objectivity found its zenith in the late nineteenth century, supported through the advent of photography. Figure 5.18 shows how all the salient details, formations, and peculiarities of a snowflake are captured.

Figure 5.17 *Arnica Montana*. (By Dr. Thomé, 1885.)

Figure 5.18   Snow crystal. (By Wilson Bentley, 2006.)

A radical break with this kind of understanding of objectivity can be observed in Figure 5.19 that implies a "trained judgment" form of objectivity. Figure 5.19 is a heat map of a wolf that has been altered with colors highlighting specific areas of the image for better comprehension, pulling out the most important aspects of the image. In order to create this image, decisions have to be made on what color shades to apply and where. This twentieth century handling of objectivity places much more responsibility on both the scientist who creates the image and the audience that receives it. The scientist "smoothens" the data. He "cleans" it, as modern data scientists would say (i.e., he applies color with meaning to the wolf). And then he leaves the audience with the challenge of making sense of this new kind of reality.

This twentieth century invitation of scientists to thus construct objectivity is equally reflected in

Figure 5.19   Thermal imagery of wolves. (From U.S. Geological Survey, 2010.)

Figure 5.20   *Guitar and Clarinet* by Juan Gris, 1920.

the modern art of this time as well as in contemporary management. The artistic works of the Cubistic period for example (see Juan Gris' *Guitar and Clarinet* in Figure 5.20) expect a considerable maturity and thinking ability from their spectators who have to reconstruct the image and attribute meaning to it. Any business analyst or modern manager that is confronted with piles of data, graphics, and statistics on his business will be able to recognize that looking at this data input produces a similar reaction than looking at a cubist painting: In both situations the recipient of the information is asked to construct reality and judge on what he deems most relevant.

The twentieth century way of creating objectivity can be regarded as an achievement, because people are challenged to create it themselves. Nobody provides them with an ideal or mechanical truth. Instead they have to train their own judgment, a process by which they "mature" (as Kant would say).

NOTE

1. Lorraine Daston and Peter Galison, *Objectivity* (New York: Zone Books, 2007).

would find a lot of stuff we need and a lot of stuff we do not need and thereby explore and learn. This does not mean that we need to go back to Stone Age library practices though. Search engines could offer us many filters that help us to order and classify search results and play around ourselves with what we get and what we are looking for. The distribution of the relevant could mean that the machine helps the user to identify the relevant and to train his or her judgment.

Teaching us "trained judgment" is not the way the web works so far. To speak with the terminology of objectivity illustrated in Box 5.3, today's web services throw us back into nineteenth century's "mechanical objectivity," because they mechanistically personalize digitally available content from a certain predetermined perspective without leaving us the possibility to engage in trained judgment on what is relevant and what is not.

The reason why we should use tools to determine for ourselves what is important and what is not has a lot to do with flourishing. Scholars like Aristotle, Wiener, and Terrell Bynum all agree: the purpose of human life has a lot to do with information processing. People need to engage in a diversity of information processing, organizing, remembering, inferring, deciding, planning, and acting in activities. Even if our machines were able to do all of this for us, the question is to what extent they should. The right balance must be found that people can still do the final selections. This is important for them to grow and be happy but also because they can carry responsibility for their information use.

5.6 Ethical Uses of Information and Knowledge, a Privacy Issue

A large part of the information and knowledge we gain in the future through our IT systems will somehow be based on data about people. Video cameras, drones, enhanced video and AR glasses, and so forth monitor individual and social interactions in public space. Sensors on our body or integrated into our homes as well as the public infrastructure will potentially measure our movements, electricity consumption, communications, noise levels, and so on. As soon as our IT systems directly or indirectly monitor us in this way and subsequently use the data for more than the initial collection purpose, then privacy issues emerge.

In 2014, Microsoft published a study together with the World Economic Forum titled "Rethinking Personal Data: Trust and Context in User-Centred Data Ecosystems" (World Economic Forum 2014). In this study the company investigated what constitutes the acceptable use of personal data from the individual's perspective. Almost 10,000 people were interviewed in eight countries from Europe, Asia, and the Americas. They indicate that the way data is collected from people and the way data is used influence acceptability. The collection method investigated in the study referred to various levels of control people could have over the collection process. This control turned out to be the most important factor for respondents' service acceptance (31%–34% of overall data-use acceptability is driven by the control construct). The finding reemphasizes the importance of ethical

data collection practices, the necessity to give people choices, ask them for consent, and build perceptions of control (compare Figure 5.9).

The second most important driver of acceptability is how the data is then used. "When data is actively collected, users prefer scenarios where the use of the data is consistent with what they originally agreed" (World Economic Forum 2014, p. 8). Purposes of data use can be explicitly agreed upon when data is initially collected from users or by giving them preference options that can be revised dynamically at any time (see, for example, today's Facebook privacy settings). However, a lot of the information we exchange in our everyday communication is based on implicit agreements on data sharing. We typically expect that the information we share in a specific role and context is treated according to the ethical information norms of the respective communication context. For example, in the role of being a patient we share details of our health with a doctor. Implicitly we expect the doctor to respect the information norm to keep our health information confidential. Helen Nissenbaum (2004) has called this kind of implicit agreement a respect for the "contextual integrity" of information use.

5.6.1 Contextual Integrity

Contextual integrity recognizes that societies have developed norms of information flow and use. Two types of norms are particularly important: The first set of norms regulates the "appropriateness" of information flows. Appropriateness means that within a given context the type and nature of information exchanged may be allowed, expected, or even demanded. It is allowed, expected, and demanded to share medical information with one's doctor for example. The second set of norms relates to the distribution of data, the "movement, or transfer of information from one party to another or others" (Nissenbaum 2004, p. 122). For example, when a medical professor working at a university clinic collects information about the personal health status of his patients, then we probably accept that this professor (who is not a regular doctor) uses the information not only to have a history of our health development but to potentially compare our case to other cases. These are expected and demanded uses. We are likely to also accept that as a university professor he wants to integrate patient cases into his research. The university context of the professor allows for this kind of data use. What we would not find appropriate is that patient data is also distributed outside of the university context, that is, the hospital sells the health records to insurance companies or international data brokers. Such a use of data would probably be judged as inappropriate and as breaching norms of distribution. Figure 5.21 visualizes contextual integrity and its potential breaches.

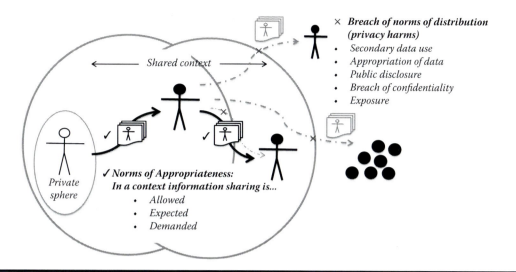

Figure 5.21 Contextual integrity and data sharing norms.

5.6.2 Privacy Harms

The respect of contextual integrity in data and information use has found widespread acceptance by governments and industry. The World Economic Forum (2012) has pointed to the need for context-aware usage of data as a key outcome of global dialogs it had conducted as part of its "Rethinking Personal Data Initiative" (Nguyen et al. 2013). Equally, the U.S. Federal Trade Commission and the Whitehouse have embraced the concept (The White House 2013). In his development of a "taxonomy of privacy" Daniel Solove (2006) discusses privacy harms resulting from breaches of contextual integrity. These are unforeseen and unwanted secondary uses of data, appropriation of personal data, public disclosure, breach of confidentiality, and exposure.

Secondary data use involves using our information in ways that we do not expect and might not find desirable. Sex sites on the web sharing category preference data with third parties is an example for such an unwanted secondary data use that is at the same time a breach of privacy. When a transmission of personal data furthermore involves monetary transactions that we ourselves do not benefit from, then we can even speak of appropriation. *Appropriation* means that our data is used to serve the aims and (potentially monetary) interests of others. Solove calls this "exploitation" (2006, p. 491). In the scenarios in Chapter 3 I give the example of United Games Corp. selling its VR customers' gaming data to third parties. The data leaves the gaming context in which it was collected and United Games benefits from the sale without its customers sharing in the profit. From a privacy perspective the data sale may not necessarily be harmful. Yet, contextual integrity is breached when the data is used.

Public disclosure of private matters means that data is made public, which is not of legitimate concern to the public and which is at the same time offensive to a reasonable

data subject. In an example from the scenarios, "Jeremy had filmed one of his teachers with his new AR glasses when she made a mistake in front of the class, and he had then published this mistake on YouTube." Public disclosure focuses on the content of a message being disclosed; in this case on the teacher making a mistake. The harm caused by public disclosure typically involves damage to the reputation of an individual. This damage is caused by the dissemination beyond context boundaries and to a larger group. In the case of Jeremy, the teacher's mistake should have stayed as information within the boundaries of the class. Solove (2006) notes that public disclosures bear risks for people's long-term reputation. Employers may base decisions on information they learn from some other context. With easily accessible digital information people can become "prisoners of (their) recorded past" (Solove 2006, p. 531).

Unlike the tort of public disclosure, the tort of *breach of confidentiality* does not require that a disclosure be "highly offensive." A breach of confidentiality occurs not only between friends and acquaintances, but also in fiduciary relationships, such as with doctors, employers, bankers, or other professionals with whom we engage. In the future work scenario in Chapter 3 employees hold private keys to their transaction data. Still the human resources department has used a "cut-ties algorithm" to suspect Carly of wanting to leave the company. Obviously there must have been a breach of confidentiality by the employer, seen that HR could apply the algorithm to her data without her consent. Breaches of confidentiality violate the trust in a relationship, because information that should be kept between parties is passed on.

Finally, *exposure* is one of the strongest forms of breach of contextual integrity, because there is certain information about us that we want to keep in our immediate private sphere, such as certain physical and emotional attributes (denoted in Figure 5.21 by the circle around the data subject).

These are attributes that we perceive as deeply primordial, and their exposure would create embarrassment and humiliation. "Grief, suffering, trauma, injury, nudity, sex, urination, and defecation all involve primal aspects of our lives—ones that are physical, instinctual, and necessary. We have been socialized into concealing these activities" (Solove 2006, p. 533). If such information about us is exposed to others, then it rarely reveals any significant new information that can be used in an assessment of our character or personality. Yet, the exposure creates injury because we have developed social practices to conceal aspects of life that we find animal-like or disgusting. "Exposure strips people of their dignity," writes Solove (2006, p. 535). Reports in the press of teenage girls stripping on Skype for their boyfriends, who then publish these private videos via e-mail or Facebook are examples for the tort of exposure.

5.6.3 Computer Bias and Fairness

I have outlined how objectivity is important for us in accessing knowledge. A special form of lack of objectivity that undermines our dignity, honor, and potentially self-esteem is when machines treat us with bias. In line with Batya Friedman and Helen Nissenbaum (1996) I use the term *bias* to refer to "computer systems that systematically and unfairly discriminate against certain individuals or groups of individuals in favor of others" (p. 332). Discrimination means that due to a machine judgment we are denied certain opportunities or goods, or confronted with some undesirable outcome. Information about us is used against us. For example, when a machine decides that we are not credit worthy and as a result we are denied a loan or confronted with a prohibitively high interest rate. Alternatively, the machine may decide that we are rich enough to pay for higher priced flight tickets. Certainly such forms of discrimination are undesirable by the people who are impacted. Figure 5.22 summarizes the concept of machine bias.

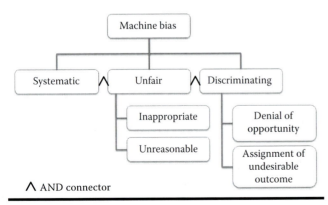

∧ AND connector

Figure 5.22 Conceptual analysis of machine bias according to Friedman and Nissenbaum (1996).

Yet, let's think about the flight ticket example again. Couldn't it be considered fair that rich people pay higher flight ticket prices than poor people? Don't we have some forms of price discrimination all along that distinguish between rich and poor and that are considered fair? Senior tariffs for public transport or free entry for kids are examples. Unfairness perceptions arise according to Friedman and Nissenbaum (1996) when the behavior of machines toward us is "inappropriate" or "unreasonable." They specify the loan example described earlier: If a person is denied a loan because she has continuously failed to pay her bills in the past, then it seems appropriate and reasonable to deny her the loan. There is no bias. Note that bias is only created when the access is "systematically" denied. If a loan is not granted once because a person has recently not paid her bills, then this is not a bias. Only if from now on the person systematically does not get a loan anymore (even though she does start paying her bills again), then this is a bias.

What is fair information use as opposed to biased information use? Synthesizing a number of influential fairness theories I broadly want to distinguish between opportunity-based fairness perceptions and equality-related fairness perceptions. By "opportunity" I mean a person's possibility to influence the outcome of events. By equality-related fairness perception I embrace the fact that people compare themselves to others and build fairness perceptions when they are treated similar to those whom they compare themselves.

Opportunity-based fairness can be built by companies through procedural justices and desert-based distributive justice. Procedural justice is created through (1) transparency of the processes (procedure) by which information and knowledge about us is used and (2) an opportunity to influence this process (Lind and Tyler 1988). Take the example of Jeremy who suspects that he does not get into university because of his outdoor times. Denying access to a student on such inscrutable ground would not be perceived as fair or procedurally just. In contrast, if it was publicized in due time that only those students get access to a certain university if they have spent sufficient time outdoors, then applicants could at least adjust their behavior in advance and have the opportunity to get accepted (even though the discrimination might still be unfair). That said, a challenge for future societies will be that Big Data analysis will suggest all kinds of beneficial and detrimental behaviors that we humans should try to live up to or avoid. Meeting all the requirements that follow out of such analyses may be procedurally just and fair (such as a university requiring pupils to have been outdoors). Yet, the number of requirements machines will figure out to be useful may overwhelm us humans. Procedural justice may have the high price of people being enslaved by their "information CVs" if people do not get the opportunity to influence the decision-making process about them.

Another set of theories related to fairness perception is dealing with "distributive justice." Distributive justice is a philosophical concept that has informed political thinking of how economic benefits and burdens should be distributed in a society. Strict egalitarians call for the equal allocation of material goods and services to all members of a society. On such philosophical grounds, machines could have a lot of information and knowledge about people, but the use of this knowledge would not be allowed to lead to any differential treatment or allocation of goods. Let's say a bank would know about the different degrees of credit worthiness of people. According to egalitarian distributive justice, the bank would not be allowed to adjust loan terms accordingly.

In his *Theory of Justice* (1971), John Rawls introduced a slightly different theory of distributive justice called the "difference principle" (Rawls 1971). The difference principle corresponds more to what I call *equality-related fairness perception*. Rawls would argue that differential unequal treatment can be fair, but only if it leads to the benefit of the least advantaged in a society. So, let's say, a bank knew that some of its customers are rich and others are poor, and would then use this knowledge to make the rich pay more for a loan than those who are poor and redistribute the gains to also give poor people a credit.

Finally, distributive justice has also been based on merit (Lamont 1994), which is more of an opportunity-based fairness approach in the sense that people earn access to resources; they deserve them.

Whereas philosophers' reflection on justice is rooted in political thinking about how to optimally distribute goods in societies, psychologists have studied the concrete construction of fairness perceptions in people as a response to the environment. The procedural justice described earlier was shown to psychologically influence fairness perceptions in various contexts (see, for example, Cox 2001 and Campbell 1999). Equally, equity theory (Adams 1963) outlines how people build up fairness perceptions. Equity theory posits that individuals who are similar to one another gauge fairness (or equity) of an exchange by comparing the ratios of their contributions and returns to that of peers in their direct reference group. Let's take again Jeremy's denial of access to university. Equity theory posits that Jeremy will compare himself to other pupils in his school who he feels similar to. If they have worked similarly hard in school and are of similar intelligence, then Jeremy would find it unfair that they get access to university while he does not. Perceived inequity leads to social tension that people strive to reduce. Such a reduction of tension can take place by changing one's reference group. Instead of comparing

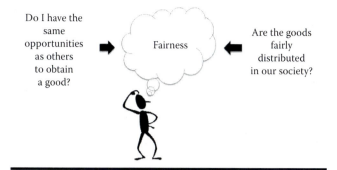

Figure 5.23 Broad overview of approaches to fairness judgments.

Jeremy to other students of similar intelligence, Stanford Online could inform him that the reason for nonadmittance was his behavior in school and his track record of breaching his teacher's privacy. They could let him know that all pupils with similar behaviors are not admitted to Stanford Online. Jeremy could then compare himself to those pupils who committed similar torts. Seeing this comparison, Jeremy would then not see his dismissal as unfair any more. This example shows that following equity theory online companies are well advised to communicate to customers their reference group. Figure 5.23 gives an overview of two approaches to fairness.

5.7 Summing Up: Ethical Knowledge Management

Knowledge is an intrinsic value, a recognized primary good to societies across the world. Esther Dyson (1994) wrote: "Cyberspace is the land of knowledge, and the exploration of that land can be a civilization's truest highest calling." For titling this chapter I have used the term "ethical" knowledge though. Ethical knowledge is what we create if we embrace and respect the intrinsic value characteristics outlined in this chapter.

At the basis of knowledge is the data and observations about the world and about ourselves. Collecting this data without the control and informed consent of the people is not advisable. Not only is consenting a legal necessity, consent and control also ensure the long-term availability of data. If people lose trust in the data collection infrastructure and feel excluded and out of control, it is likely that they quit. Psychologists know that humans avoid environments in which they are out of control (Mehrabian and Russell 1974). If they are forced to stay, they physically suffer (Langer 1983). Both of these reactions are certainly not an aim of the knowledge society we want to build.

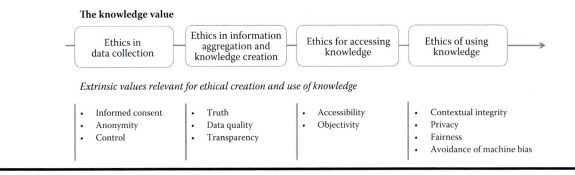

Figure 5.24 Summary of ethical knowledge creation and values supporting it.

Information aggregation and knowledge creation must withstand the expectations of truth. Without truth there is no knowledge. And truth at all times needed discourse and challenge. If we do not allow for transparency in our knowledge creation processes and have our information aggregations and conclusions regularly challenged, then we are likely to hamper progress. We will create suboptimal half-truths on the basis of which neither science nor management nor economics can operate efficiently in the long term. Current data quality problems in companies are a warning precursor of such developments.

When it comes to knowledge accessibility, an ethical vision of society as a free-minded and mature community of individuals implies that as many people as possible should be granted the possibility to learn from and access created knowledge and information. Learning about the world should however not be predetermined by simplistic filter bubbles. Human beings have a high capacity for filtering reality themselves and according to their own needs. Machines should respect this human capacity and give people multiple filter technologies that allow them to find the objective truth for themselves.

Finally, knowledge is a treasure and an asset if it is produced ethically. Societies should treat it as such as they use it. Using knowledge against the people, breaching their privacy, undermining fairness, and establishing ubiquitous machine bias will undoubtedly be a short-term strategy on which a knowledge society cannot flourish.

Figure 5.24 summarizes all those extrinsic values that are relevant along the process of managing data, information, and knowledge.

A short disclaimer is necessary at this point: In this chapter I have exclusively talked about knowledge that we gain through the use of machines. Thereby I have understated that a great extent of knowledge is actually gained without the help of machines, and should be so in the future. Maslow once wrote: "Science is only one means of access to knowledge of natural, social and psychological reality. The artist,

the philosopher, the literary humanist, or for that matter, the ditch digger, can also be the discoverer of truth, and should be encouraged as much as the scientist" (1970, p.8).

EXERCISES

1. From the scenario on the future of university education in Chapter 3, identify scenes where the ethical use of knowledge is at stake. Align these incidents with Figure 5.24. What extrinsic values are impacted? Explain your reasoning.

2. Identify relevant values at stake if universities select their students based on Big Data analysis. Then focus on those values you identify and which correspond to the values described in this chapter on ethical knowledge. Break down each value with the help of the conceptual schemata presented in this chapter. Use this conceptual analysis to think broadly about technical and governance measures that Stanford Online might consider improving to ensure that its student selection process is ethical.

3. In-class debate: The existence of software that uses agents to protect against data collection could create a digital divide between people who can afford privacy protection and those who cannot. In the retail scenario in Chapter 3, I describe how rich people can forgo economic incentives and stay anonymous, while those who need the economic incentives must reveal their personal data. Economic benefits are traded for personal data. Debate in class whether it is ethically correct to trade privacy for economic benefits.

4. In-class debate: Should we have a right to be forgotten?

5. Think of a contemporary IT application and outline whether, how, and to what extent it respects the principle of contextual integrity for the use of its customer data.

References

Adams, J. 1963. "Toward an Understanding of Inequity." *Journal of Abnormal Psychology* 67:422–436.

Aristotle. 1984. "Physics." In *The Complete Works of Aristotle, Volumes I and II*, edited by Jonathan Barnes. Princeton, NJ: Princeton University Press.

Article 29 Working Party. 2013. Working Document 02/2013 providing guidance on obtaining consent for cookies.

Averill, J. R. 1973. "Personal Control Over Aversive Stimuli and Its Relationship to Stress." *Psychological Bulletin* 80:286–303.

Campbell, M. 1999. "Perceptions of Price Unfairness: Antecedents and Consequences." *Journal of Marketing Research* 36(2): 187–199.

Casassa Mont, M., S. Pearson, and P. Bramhall. 2003. Towards Accountable Management of Identity and Privacy: Sticky Policies and Enforceable Tracing Services. HP Laboratories Bristol.

Cohen, J. E. 2012. "What Privacy Is For." *Harvard Law Review* 126:1904–1933.

Cox, J. 2001. "Can Differential Prices Be Fair?" *The Journal of Product and Brand Management* 10(4):264–276.

Cranor, L. 2012. "Necessary But Not Sufficient: Standardized Mechanisms for Privacy Notice and Choice." *Journal of Telecommunications and High Technology Law* 10(2):273–308.

Cranor, L. F., B. Dobbs, S. Egelman, G. Hogben, J. Humphrey, and M. Schunter. 2006. "The Platform for Privacy Preferences 1.1 (P3P1.1) Specification—W3C Working Group Note 13 November 2006." World Wide Web Consortium (W3C)—P3P Working Group 2006 [cited July 17, 2007]. http://www.w3.org/TR/P3P11/.

Cranor, L. F., K. Idouchi, P. G. Leon, M. Sleeper, and B. Ur. 2013. "Are They Actually Any Different? Comparing Thousands of Financial Institutions' Privacy Practices." Paper presented at The Twelfth Workshop on the Economics of Information Security (WEIS 2013), Washington.

Danezis, G., M. Kohlweiss, B. Livshits, and A. Rial. 2012. "Private Client-Side Profiling with Random Forests and Hidden Markov Models." Paper presented at 12th International Symposium on Privacy Enhancing Technologies (PETS 2012), Vigo, Spain.

Daston, L. and P. Galison. 2007. *Objectivity*. New York: Zone Books.

European Commission. 2001. Promoting a European Framework for Corporate Social Responsibility. Luxembourg: Office for Official Publications of the European Commission.

European Parliament and the Council of Europe. 1995. "Directive 95/46/EC of the European Parliament and of the Council of 24 October 1995 on the Protection of Individuals with Regard to the Processing of Personal Data and on the Free Movement of Such Data." *Official Journal of the European Communities* L281:31–50.

Federal Trade Commission (FTC). 2000. Fair Information Practice Principles.

Floridi, L. 2005. "Is Semantic Information Meaningful Data?" *Philosophical and Phenomenological Research* 70(2):351–370.

Friedman, B. and H. Nissenbaum. 1996. "Bias in Computer Systems." *ACM Transactions on Information Systems* 14(3): 330–347.

Friedman, B., E. Felten, and L. I. Millett. 2000. "Informed Consent Online: A Conceptual Model and Design Principles." Technical report. University of Washington, Seattle.

Gentry, C. 2009. "Fully Homomorphic Encryption Using Ideal Lattices." Paper presented at 41st ACM Symposium on Theory of Computing (STOC '09), Bethesda, Maryland, May 31–June 2.

Ginzburg, C. 1976. "High and Low: The Theme of Forbidden Knowledge in the Sixteenth and Seventeenth Centuries." *Past and Present* 73:28–41.

Greenleaf, G. 2011. "Global Data Privacy in a Networked World." In *Research Handbook of the Internet*. Cheltenham, UK: Edward Elgar.

Guenther, O. and S. Spiekermann. 2005. "RFID and Perceived Control: The Consumer's View." *Communications of the ACM* 48(9):73–76.

Halpern, S. 2011. "Mind Control & the Internet." *The New York Review of Books*, June 23.

Hastak, M. and M. B. Mazis. 2011. "Deception by Implication: A Typology of Truthful but Misleading Advertising and Labeling Claims." *Journal of Public Policy & Marketing* 30(2):157–167.

Hess, C. and E. Ostrom. 2006. *Understanding the Knowledge Commons*. Cambridge, MA: MIT Press.

Jaspers, K. 1973. *Philosophie I: Philosophische Weltorientierung*. Berlin: Springer Verlag.

Kant, I. 1784/2009. *An Answer to the Question: "What Is Enlightenment?"* London: Penguin Books.

Kaptein, M. 2004. "Business Codes of Multinational Firms: What Do They Say?" *Journal of Business Ethics* 50(1):13–31.

Kaye, J., E. A. Whitley, D. Lund, M. Morrison, H. Teare, and K. Melham. 2015. "Dynamic Consent: A Patient Interface for Twenty-First Century Research Networks." *European Journal of Human Genetics* 23:141–146.

Khabsa, M. and C. L. Giles. 2014. "The Number of Scholarly Documents on the Public Web." *PLoS One* 9(5):e93949.

Lamont, J. 1994. "The Concept of Desert in Distributive Justice." *The Philosophical Quarterly* 44(174):45–64.

Langer, E. 1983. *The Psychology of Control*. Beverly Hills, CA: Sage Publications.

Langheinrich, M. 2003. "A Privacy Awareness System for Ubiquitous Computing Environments." Paper presented at 4th International Conference on Ubiquitous Computing, UbiComp2002, Göteborg, Sweden, September 29–October 1.

Langheinrich, M. 2005. "Personal Privacy in Ubiquitous Computing: Tools and System Support." Institut für Pervasive Computing, ETH Zürich, Zürich, CH.

Lind, A. E. and T. R. Tyler. 1988. *The Social Psychology of Procedural Justice*. New York: Plenum Press.

Maslow, A. 1970. *Motivation and Personality*. 2nd ed. New York: Harper & Row Publishers.

Mehrabian, A. and J. A. Russell. 1974. *An Approach to Environmental Psychology*. Cambridge, MA: MIT Press.

Menéndez-Viso, A. 2009. "Black and White Transparency: Contradictions of a Moral Metaphor." *Ethics and Information Technology* 11(2):155–162.

Meyer, B. 2007. "The Effects of Computer-Elicited Structural and Group Knowledge on Complex Problem Solving Performance." Mathematics and Natural Science Faculty, Humboldt University Berlin.

Nguyen, C., P. Haynes, S. Maguire, and J. Friedberg. 2013. "A User-Centred Approach to the Data Dilemma: Context, Architecture, and Policy." In *The Digital Enlightenment Yearbook 2013*, edited by Mireille Hilebrandt. Brussels: IOS Press.

Nissenbaum, H. 2004. "Privacy as Contextual Integrity." *Washington Law Review* 79(1).

Nonaka, I. and H. Takeuchi. 1995. *The Knowledge Creating Company: How Japanese Companies Create the Dynamics of Innovation*. London: Oxford University Press.

Organisation for Economic Co-operation and Development (OECD). 1980. OECD Guidelines on the Protection of Privacy and Transborder Flows of Personal Data.

Organisation for Economic Co-operation and Development (OECD). 2015. "Understanding Data and Analytics." In *Toward Data-Driven Economics: Unleashing the Potential of Data for Growth and Well-Being*, chap. 3. Paris: Organisation for Economic Co-operation and Development (OECD).

Pariser, E. 2011. "Beware Online Filter Bubbles." TED.com.

Pariser, E. 2012. *The Filter Bubble: How the New Personalized Web Is Changing What We Read and How We Think*. London: Penguin Press.

Plato. 2007. *The Republic*. Translated by H. D. P. Lee and Desmond Lee. London: Penguin Classics.

Popper, K. R. 1974. "Selbstbefreiung durch Wissen." In *Der Sinn der Geschichte*, edited by Leonhard Reinisch. München, Germany: C.H. Beck.

Ragnedda, M. and G. W. Muschert. 2013. *The Digital Divide: The Internet and Social Inequality in International Perspective*. London and New York: Routledge.

Rawls, J. 1971. *A Theory of Justice*. Oxford: Oxford University Press.

Redman, T. C. 2013. "Data's Credibility Problem." *Harvard Business Review*, 2–6.

Scannapieco, M., P. Missier, and C. Batini. 2005. "Data Quality at a Glance." *Datenbank-Spektrum* 14:6–14.

Seneviratne, O. and L. Kagal. 2014. "Enabling Privacy Through Transparency." Paper presented at IEEE Conference on Privacy, Security and Trust, Toronto, Canada, July 23–24.

Solove, D. J. 2001. "Privacy and Power: Computer Databases and Metaphors for Information Privacy." *Stanford Law Review* 53:1393–1462.

Solove, D. J. 2002. "Conceptualizing Privacy." *California Law Review* 90(4):1087–1156. doi: 10.2307/3481326.

Solove, D. J. 2006. "A Taxonomy of Privacy." *University of Pennsylvania Law Review* 154(3):477–560.

Spence, E. H. 2011. "Information, Knowledge and Wisdom: Groundwork for the Normative Evaluation of Digital Information and Its Relation to the Good Life." *Journal of Ethics in Information Technology* 13:261–275.

Spiekermann, S. 2007. "Perceived Control: Scales for Privacy in Ubiquitous Computing." In *Digital Privacy: Theory, Technologies and Practices*, edited by A. Acquisti, S. De Capitani, S. Gritzalis, and C. Lambrinoudakis. New York: Taylor & Francis.

Stehr, N. 1994. *The Knowledge Societies*. London: Sage Publications.

Tavani, H. 2012. "Search Engines and Ethics." In *The Stanford Encyclopedia of Philosophy*, edited by Edward N. Zalta. Stanford: The Metaphysics Research Lab.

Truncellito, D. A. 2015. "Epistemology." In *The Internet Encyclopedia of Philosophy*. http://www.iep.utm.edu/epistemo/ (accessed August 11, 2015).

Turilli, M. and L. Floridi. 2009. "The Ethics of Information Transparency." *Ethics and Information Technology* 11(2): 105–112.

van den Hoven, J. and E. Rooksby. 2008. "Distributive Justice and the Value of Information." In *Information Technology and Moral Philosophy*, edited by Jeroen van den Hoven and John Weckert. Cambridge, UK: Cambridge University Press.

van Dijk, J. A. 2013. "A Theory of the Digital Divide." In *The Digital Divide: The Internet and Social Inequality in international Perspective*, edited by Massimo Ragnedda and Glenn W. Muschert, 29–51. London and New York: Routledge.

Wang, R. Y. 1998. "A Product Perspective on Total Data Quality Management." *Communication of the ACM* 41(2):48–65.

Weiser, M. 1991. "The Computer for the 21st Century." *Scientific American* 265(3):94–104.

The White House. 2009. "Transparency and Open Government: Memorandum for the Heads of Executive Departments and Agencies." Washington, D.C.

The White House. 2013. "Consumer Data Privacy in a Networked World: A Framework for Protecting Privacy and Promoting Innovation in the Global Digital Economy." Washington, D.C.

World Economic Forum. 2012. "Rethinking Personal Data: Strengthening Trust."

World Economic Forum. 2014. "Rethinking Personal Data: Trust and Context in User-Centred Data Ecosystems."

Zimbardo, P. G. and R. J. Gerrig. 1996. *Psychologie*. 7th ed. Berlin: Springer Verlag.

Chapter 6

Preserving Freedom in Future IT Environments

"When a man is denied the right to live the life he believes in, he has no choice but to become an outlaw."

Nelson Mandela (1918–2013)

Freedom is one of the most cherished values in today's democratic societies. It is embedded in most constitutions and international conventions. In many respects, machines have given us greater autonomy in the way we live. For example, we can now work remotely from almost any geographic location and delegate many tedious or time-consuming activities to machines. However, the personal liberty we gain in some parts of our life is offset by quite a lot of control we delegate to machines (Brey 2004; Spiekermann 2008). When machines exercise control on our behalf, they can also infringe on our freedoms. For example, machines may force us into behaviors, deny us access to locations, and tell us how and where to drive. Therefore we must carefully design machines with the goal to strike a fine balance between delegated tasks and tasks kept by humans.

Before delving into how to achieve this balance, I first want to clarify what freedom, liberty, and autonomy actually are as constructs. Do we need to distinguish between the terms *freedom* and *liberty*, for instance? Pitkin (1988) makes a distinction between freedom and liberty due to the different etymological heritages of the two words. "Free" comes from the Indo-European adjective *priyos*, which means something like "one's own," "dear," or "the personal," with a connotation of affection or closeness. So freedom has been associated more with an inner state (similar to *positive liberty* introduced in Section 6.2). An abuse of freedom means a threat "to engulf the self, to release to uncontrollable and dangerous forces" (Pitkin 1988, p. 543). In contrast, "liberty" stems from the Indo-European verbal root *leudh*, which means "to grow" or "to develop" in the face of external controls. It has been more associated with external

states, like a political system in the face of which someone is permitted to grow (similar to the concept of *negative liberty* explained in Section 6.1). Despite these different linguistic roots many languages (surprisingly) do not separate the two terms. For example, Germans use the word *Freiheit* (similar to freedom), while French use the word *liberté* (similar to liberty) to denote the same thing. Most political and social philosophers have also used the two terms interchangeably (Carter 2012). Therefore, hereafter, I will not discern them.

David Hume (1711–1776) defined liberty (or freedom) as the "power of acting or of not acting, according to the determinations of the will; that is, if we choose to remain at rest, we may; if we choose to move, we also may" (Hume 1747/1975, p. 95). But already Hume recognized that choice alone does not lead to concrete action. After we make a choice, we need to be able to carry out what we choose and, unfortunately, our inner and outer environment sometimes impedes the execution of our choice. Therefore, we must distinguish two constructs: freedom of action and freedom of will.

Closely related to this distinction between freedom of action and freedom of will are the concepts of negative and positive liberty. Negative liberty looks at external obstacles and is therefore close to the concept of freedom of action. In contrast, positive libertarians recognize that not all constraints on freedom come from external sources. Instead, positive libertarians emphasize internal constraints, such as irrational desires, addictions, fear, and ignorance. They argue that to be free means to be independent from too much external influence.

6.1 Negative Liberty and Machines

Negative liberty is defined as "the absence of obstacles, barriers or constraints … (meaning) to be unprevented from doing whatever one might desire to do" (Carter 2012). Some authors have called this perspective on liberty the "republican view" (Pettit 1979). According to this view, liberties include freedom of movement, freedom of religion and freedom of speech. Berlin (1969) described negative liberty as a kind of free space in which people are sovereign: "the area within which the subject—a person or a group of persons—is or should be left to do or be what he is able to do or be, without interference by other persons" (p. 2).

Technology can be designed as an external obstacle to our freedom of action. In Chapter 3, I described several ways in the scenarios in which this might be done. A very subtle form of interference with our freedom is when objects act autonomously and make decisions for us without us (their owners) being in the loop:

> As he turns around in his bed, he knows that his bracelet has now signaled to the coffee machine to prepare his morning café latte—a friendly nudge to get up and get going. But Stern does not feel like it at all today … Stern slowly walks up to the kitchen. His café latte is not as hot as he likes it, and the machine has not put as much caffeine as usual into his cup due to his raised emotional arousal. But never mind.

This example is one for *direct negative liberty infringement*, or, as I have coined it elsewhere, "technology paternalism" (Spiekermann and Pallas 2005). *Technology paternalism* involves autonomous actions of machines that interfere with peoples' liberty when humans cannot overrule machines. We are already regularly confronted with technology paternalism. Sometimes sensor enabled (doors) do not open in public buildings, because people are supposed to walk different paths. University rooms cannot be locked from the inside, because they are supposed to be accessible from the outside at any time. Cars do not start when sensors in the car recognize alcohol consumption by the drivers. They beep and force us to fasten the seatbelt, no matter what situation the driver is in. An even stronger form of technology paternalism will occur when whole infrastructures and processes make us incur extra time and money. An example is the Halloville Mall in the retail scenario:

> [P]eople who are on a truly anonymous scheme can use a separate smaller mall entrance on the east side of Halloville that does not track any data … However, when people go through that entrance, they don't receive the 3% discount he gets and have to pay for parking, a luxury that Roger cannot afford.

Another form of negative liberty infringement is of an *indirect* nature. The infringement is indirect because the controlling machine entity or entities interfere with our activities without revealing their identity or the identities of their operators. Take the education scenario, where Big Data analysis is done by an unspecified entity, which then determines that Jeremy is not allowed to join Stanford's online university program:

> "I guess they think I can't do any better because of my outdoor times." … "What do you mean by outdoor times?" Roger asks … Jeremy doesn't know whether it's true, but a whistleblower software agent told him that Big Data analytics found that individuals' intelligence were highly correlated with their average time outdoors over the past 10 years. Since Jeremy had stayed indoors a lot when he played the *Star Games* VR, his average outdoor 10-year rating was probably pretty low. And he now suspected that this data was being used to predict applicant performance.

We sometimes face a similar situation today, when credit scores are used against us and impede us from getting a loan or an attractive interest rate. Some people argue that it is appropriate to use technology in this way, and therefore the technology does not infringe on freedom. They would probably say that it was Jeremy's choice to not exercise and spend more time outdoors. They would point out that if he had known about the outdoor expectations of Stanford Online and if those expectations were stable over time, he would have complied with them. They would also point to the fact that Jeremy is free to go to another school. Perhaps his desire to go to Stanford is simply too ambitious. Philosopher Ian Carter (2012) replies to this line of argument as follows: "If being free meant being unprevented from realizing one's desires, then one could, again paradoxically, reduce one's unfreedom by coming to desire fewer things one is unfree to do. One could become free simply by contenting oneself with one's situation. A perfectly contented slave is perfectly free to realize all of her desires."

6.2 Positive Liberty and Machines

"Positive liberty is the possibility of acting … in such a way as to take control of one's life and realize one's fundamental purposes" (Carter 2012). Positive freedom does not focus on the content of desires. Instead, positive freedom focuses on the ways in which desires are formed and whether they are the result of an individual's reflection and choice or the result of pressure, manipulation, or ignorance (Christman 1991). There are four kinds of challenges to positive liberty in the machine age that I will describe: manipulation, addiction, denial of autonomy, and allocation of attention.

I start with the danger of manipulation. Remember the retail scenario, where I describe the user dynamics of the Talos suit:

> The suit tracks all body functions and analyzes his moves and progress. Unlike some of his neighbors, Roger does not think that he will get paranoid about the suit. Many of his friends have gone crazy. The textiles transmit everyone's activity data to a regional fitness database that displays everyone's performance. So, many of his peers became preoccupied about their physical condition when seeing how they perform in comparison to their peers. They feel like they have to meet at least the average performance standard in the region, which is pretty high. One of his friends was so thrilled by the Talos force that he exhausted himself in a 12-hour run in the woods. He later had to be hospitalized for his exhaustion.

Machine feedback—as the Talos case shows—may manipulate our will in good and bad ways. In the example, Roger's neighbors become pressured to improve their fitness to a level that is not necessarily healthy. But they are also motivated to exercise more. No matter the positive or negative effects, we must note that machines (like the Talos) do influence our will. Some philosophers have claimed that the will is always free. Descartes famously wrote, "the will is by nature free in such a way that it can never be constrained" (1649/1989, Article 41). But the majority of scholars agree that there are many situations where the will is not free, which is true for the digital world just as much as for the physical on. Physical, biological, and social factors influence how we think and act. Machines do, however, amplify this influence. They amplify, because they constantly access our consciousness. They make comparative factors (such as peers' performance) or behavioral rules more visible, and nudge us to behave in a certain way.

Positive liberty is not only about the way in which we form our desires but also about our active commitment to them. Harry Frankfurt (1971) distinguished between two kinds of desires: "First-order desires" are those we share with animals. We feel that we want something, for example, a cake. In contrast, "second-order desires" reflect on first-order desires. And here we may decide to not follow our first-order desires. For example, we may decide not to eat the cake because we do not want to get fat. Frankfurt argues that we act freely when we are able to act on our second-order desires. These desires are the ones that we actually identify with and that actually satisfy us. The second-order desires reflect the true self (Frankfurt 1971).

Frankfurt's distinction allows us to distinguish free people from addicts. Addicts can only follow their first-order desires. They are not free because their ability to follow second-order desires is impaired. The scenarios mention the potential for addiction in machine environments:

> On average, players spend 3 hours in the game per day, with 10% at 6 hours a day. And—good for Stern—the hours are growing. The game is really addictive, or, as Stern would put it, "compelling."

In the Chapter 7 on health, I will return to the problem of addiction to machines.

One challenge relating to positive liberty is how we can guard our *autonomy* vis-à-vis our intelligent machines. Human autonomy is regarded as one of the cornerstones of social enlightenment. Kant's classical definition of autonomy is that it is "the property of the will by which it is a law to itself" (Kant 1785/1999, 4:440). In Kant's perception, autonomy means that we are sovereigns of our own will. This does not mean that everyone can do what he or she wants. We do have the *duty* to act such that we respect the freedom of others as well, which again limits our own actions or autonomy to act. But Kant believed that as individuals we must reach a state of internal and external autonomy in order to be able to consciously and freely take action in favor of a universal good. Autonomy in Kant's interpretation is hence a form of positive liberty and negative liberty, which is why I position it between the two forms of freedom in Figure 6.1.

In the future, it might be difficult to guard our sovereignty and autonomy in the way we still have it today. This becomes clear when I describe the intimate relationship between Sophia and her agent Arthur:

> Sophia chats with her 3D software dragon Arthur, who gives her advice on what products and shops to avoid for bad quality and where to find stuff

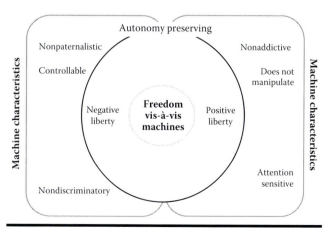

Figure 6.1 Machine characteristics protecting our freedom.

she likes and needs. Sophia almost cannot live without Arthur's judgment anymore. She really loves him even though he recently started to criticize her sometimes; for example, when she was lazy or unfair to a friend.

Arthur is described as a highly intelligent machine being that influences Sophia's thinking and behavior to an extent where she "almost cannot live without Arthur's judgment anymore." Consequently, agent Arthur is crucial to Sophia's autonomy and positive liberty. Note again Berlin's (1969, p. 122) definition of positive liberty: "What, or who, is the source of control or interference that can determine someone to do, or be, this rather than that?" Kant would ask whether Sophia (and her parents) have autonomously determined that Arthur's judgments are beneficial. If they have had the choice to use competitive agent software or if they were able to go without an agent altogether, their autonomy in Kant's understanding would be guarded. But this implies, of course, that the sources and interests underlying Arthur's functioning are transparent to Sophia and her parents; so that they can judge the agent's trustworthiness. Berlin's and Kant's clear reflections on what autonomy really means highlight the special care we must take in the transparent design of future decision-support systems.

Finally, a subject that has not yet been associated with positive liberty infringement is the effect that machines have on our attention allocation. Already, our information technology (IT) systems channel a lot of our attention, which in turn controls what we see and do at a given time. The systems force us to attend to things other than what we have actually chosen to do at that moment. For example, machines capture our attention when we receive a phone call or message while in deep conversation with a friend. As a result, we are often not in control over what to look at and attend to. Of course, some people would argue that we do not have to pick up a ringing phone, and we do not need to look at the ads we receive. This argument is, however, true only to the extent that the incoming signals can be ignored. Depending on the design of interruptions, we sometimes can ignore them. We can set, for instance, our handsets to be silent and we can determine that they do not ring at certain hours. Often however we cannot ignore the machines surrounding us. This is the case, for example, when incoming messages are highly salient, moving, or popping up in front of us hindering ongoing work and inhibiting free wondering of the mind. I will detail later how systems can be better designed to be less intrusive so that they do not infringe on our attention priorities and hence our positive liberty.

Figure 6.1 summarizes the various machine characteristics that this chapter covers as issues in our struggle to maintain positive and negative liberty in the machine age.

6.3 Technology Paternalism and Controllability

In his famous article on the computer of the twenty-first century, Marc Weiser (1991) wrote, "The [social] problem [associated with Ubiquitous Computing], while often couched in terms of privacy, is really one of control." Weiser was working to embed computer power into objects. He thought that we will weave digital functionality into most of our ordinary objects and that clumsy computers would be replaced by machines that exercise power invisibly and through our objects. Some people have started to call this vision "The Internet of Things." Stern's bracelet is an example of the new machine. Because it knows when Stern usually wakes up and recognizes that Stern is moving in his bed, it prepares Stern's morning coffee. It also knows, based on Stern's pulse and skin conductivity, that he is relatively stressed. As a result, the bracelet signals the coffee machine to not only prepare regular coffee but to reduce the caffeine level so that nutrition is optimized for Stern's body state. Much of this scenario addresses paternalism.

How and why is Stern's coffee machine paternalistic? Find what is technology paternalism anyways? "[Paternalism is] a system under which an authority ... regulates the conduct of those under its control in matters affecting them as individuals as well as in their relations to the authority and to each other" (*Merriam-Webster's Collegiate Dictionary* 2003). The goal is "protecting people and satisfying their needs, but without allowing them any freedom or responsibility" (*The Longmans Dictionary of Contemporary English* 1987). Together with Frank Pallas, I extensively discussed and developed the concept of technology paternalism in an earlier publication (Spiekermann and Pallas 2005). Against the background of radio-frequency identification (RFID) and sensor technology, and with the help of several focus groups, we identified the main conditions under which we can talk about paternalist machines and what should be done to avoid them.

The first trait of a paternalist machine is that it starts acting autonomously. It is independent and out of the control of the machine owner. Because the coffee machine is not controlled, it cannot be overruled. Stern cannot stop his bracelet from ordering coffee even though he knows that morning that he needs more time in bed. The result of the action cannot be disregarded: the coffee is there. And this activity might limit or infringe on freedom. Stern would have liked to have his coffee later and with the usual caffeine level, but the machine does not react to that desire or even give him an option. Instead, the lower caffeine level is legitimized by the argument that Stern's body is better off with less caffeine. Figure 6.2 summarizes these traits of paternalistic systems.

How can we avoid such paternalist machines? A major result from our empirical research showed that owners should be able to overrule machines. "Decisions made by technology

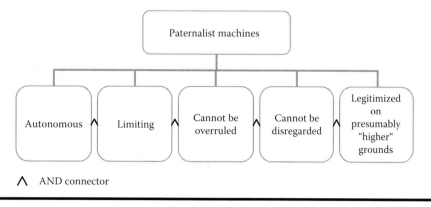

Figure 6.2   Traits of paternalistic machines.

and any exceptions from this should be considered very carefully. People should always have the last word!" one of the study participants said (Spiekermann and Pallas 2005, p. 9). In Stern's scenario, Stern could have the last word if his bracelet only signaled that it was ready to order coffee. Stern could confirm the order before it is placed. In a more sophisticated version of the bracelet, Stern may possess an agent as Sophia's dragon Arthur. Arthur could ask him whether he wants his coffee prepared and whether he wants it with less caffeine. This version of the future sounds much more promising. The example shows that simple system design elements that determine how control is allocated between men and machines can alter the whole relationship. A freedom-depriving scenario, where coffee is prepared without request in an undesired way, is turned into a freedom-enhancing one where coffee is prepared for us by a machine exactly when and how we want it.

Optimal control allocation (often called "function allocation") is at the core of a scientific field investigating "human-centered automation" (Billings 1991). In this field, a traditional approach to understanding optimal function allocation between humans and machines was Fitts's "Men-Are-Better-At/Machines-Are-Bettter-At" (MABA–MABA) list (Fitts 1951). Fitts suggested allocating tasks to humans and machines in accordance with their relative strengths and weaknesses.

Fitts's list noted humans being better than machines in

- Detecting small amounts of visual, auditory, or chemical energy
- Perceiving patterns of light or sound
- Improvising and using flexible procedures
- Reliably storing information for very long periods of time and recalling appropriate parts
- Reasoning inductively and at exercising judgment

In contrast, machines were found to be better at

- Responding quickly to control signals
- Applying great force smoothly and precisely

- Storing information for some time and erasing it completely
- Reasoning deductively

Even though these relative strengths of men and machines were formulated over 50 years ago, most of them still hold some truth. Machines today can detect and perceive quite a lot of obvious, bold, and repeated patterns. Truly *understanding* a situation though beyond what is encodeable is an exclusive human skill. Although reliable storage of information is often identified as a clear machine advantage, in truth the preservation of digital information is a challenge (Berman 2008). And decay of what is digitally stored is unfortunately independent of information relevance. Human beings can, in contrast, often remember at least the most important parts of a story. When they are trained they can well remember details.

That said, the relative skills of men and machines are rapidly evolving. Machine capabilities progress extremely rapidly, whereas human capability needs long training and practice. Consequently, no clear and long-term guidelines on how functions should generally be shared between humans and machines are available. In contrast, Thomas Sheridan (2000), a pioneer in questions of automation design, described any attempt to develop such guidelines as "alchemy."

What is clear, however, is that we must carefully consider our options for allocating function control. Sheridan himself identified eight levels of relative control between humans and machines (Sheridan 1988, 2002). These levels range from one extreme, where a computer does everything and people have no control, to the opposite extreme, where humans have full control. Figure 6.3 summarizes this control manipulation scale and demonstrates how it could be applied to the scenario of Stern's bracelet interacting with the coffee machine.

Although engineers benefit from knowing the different options for fine-grained control, they must base their decision on the ultimate goal of a machine service. Is it efficiency and productivity? Or are human freedom, dignity, growth,

8–10 Stages	Automation and control allocation between machine and human	Example: Smart bracelet (SB) and coffee machine (CM)	Degree of paternalism
1	M does not offer assistance; H must do the task completely herself.	Stern gets up and brews coffee himself. SB and CM do nothing (status quo).	
2	Upon request, M shows all alternative options to do a task. H executes.	Stern pulls up CM menu (i.e., on his mobile device or bracelet display) or starts agent. CM menu or agent state *all* order options (time, caffeine level, etc.). Stern manually chooses one *and* presses order button.	
3	M recommends specific way to do the task. H has to execute or not execute recommendation.	Stern pulls up CM menu (i.e., on his mobile device or bracelet display)/starts agent. CM menu or agent states all order options (time, caffeine level, etc.) and *recommends one* in particular. Stern chooses. His choice is *automatically interpreted as an order* and is executed by CM.	
4	M recommends a specific way to do a task and it executes upon H's approval.	SB or agent signals an option to order coffee. When Stern approves, CM executes.	
5	M recommends a specific way. Allows H a restricted time to veto before automatic execution.	SB or agent signals an option to order coffee and recommends less caffeine. If Stern does not veto within a certain time frame CM brews coffee automatically.	
6	M executes automatically and informs H about action taken.	CM brews coffee automatically when receiving signal from SB. SB/agent informs Stern that coffee is ready.	
7	M executes automatically and informs H only if asked to.	CM brews coffee automatically when receiving signal from SB. Stern can consult SB/agent whether coffee is ready.	
8	M selects the method and executes task. H is out of the loop.	CM brews coffee automatically when receiving signal from SB. Stern has no information. Goes to the kitchen to get the coffee that was chosen for him (scenario status).	

Figure 6.3 Levels of output automation as distinguished by Sheridan. (From Sheridan, T. B., 1988, "Task Allocation and Supervisor Control," in *Handbook of Human–Computer Interaction*, edited by M. Helander, 159–173, Amsterdam, Netherlands: Elsevier Science. With permission.)

and emotional well-being more important? In classical automation environments like factories, where productivity, efficiency, and safety have been the main design priorities, control has regularly been delegated to machines. For some reason, the notion that more automation is always better persists for these settings. But in the coming machine age, where man and machine will interact ubiquitously in daily life, efficiency and productivity may not necessarily be the best IT design goals. The idea of full automation—some authors talk about "fully autonomous agents"—will probably need to make room for a more balanced man-machine vision. This is because "communication and control belong to the essence of man's inner life," as Norbert Weiner (1954, p. 18) once said. Humans' emotions and behavior are strongly determined by the degree of control they have over their environments. In the 1970s, Mehrabian and Russell (1974) found that perceived control ("dominance") over an environment lead people to approach that environment. In contrast, when people are deprived of control, they avoid environments, show reactance (Brehm 1966), feel

helpless (Seligman 1975; Abramson, Seligman, and Teasdale 1978), are unhappy (Thompson and Spacapan 1991), and even die earlier (Langer and Rodin 1976). It is therefore not surprising that early studies on control perceptions and Internet use found that people are more motivated to use e-commerce sites over which they have control (Novak, Hoffman, and Yung 2000). Control allocation in favor of the machine is therefore a less obvious decision in consumer IT markets than it has been in enterprise systems. A simple illustration of how "the pleasure of control" can play out in tech-driven consumer markets is the continued use of stick-shift cars in Europe, where, as of 2012, over 75% of the cars still have manual transmission.*

When engineers decide for more automation, they must consider how to provide user feedback. Human–computer

---

* For transmission-type statistics, see "Projected EU Light Vehicle Production in 2012 and 2015, by Transmission Type," The Statistics Portal, http://www.statista.com/statistics/309940/transmission-type -of-light-vehicles-produced-in-europe/ (commercial source).

interaction (HCI) scholars outline how systems' feedback can be used to effectively foster perceptions of control; see, for example, the following six feedback rules for system designers below*:

1. Feedback should correspond to a user's goals and intentions.
2. Feedback should help evaluate a user's goal accomplishment.
3. Feedback should be sufficiently specific to control user activity.
4. Feedback should help develop accurate mental models of the system.
5. Feedback should fit the task representation (verbal and visual).
6. Feedback should fit the type of behavior (controlled, automatic).

In industrial environments that are highly automated, people have been observed to often not know what is going on. For instance, pilots in cockpits of highly automated modern planes frequently ask questions such as *what is it doing, why is it doing that, what will it do next, and how did it ever get into that mode* (Woods 1996). Similar problems now arise in everyday life. People regularly "feel stress due to subjectively unpredictable behavior of technical systems" (Hilty, Som, and Köhler 2004, p. 863). For example, modern cars sometimes brake autonomously, even when on a motorway. People become stressed in such cases due to a lack of "situational awareness" (Endsley 1996). They do not know the mode a machine is in. Often they do not know or forget about underlying machine mechanisms at work (Endsley 1996). Poor system design contributes to such a lack of situation awareness. Too often a discrepancy can be observed between an engineer's "conceptual model," which determines how a machine acts, and the "user's mental model," which determines how humans understand this action (Norman 1988; Scerbo 1996). Work by Donald Norman (2007) and others on control affordances in the design of future machines is therefore vital. Intelligent machines should meaningfully interact with users, provide reasons for suggestions, allow users to easily pause and resume activity, only gradually advance to take over decisions for humans, and respect that people have very different predispositions for how much control they want to delegate (Maes and Wexelblat 1997).

Coming back to Stern's coffee example: If we used stage 4 of the automation scale as shown in Figure 6.3, we would see that Stern's bracelet gives him the option to order coffee. Only when Stern approves this order, does the coffee machine actually start brewing. If we implement the machine in this way, we do not have a problem of liberty infringement or technology paternalism, and Stern is aware of what is going on. But let's think of a case that is ethically more ambiguous. Imagine that Stern had a heart attack in the past. He loves coffee, but it is not good for him. He always fails to comply with his own wish to drink less caffeine. His second-order desire is hence constantly undermined. Also, his health insurance company does not want him to drink coffee because drinking coffee increases his risk of another heart attack. The coffee machine is therefore set to not put caffeine in the brewer. Should this rule be a default that Stern cannot override? To ethically judge this question on machine design, we need to consider another construct related to positive liberty; that is how humans' *autonomy* can be guarded (or lost) in the face of machines.

6.4 Autonomy vis-à-vis Machines

"Autonomy … refers to the capacity to be one's own person, to live one's life according to reasons and motives that are taken as one's own and not the product of manipulative or distorting external forces. Autonomy might be defined as the freedom to make self-regarding choices, in which a person expresses his/her authentic self" (Koopmans and Sremac 2011, p. 177). Taking this definition of autonomy, we immediately see the concept's close link to positive libertarian thinking. What is key to Stern's case is to consider *who* decided to put less caffeine in the coffee. To speak with Berlin's words (1969, p. 122): "What, or who, is the source of control or interference…?" Consider these potential sources of control:

1. Coffee machines in the future will generally not put caffeine in coffee at all. Some political entity has decided that the health risks are too high for everyone, and so by default all coffee machines comply with the zero-caffeine rule.
2. The manufacturer of the coffee machine has decided that the company wants to compete based on healthy coffee and therefore markets most of its machines with the zero-caffeine default.
3. The health insurance company has asked Stern for permission to monitor his caffeine consumption by obtaining the coffee machine's usage data and will deny him insurance for another heart attack if he drinks any caffeine.
4. The machine is flexible, and Stern can overrule it and brew his coffee however he wants it.

---

* Source: Te'eni, D., J. Carey, and P. Zhang, 2007, *Human–Computer Interaction: Developing Effective Organizational Information Systems*, New York: John Wiley & Sons, p. 211. With permission.

With options 1 and 3, Stern effectively loses his autonomy: he is not in control. Instead, the regulator (option 1) or the insurance company (option 3) has taken over and infringed on his liberty. To quote Kant, Stern is "constrained by another's choice" (1785/1999, 6:237). With option 2, Stern has a bit more autonomy as long as there is competition in the market for coffee machines. He can still purchase from a vendor who gives him more freedom. Finally, option 4 respects Stern's freedom. He can choose each day how much caffeine he wants to drink. Here, theoretically, Stern is autonomous. Yet, if he has this freedom, then he is tempted each day to have a little real coffee and so to forgo his second-order desire to remain healthy and avoid another heart attack. So we can consider a fifth design option, one that optimizes both his freedom in terms of autonomy and his ability to realize his second-order desire:

5. The vendor sets a zero-caffeine default in the coffee machine. Stern can easily override this default. But when he does so, he sees a little signal on the machine that shows an unhappy smiley face that seems to say: "Given your health status, do you really want so much caffeine?"

This option of setting machine defaults to protect individuals while leaving them meaningful room to make a different decision has been called "nudging" (Thaler and Sunstein 2009). Nudging has recently found strong resonance in political practices. Of course, "by choosing their actions, one by one, humans continually create and adjust their own ethical characters—and their own lives and personal identities as well" (Bynum 2006, p. 160). Nudging people interferes with this continued identity construction. Yet, from psychology and behavioral economics, we know that humans often have trouble making rational decisions (Kahneman and Tversky 2000). Many of us cannot effectively judge short-term and long-term risks. We are often tempted by immediate gratification (O'Donoghue and Rabin 2000). We tend to hyperbolically discount long-term risks (Laibson 1996; O'Donoghue and Rabin 2001). The last option, where the machine carefully nudges a person to his true advantage, is an example of a machine helping us to follow our second-order desires while preserving our liberty to choose. This, of course, presumes that a machine or rather a machine's designer knows about our second-order desires. If such nudging is then not done too intrusively, but is transparent and can easily be countermanded by the user, machines may support our positive liberty. But note, continuing in the line of positive libertarian thinking, it is crucial for a nudging machine to reveal the source of its defaults: Why was the default set, who set it, and how can it be changed? This information should be easily accessible. At the very least it should be part of the machine's manual.

In the case of the coffee machine, giving this kind of information seems feasible. But think of the autonomous Alpha1 robots or agent Arthur. These extremely advanced machines may be called "agents" because they display three key characteristics: interactivity, autonomy and adaptability. Box 6.1 details these machine characteristics. It will be much harder to remain autonomous in the face of these systems.

BOX 6.1   AUTONOMOUS AGENTS: A CHARACTERIZATION

Some scholars have proposed that machines qualify as agents if they are interactive, autonomous, and adaptive (Allen, Varner, and Zinser 2000; Floridi and Sanders 2004). These scholars define these three traits as follows.

Interactivity means that machine agents react to input from their environment. For example, they know where they are based on geocoordinates. They can sense their environments, receive and interpret video streams, and use this data input to react. Interactivity may be realized with a simple "if <some external state> = x, then do a."

Machine *autonomy* means that the system can trigger an action on its own. It does not necessarily need an external stimulus or command. It can perform internal transitions to change its state. For example, a machine can contain an internal clock that measures its lifetime. After 2 years, the internal clock tells the machine to stop functioning. Autonomy may be realized with a simple "if <some internal state> == x, then do b."

Adaptability means that a machine seems to learn. Through its interactions, it can change the rules by which it changes state. Thus the machine takes decisions based on factors; the combination and sequence of actions cannot be perfectly predicted even by a machine's designer. The lifetime algorithm of it depends on too many internal and external states. Take the example of a house cooling system, which may start with an initial rule to balance temperature at 18°C (64°F) based on inhabitants' sensed body heat. When three out of four people in the house fall below the recommended body temperature, the system increases the house temperature to 20°C (68°F). Later, the machine might detect that body temperature is a little too high for some of the house's inhabitants. Instead of decreasing the temperature to 19°C (66°F), it adds perspiration as an additional indicator for bodily health to its "optimal temperature setting" algorithm; it hence effectively changes (extends) its initial rule. The system might later determine that maintaining a temperature of 20°C (68°F) is best because this temperature guarantees a good balance between body heat and transpiration. Alternatively, by retrieving the newest research from the web, the house temperature system may learn that for people beyond 80 years of age, the optimal room temperature is 23°C. It recognizes that one lady in the house has just turned 80 and it therefore sets the room that she is in to 23°C.

NOTES

1. F. Luciano and J. W. Sanders, "On the Morality of Artificial Agents," *Minds and Machines* 14 (2004): 349–379.
2. Colin Allen, Gary Varner, and Jason Zinser, "Prolegomena to Any Future Artificial Moral Agent," *Journal of Experimental and Theoretical Artificial Intelligence* 12 (2000):251–261.

A key challenge for *human autonomy* in the face of autonomous agents will be to have the agents correctly model users and their decision environments in real-time. Vendors will not always be able to set simple defaults correctly. In scenarios where humans interact with agents like Arthur, adaptive machines regularly make decisions based on some internal or external stimuli, states, and more or less comprehensive reasoning. The question is how these future machines will learn from us and "adapt" to us. How will we communicate our desires to our machine agents, and how strongly will our agents' algorithms then respect these desires when they make decisions for us? As Batya Friedman and Helen Nissenbaum outline in their paper "Software Agents and User Autonomy": "A lack of technical capability on the part of the software agent—to be able to accurately represent the user's intentions—can lead to a loss of autonomy for the user" (1997, p. 467). The same is certainly true when it comes to the capturing of second-order desires that precede intentions.

When machine agents do not possess the right level of capability and make the wrong decisions for us, they undermine our autonomy and may become a nuisance for us as users (as some machines are already today). Take the cooling system example from Box 6.1: What if the old lady in the house has always loved to sleep in a cold room below 17°C (62°F)?

A way to support machines' capability is to design them in a fluid way. *Machine fluidity* means that agents can adjust to changes in users' goals, even smaller subgoals (Friedman and Nissenbaum 1997). Returning to the coffee machine example: If Stern's health improves, he may be able to have

real coffee from time to time without negative health effects. The coffee machine needs to be able to accommodate this change of Stern's goals or spontaneous deviation from his default. Machines are more fluid when users can influence a machine's inner workings. The fluidity implies that users should be able to access machines' settings and rules. They should be able to alter them and have insight into and choice over the behavioral options available. This requirement seems like an irrational call from the past. As Floridi and Sanders (2004) note, "The user of contemporary software is explicitly barred from interrogating the code in nearly all cases." Does this need to be the case? And is this even desirable? Some people would argue that people are lazy and that they are not interested in manipulating anything nor are they qualified. They can do more harm than good. On the other hand, the history of technology is paved with evidence that people do manipulate their machines. Take the example of the automobile. People not only became innovative around machines, thinking about the most eccentric uses (Figure 6.4), but many spent parts of their lives understanding them, repairing them, and ramping them up (Figure 6.5). This is not different in the computing era. A whole youth culture has evolved around the manipulation of computers, testing out their limits and capabilities.

This last point of *Machine Accessibility* is one that I will take up again in more detail in Chapter 11, where I discuss the benefits of free and open source software. For people to not lose freedom and make decisions in cooperation with their machines, they need to be able to control the code of their machines. Some scholars will firmly contest this claim. We do not know all of the details about how our cars work either, but we can still use it and feel very free as we speed up

Figure 6.4 People putting their machines to new uses, for example, a Ford Model T converted for farm use between 1925 and 1935. (From Library and Archives Canada.)

Figure 6.5 People manipulating their machines. (© CC BY 2.0, Bitnar 2009.)

and down the highway. Furthermore, many machines now come with "software as a service" architectures. Economic reality is that the code is often black-boxed on some remote server that is not accessible to users. An open question for machine's future design is however *how much* insight we do need in order *to feel* in control over the machines we use. And at what level should we then effectively be allowed to manipulate them? Only at the application layer? Or also at lower levels of a machine's design?

From a positive liberty perspective, it seems to be important for a machine owner to determine and change "the source" of recommendations coming from the machine. For example, Sophia should be able to control and choose whether Arthur's recommendations are based on the mall's advertisement system, the Global loyalty card system, or some nongovernmental organizations (NGOs).

> Sophia receives less information, but the information she receives is of higher quality and more tailored to her preferences. She can let Arthur know from what sources he should retrieve recommendations. She trusts Playing the World and believes that Arthur respects her orders, looks after her privacy, and recommends what is best for her.

Even if the influence of a machine's sources and defaults remains unknown for most users, who are neither willing nor capable of digging into machine details, the potentiality of that *freedom to access* and change positively influences

liberty at the societal level. As I will argue later—and many scholars have argued before me—open and free code is vital for freedom.*

If individual users do not manipulate their machines at a deep level but want to access and control their agents to some degree, they should be able to do so on the application layer at least. This means of course more work for designers who have to think about how to make the logic of applications comprehensible. Users must be able to easily understand the application layer interface. The application layer interface should fully represent how machine states are determined and how they can be altered. Most important, users should be able to make changes easily and at minimum transaction cost. Some authors have noted critically that "in some instances, software agents may supply users with the necessary capability to realize their goals, but such realization in effect become impossible because of complexity. That is, the path exists to the state the user desires to reach, but negotiating that path is too difficult for the user" (Friedman and Nissenbaum 1997, p. 467).

Against this background, it would make sense for system designers to embrace progressions in the field of end-user programming and development (Ko et al. 2011). "People who are not professional developers can use End User Development (EUD) tools to create or modify software artifacts (descriptions of automated behavior) and complex data objects without significant knowledge of a programming language."† I have pointed to this kind of user control over future machines in the scenario when I describe how the 8-year-old Sophia is able to access and manipulate her agent Arthur:

> One cool thing about the game is the fostering of players' creativity. Anyone can design his or her own game characters with an easy-to-use programming tool and have these characters accompany them or engage with other people and their characters … [Sophia] has configured Arthur to run on top of her personal data vault. She could do so because of an agreement between the game company Playing the World and her personal data vault provider.

In my sci-fi cases, Sophia is smart enough to manipulate her agent Arthur. If people are anxious about changing

---

* Machine manufacturers will argue, of course, that deep tampering with machines is dangerous and creates liability issues if a machine fails. I would argue that tampering below the application layer could shift liability to the user. This liability is no different from the liability of people who customize or tinker with their cars: When those machines break, the driver is often responsible. Digitally enforced logging of code changes in future machines could ensure that liability goes where it belongs.

† "End-User Development," *Wikipedia*, accessed July 20, 2014, http://en.wikipedia.org/wiki/End-user_development.

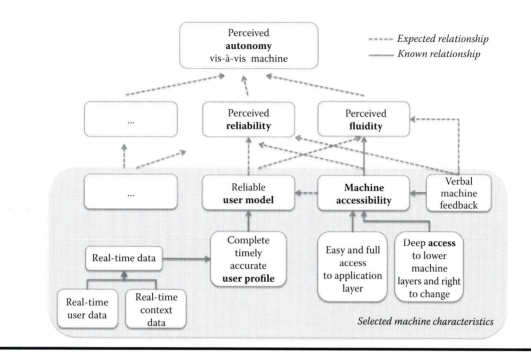

Figure 6.6 **Factors expected or known to influence a machine's perceived autonomy.**

how their agent works, then at the very least machines should explain why they act as they do. Petty Maes (1994), a software agent pioneer, wrote that "the particular learning approach adopted allows the agent to give explanations for its reasoning and behavior in a language the user is familiar with, namely in terms of past examples similar to the current situation. For example, "'I thought you might want to take this action because this situation is similar to this other situation we have experienced before, in which you also took this action' or 'because assistant Y to person Z also performs tasks that way, and you and Z seem to share work habits…'" (pp. 32–33). When designers account for accessibility, they must match users' abilities to what the machine assumes the user is capable of (Friedman and Nissenbaum 1997).

Finally, machine agents need to be *reliable*. If machine agents use inaccurate or false information, people cannot trust them. And if people then still have to rely on the agents, they will naturally feel out of control. At first sight, this requirement sounds easy to meet. Of course machine agents need to work with correct information! However, much machine judgment and feedback is based on probabilities today. The responses we receive from machines are based on what the machines think is correct. To date, machines do not reveal that their feedback is just a statistical probability and cannot be taken for granted. For example, timely ad networks often make projections of people's likely traits and interests. The networks do so based on observed and probabilistic behavioral patterns and demographics. In many cases, the resulting judgments are outdated or false though. Future

machines cannot be based on such suboptimal user knowledge (see Chapter 5). If we want to build autonomous machines that gain our trust and do not undermine our autonomy, then users and (software) agents must cooperate to ensure that agent behaviors are based on timely and accurate *user profiles* available in real-time. We need excellent and reliable *user models* (Kobsa 2007). Again, this data collection needs to happen with the consent of users, as was outlined in Chapter 5.

Figure 6.6 summarizes the factors that influence how we will perceive our autonomy vis-à-vis machines when interacting with them. It shows the technical factors this perception depends on. The empty boxes in the figure indicate that other factors not covered here might equally influence our perception of autonomy.

6.5 Attention-Sensitive Machines

"What information consumes is rather obvious: it consumes the attention of its recipients. Hence a wealth of information creates a poverty of attention, and a need to allocate that attention efficiently among the overabundance of information sources that might consume it," according to Simon (1971). Herbert Simon foresaw the current explosion of information: Every day, consumers are confronted with 2500 to 5000 advertising messages (Langer 2009). In addition, employees receive an average of 120 e-mail messages per day (Nuria, Garg, and Horvitz 2004; Fischer 2012; Radicati 2014), leading a typical office employee to check his or her messages around 50 times a day (Robinson 2010). As a result, attention spans are shrinking. A typical office worker is interrupted every 4 to 12 minutes (Dabbish, Mark, and Gonzàlez 2011). As hardly any room is left for concentrated efforts, are we still free masters of our attention? Or are we addicted to machines, pushed by them into attending them?

> First (the HR representative) chatted about Stern's recent attention scores. The company's attention management platform had found that Stern's attention span to his primary work tasks as a product manager was below average. "You seem to be interrupting yourself too often," the HR representative had said. "But what could I do?" thought Stern. There are simply too many messages, e-mails, social network requests, and so forth that would draw on his attention. So he obviously did not match the 4-minute minimum attention span that the company had set as a guideline for its employees. Employees' attention data was openly available to the HR department and management in order to deal with people's dwindling capability to concentrate.

Stern's story suggests that the problem of attention allocation may force companies to set minimum requirements for employees' concentration capacity and monitor employees for compliance. We might even see a new digital divide between those who can concentrate and focus on primary tasks and those who do not have the willpower to do so.

William James defined attention as "the taking possession by the mind in clear and vivid form, of one out of what seem several simultaneously possible objects or trains of thought … It implies withdrawal from some things in order to deal effectively with others" (1890/2007, pp. 403–404). James's definition of attention suggests a link between attention allocation and positive liberty. He talks about a "taking possession by the mind," but we must also consider who might take the mind over. Do we take possession of our own minds? Is Stern free, since he often consciously decides to interrupt himself and turn to another task in less than 4 minutes? Or is he not free, seen the number of external entities that ping him constantly and so aggressively that he cannot avoid the intrusion?

We know that the external environment causes the interruption in at least 51% of attention switches between knowledge work tasks. In 49% of the cases, people initiate the switch themselves (González and Mark 2004). Given this high number of external interruptions, we must build machines that are less intrusive. Attention is our scarcest and most valuable human resource. What we attend to defines who we are. And if our daily work environment, where we spend over 50% of our waking time,* turns us into creatures whose attention is externally manipulated by machine signals, then we risk losing a considerable part of our autonomy.

Do machines need to interrupt us in such an intrusive way that we lose control over our attention allocation? Research in psychology, computer science, and human–computer interaction shows that they do not. If machines better understand interruption situations and are sensitive to our natural attention allocation habits, they can be much less intrusive.

6.5.1 Attention-Sensitive Interruption

Interruption can be defined as an event where a stimulus effectively redirects an individual's attention away from an ongoing *primary task* and shifts that attention toward a secondary task. Examples of interruptions include an e-mail notification popping up at the side of a screen or an ad on the border of a website. McFarlane (2002) distinguishes between "negotiated timing" of interruptions and "immediate timing." In negotiated interruption timing, users have some control. They are notified of an incoming message, but they can determine whether and when to view it. For example, today's e-mails or instant messages (IMs) may be announced by the appearance of a small window, but the messages stay in the window or inbox until the user chooses to view them. In contrast, immediate interruption timing gives users no control and deprives them of freedom. The system enforces immediate attention. Full-screen pop-ups or screen freezing are examples of immediate interruptions, as are emergency messages in the car that freeze the radio show.

A third option is to not announce secondary tasks at all. Instead, users can "pull" information when they want to view it; an example is opening a mail program when we want to check what is there. The negotiated and immediate message delivery designs are both "push" strategies. Strictly speaking, they both reduce our liberty because they directly or indirectly force us to attend to incoming information. Our minds are not free to choose what to look at. Only interruption delivery strategies that employ the pull strategy give users full positive liberty to decide whether and when to retrieve information (see Figure 6.7). If the pull strategy is fully thought-out it means that only people's intention to do something, to search for something or to buy something drives their action. Doc Searls (2012) called this potential avenue of technology and economics "the intention economy."

Consider a system design that involves *negotiated* notification and hence some liberty infringement. How can this

---

* U.S. Bureau of Labor Statistics, Time Survey 2011, accessed July 22, 2014, http://www.bls.gov/news.release/pdf/atus.pdf.

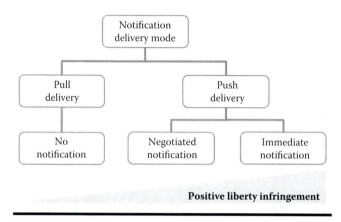

Figure 6.7   Liberty infringement can be limited by message pull delivery.

notification be delivered so that users feel minimal intrusion on their freedom? Note that intrusion is also a privacy harm. Solove (2006) defined intrusions as "invasive acts that disturb one's tranquility or solitude" (p. 491). Warren and Brandeis (1890), pioneers of privacy research, talked about "the right to be let alone."

McFarlane's (2002) study of interruption coordination techniques found that users who can negotiate their attention to an interruption normally take about 10 times longer to attend to it than when systems push notifications and people have to attend to them immediately. This lengthy delay before attending to secondary tasks is due to task chunking behavior. Humans naturally wait to switch to another task until a current task is completed and mental workload is low enough to accommodate something new (Buxton 1986). An ideal moment for interrupting someone is therefore after such a task chunk, at breakpoints between tasks (Salvucci and Bogunovich 2010). That said, when we are immersed in evaluative or perceptual tasks, fewer such breakpoints are available. This is the case, for example, when people read, learn, or write. The perceptual and nonrepetitive nature of knowledge work tasks makes them harder to interrupt in a nonintrusive way than executive tasks with multiple breakpoints (Brumby, Howes, and Salvucci 2009). Knowledge tasks are characterized by unfamiliar situations, objects, or texts that we need to learn about. Because we do not have rules of control or "production rules" for them yet (Rasmussen 1983), we need our full cognitive capacity to complete them. While people are engaged in knowledge tasks, systems should therefore protect people's attention resource and withhold notifications until the knowledge task is completed or people choose to stop working themselves. Only then should notifications be delivered.

Waiting for the end of a task or the next subtask boundary is, of course, not always possible. In emergency cases, an immediate notification may be justified to ensure immediate user attention and reaction. Some primary tasks, especially

those that are skill based, may also be interrupted without waiting for breakpoints. Rasmussen (1983) distinguishes between skill-, rule-, and knowledge-based tasks. Skill-based tasks, such as sensory-motor skills, need little conscious control because they have become a kind of habit. These tasks do not require much cognitive capacity and hence can more easily integrate interruption or even a secondary task. For example, many people can drive a car and talk on the phone. Although the ability to focus on driving might be impaired, we do not view the phone call as a liberty infringement.

Primary task types and breakpoints are not the only factors available to minimize intrusion. The design and relevance of a notification are also important (see Figure 6.8). The interruption design, which includes colors, size, vividness, and motion, influence how strongly people's attention is captured by it (Taylor and Thompson 1982; Beattie and Mitchell 1985). In particular, movement in notifications deprives people of the ability to avoid them (Bartram, Ware, and Calvert 2003; McCrickard and Chewar 2003), because humans' innate "orientation reaction" forces them to react to unexpected motion (Pavlov 1927; Diao and Sundar 2004). Another way to make a message pass is by choosing the right modality. Modality is the sensory channel used for information transmission: visual, tactile, or auditory. The modality of a notification can be identical to or different from the modality of the primary task. For example, a notification that a new e-mail message has arrived can be delivered in a visual modality (form) by using an on-screen pop-up window or in the auditory modality by using an alert sound. When the notification comes in the same modality as the primary task, the two interfere because they use the same perceptual

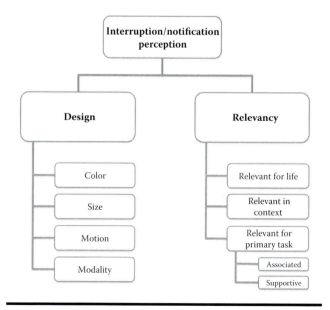

Figure 6.8   Parameters determining the positive or negative perception of an interruption.

resources (Storch 1992; Wickens 2002). People are then maximally disturbed. Interruptions of the same mode as the primary task should therefore be avoided if possible.

Modality configuration has challenging consequences for voice-based human–agent interaction. If the same modality must be used, one way to optimize human–agent interaction is to have people systematically initiate the conversations with agents. In situations where this is not possible, agents should only interrupt users when the incoming information is relevant to the user. We refer to relevant situations as those where mutual task relevance is given. Mutual task relevance exists when the notification or secondary task contains information that is topically associated with the primary task or in the same domain. The primary and secondary tasks can also be mutually relevant in terms of goal utility. Goal utility is given when the secondary task is complementary to the primary task and supports completing the primary task. When notifications and the tasks announced to users are relevant, they are perceived as less intrusive. An example is the Google search engine that displays relevant ads to people corresponding to what they are looking for. Such ads that directly correspond to search terms are less intrusive than banner ads that display any kind of information. Figure 6.8 summarizes the described parameters, which determine the positive or negative perception and level of disruptiveness of an interruption.

6.6 Summing Up: Freedom and Liberty in Future IT Environments

In this chapter, I used the distinction between negative and positive liberty as a starting point to discuss several ways in which machines can undermine our freedom and how they can be built to not do so. In particular, I identified technology paternalism as a threat to our liberty. Designers can choose levels of automation to avoid coercion of our lives by machines. Automation can be tweaked in a fine-grained manner so that people retain control while reaping the benefit of machines taking over tedious responsibilities. This has the potential to free people in many respects.

If intelligent, autonomous machine agents become common, we must consider from where they receive their orders: their owners or some external source? Can we control the source? Are the machines' actions transparent? These questions will be vital if we want to remain autonomous as individuals. "Freedom of thought" is probably the major threshold of human liberty, and if our thought is increasingly manipulated by machines, we risk losing a great part of our human identity. We differentiate ourselves from animals through our capacity to direct our personal thoughts. If that capacity is lost, we risk reducing ourselves to Pavlovian dogs responding to our machines.

EXERCISES

1. Describe the difference between positive and negative liberty. Use the robot scenario to illustrate where positive and negative liberty may be infringed upon by the future robot and agent systems.
2. Think of a system in your daily life that infringes upon your negative liberty and behaves in a paternalistic way. Describe how and why it is paternalistic by structuring your reasoning according to the dimensions depicted in Figure 6.2.
3. Take the robot Martha that has become Jeremy's friend in the retail scenario. Martha would like to show Jeremy around the Halloville Mall, but Jeremy and his family have their own plans. Using Sheridan's eight levels of automation, shown in Figure 6.3, describe how the degree of automation could be varied in Martha. How would Martha behave on each of the eight levels?
4. Think of agent Hal, a recommendation and monitoring system described in the robot scenarios. Hal acts in a highly autonomous way. Describe what characteristics Hal would need to have to ensure that its users feel autonomous in their management tasks. Use Figure 6.6 to conduct the analysis.
5. Take one of your own current messaging applications and analyze how it notifies you about new incoming messages (see Figure 6.7). Think about how its message notifications could be designed to minimally distract you. How intrusive is the current interruption management system in your application? How could attention sensitivity be improved?

References

Abramson, L. Y., M. E. P. Seligman, and J. D. Teasdale. 1978. "Learned Helplessness in Humans." *Journal of Abnormal Psychology* 87(1):49–74.

Allen, C., G. Varner, and J. Zinser. 2000. "Prolegomena to Any Future Artificial Moral Agent." *Journal of Experimental and Theoretical Artificial Intelligence* 12:251–261.

Bartram, L., C. Ware, and T. Calvert. 2003. "Moticons: Detection, Distraction and Task." *International Journal of Human–Computer Studies* 58(5):515–545.

Beattie, A. E. and A. A. Mitchell. 1985. "The Relationship between Advertising Recall and Persuasion: An Experimental Investigation." In *Psychological Processes and Advertising Effects: Theory, Research, and Application*, edited by L. F. Alwitt and A. A. Mitchell, 129–155. Hillsdale, NJ: Lawrence Erlbaum.

Berlin, I. 1969. "Two Concepts of Liberty." In *Four Essays on Liberty*, edited by Isaiah Berlin. Oxford: Oxford University Press.

Berman, F. 2008. "Got Data? A Guide to Data Preservation in the Information Age." *Communications of the ACM* 51(12):50–56.

Billings, C. E. 1991. "Human-Centered Aircraft Automation: A Concept and Guidelines." NASA Technical Memorandum. Meffet Field, CA: NASA Ames Research Center.

Brehm, J. W. 1966. *A Theory of Psychological Reactance.* New York: Academic Press.

Brey, P. 2004. "Disclosive Computer Ethics." In *Readings in Cyber Ethics*, edited by Richard A. Spinello and Herman T. Tavani, 55–66. Sudbury, MA: Jones Bartlett Learning.

Brumby, D. P., A. Howes, and D. D. Salvucci. 2009. "Focus on Driving: How Cognitive Constraints Shape the Adaptation of Strategy When Dialing While Driving." Paper presented at ACM Conference on Human Factors in Computing Systems (CHI 2009), Boston, April 8.

Buxton, W. 1986. "Chunking and Phrasing and the Design of Human–Computer Dialogues." Paper presented at IFIP 10th World Computer Congress, Dublin, Ireland, September 1–5.

Bynum, T. W. 2006. "Flourishing Ethics." *Ethics and Information Technology* 8(4):157–173.

Carter, I. 2012. "Positive and Negative Liberty." In *The Stanford Encyclopedia of Philosophy*, edited by Edward N. Zalta. Stanford, CA: The Metaphysics Research Lab

Christman, J. 1991. "Liberalism and Individual Positive Freedom." *Ethics* 101(2):343–359.

Dabbish, L., G. Mark, and V. M. Gonzàlez. 2011. "Why Do I Keep Interrupting Myself?: Environment, Habit and Self-Interruption." Paper presented at Conference on Human Factors in Computing Systems (CHI '11), Vancouver, Canada, May 7–12.

Descartes, R. 1649/1989. *The Passions of the Soul.* Translated by Stephen H. Voss. Indianapolis, IN: Hacket Publishing.

Diao, F. and S. S. Sundar. 2004. "Orienting Response and Memory for Web Advertisements: Exploring Effects of Pop-Up Window and Animation." *Communication Research* 31(5):537–567.

Endsley, M. R. 1996. "Automation and Situation Awareness." In *Automation and Human Performance: Theory and Application*, edited by Raja Prasuraman and Mustapha Mouloua, 163–181. Mahwah, NJ: Lawrence Erlbaum Associates.

Fischer, G. 2012. "Context-Aware Systems—The "Right" Information, at the "Right" Time, in the "Right" Place, in the "Right" Way, to the "Right" Person." Paper presented at International Working Conference on Advanced Visual Interfaces (AVI '12), Capri Island, Italy, May 22–25.

Fitts, P. M. 1951. *Human Engineering for an Effective Air-Navigation and Traffic-Control System.* Columbus, OH: Ohio State University.

Floridi, L. and J. W. Sanders. 2004. "On the Morality of Artificial Agents." *Journal of Minds and Machines* 14(3):349–379.

Frankfurt, H. 1971. "Freedom of the Will and the Concept of a Person." *Journal of Philosophy* 68(1):5–20.

Friedman, B. and H. Nissenbaum. 1997. "Software Agents and User Autonomy." Paper presented at International Conference on Autonomous Agents '97, Marina Del Rey, California, February 5–8.

González, V. M. and G. Mark. 2004. "Constant, Constant, Multi-Tasking Craziness: Managing Multiple Working Spheres." Paper presented at Conference on Human Factors in Computing Systems (CHI '04), Vienna, Austria, April 24–29.

Hilty, L., C. Som, and A. Köhler. 2004. "Assessing the Human, Social, and Environmental Risks of Pervasive Computing." *Human and Ecological Risk Assessment* 10(5):853–874.

Hume, D. 1747/1975. *An Enquiry Concerning Human Understanding.* Edited by L. A. Selby-Bigge and P. H. Nidditch. Oxford: Clarendon Press.

Jackson, T. W., R. Dawson, and D. Wilson. 2003. "Understanding Email Interaction Increases Organizational Productivity." *Communications of ACM* 46(8):80–84.

James, W. 1890/2007. *The Principles of Psychology.* Vol. 1. New York: Cosimo Classics.

Kahneman, D. and A. Tversky. 2000. *Choices, Values, and Frames.* New York: Cambridge University Press.

Kant, I. 1785/1999. "Groundwork for the Metaphysics of Morals." In *Practical Philosophy*, edited by Mary J. Gregor and Allen W. Wood. New York: Cambridge University Press.

Ko, A. J., R. Abraham, L. Beckwith, A. Blackwell, M. Burnett, M. Erwig, C. Scaffidi, J. Lawrance, H. Lieberman, B. Myers, M. B. Rosson, G. Rothermel, M. Shaw, and S. Wiedenbeck. 2011. "The State of the Art in End-User Software Engineering." *ACM Computing Surveys* 43(3).

Kobsa, A. 2007. "Privacy-Enhanced Personalization." *Communications of the ACM* 50(8):24–33.

Koopmans, F. and S. Sremac. 2011. "Addiction and Autonomy: Are Addicts Autonomous?" *Nova Prisutnost* 9(1):171–188.

Laibson, D. 1996. *Hyperbolic Discount Functions, Undersaving, and Savings Policy.* Cambridge, MA: National Bureau of Economic Research.

Langer, S. 2009. *Viral Marketing: Wie Sie Mundpropaganda gezielt auslösen und Gewinn bringend nutzen.* Wiesbaden: Gabler Verlag.

Langer, E. and J. Rodin. 1976. "The Effects of Choice and Enhanced Personal Responsibility for the Aged: A Field Experiment in an Institutional Setting." *Journal of Personality and Social Psychology* 34(2):191–198.

Maes, P. 1994. "Agents That Reduce Work and Information Overload." *Communications of the ACM* 37(7):30–40.

Maes, P. and A. Wexelblat. 1997. "Issues for Software Agent UI." MIT Media Lab. Cambridge, Massachusetts.

McCrickard, D. S. and C. M. Chewar. 2003. "Attuning Notification Design to User Goals and Attention Costs." *Communications of ACM* 46(3):67–72.

McFarlane, D. 2002. "Comparison of Four Primary Methods for Coordinating the Interruption of People in Human–Computer Interaction." *Human–Computer Interaction* 17(1):63–139.

Mehrabian, A. and J. A. Russell. 1974. *An Approach to Environmental Psychology.* Cambridge, MA: MIT Press.

Norman, D. A. 1988. *The Psychology of Everyday Things.* New York: Basic Books.

Norman, D. A. 2007. *The Design of Future Things.* New York: Basic Books.

Novak, T. P., D. L. Hoffman, and Y.-F. Yung. 2000. "Measuring the Customer Experience in Online Environments: A Structural Modeling Approach." *Marketing Science* 19(1):22–42.

Nuria, O., A. Garg, and E. Horvitz. 2004. "Layered Representations for Learning and Inferring Office Activity from Multiple Sensory Channels." *Computer Vision and Image Understanding* 96(2):163–180.

O'Donoghue, T. and M. Rabin. 2000. "The Economics of Immediate Gratification." *Journal of Behavioral Decision Making* 13(2):233–250.

O'Donoghue, T. and M. Rabin. 2001. "Choice and Procrastination." *Quarterly Journal of Economics* 116(1):121–160.

Orwell, G. 1949. *1984.* New York: The New American Library.

Pavlov, I. P. 1927. *Conditional Reflexes: An Investigation of the Physiological Activity of the Cerebral Cortex.* London: Wexford University Press.

Pettit, P. 1979. *Republicanism: A Theory of Freedom and Government.* Oxford: Oxford University Press.

Pitkin, H. F. (1988) "Are Freedom and Liberty Twins?" *Political Theory* 16(4):523–552.

Radicati, S. 2014. *E-Mail Statistics Report, 2014–2018.* Palo Alto, CA: The Radicati Group Inc.

Rasmussen, J. 1983. "Skills, Rules, and Knowledge; Signals, Signs, and Symbols, and Other Distinctions in Human Performance Models." *IEEE Transactions on Systems, Man, and Cybernetics* 13(3):257–266.

Raz, J. 1996. *The Morality of Freedom.* Oxford: Clarendon Press.

Robinson, J. 2010. "Blunt the E-Mail Interruption Assault: If You're Constantly Checking Messages, You're Not Working." http://www.msnbc.msn.com/id/35689822/ns/business-small _business/t/blunt-e-mail-interruption-assault/.

Salvucci, D. D. and P. Bogunovich. 2010. "Multitasking and Monotasking: The Effects of Mental Workload on Deferred Task Interruptions." Paper presented at ACM Conference on Computer Human Interaction (CHI 2010), Atlanta, Georgia, April 10–15.

Scerbo, M. W. 1996. "Theoretical Perspectives of Adaptive Automation." In *Automation and Human Performance*, edited by Raja Parasuraman and Mustapha Mouloua, 37–63. Mahwah, NJ: Lawrence Erlbaum Associates.

Searls, D. 2012. *The Intention Economy: When Customers Take Charge.* Boston: Harvard Business Review Press.

Seligman, M. E. P. 1975. *Helplessness: On Depression, Development, and Death.* San Francisco: Freeman.

Sheridan, T. B. 1988. "Task Allocation and Supervisor Control." In *Handbook of Human–Computer Interaction*, edited by M. Helander, 159–173. Amsterdam, Netherlands: Elsevier Science.

Sheridan, T. B. 2000. "Function Allocation: Algorithm, Alchemy or Apostasy?" *International Journal of Human–Computer Studies* 52(2):203–216.

Sheridan, T. 2002. *Humans and Automation: System Design and Research Issues.* Santa Monica, CA: John Wiley & Sons.

Solove, D. J. 2006. "A Taxonomy of Privacy." *University of Pennsylvania Law Review* 154(3):477–560.

Spiekermann, S. 2008. *User Control in Ubiquitous Computing: Design Alternatives and User Acceptance.* Aachen, Germany: Shaker Verlag.

Spiekermann, S. and F. Pallas. 2005. "Technology Paternalism: Wider Implications of RFID and Sensor Networks." *Poiesis & Praxis: International Journal of Ethics of Science and Technology Assessment* 4(1):6–18.

Storch, N. A. 1992. "Does the User Interface Make Interruptions Disruptive? A Study of interface Style and Form of Interruption." Paper presented at Conference on Human Factors in Computing Systems, Monterey, California, June 3–7.

Taylor, S. E. and S. C. Thompson. 1982. "Stalking the Elusive Vividness Effect." *Psychological Review* 89(2):155–181.

Te'eni, D., J. Carey, and P. Zhang. 2007. *Human–Computer Interaction: Developing Effective Organizational Information Systems.* New York: John Wiley & Sons.

Thaler, R. and C. R. Sunstein. 2009. *Nudge: Improving Decisions About Health, Wealth, and Happiness.* New York: Penguin Books.

Thompson, S. C. and S. Spacapan. 1991. "Perceptions of Control in Vulnerable Populations." *Journal of Social Issues* 47(4):1–21.

Warren, S. D. and L. D. Brandeis. 1890. "The Right to Privacy." *Harvard Law Review* 4(5):193–220.

Weiner, N. 1954. *The Human Use of Human Beings: Cybernetics and Society.* 2nd ed. Da Capo Series of Science. Boston: Da Capo Press.

Weiser, M. 1991. "The Computer for the 21st Century." *Scientific American* 265(3):94–104.

Wickens, C. D. 2002. "Multiple Resources and Performance Prediction." *Theoretical Issues in Ergonomics Science* 3(2): 159–177.

Woods, D. D. 1996. "Decomposing Automation: Apparent Simplicity, Real Complexity." In *Automation and Human Performance: Theory and Application*, edited by Raja Parasuraman and Mustapha Mouloua, 3–17. Mahwah, New Jersey: Lawrence Erlbaum Associates.

Chapter 7

Health and Strength in the Machine Age

"Early to bed and early to rise, makes a man happy, wealthy, and wise."

Benjamin Franklin (1706–1790)

Health is at the core of humans' physiological needs. If we are not healthy, all of the rest of Maslow's human needs are impacted. Frederick Herzberg (1968) would probably argue that health is a "hygiene factor" of motivation, which means that when people feel sick, attempts to motivate them by appealing to higher needs or values is difficult. Norbert Wiener (1954) regarded human physiology as core to a person's information processing potential.

The World Health Organization (WHO) defines health as "a state of complete physical, mental and social wellbeing and not merely the absence of disease or infirmity" (1946, p. 100). Physical health is a state and perception of bodily well-being in which an individual can or feels that he or she can perform daily activities and duties without any problem. In contrast, mental health is a state of well-being in which the individual realizes his or her own abilities, can cope with the normal stresses of life, can work productively and fruitfully, and can contribute to his or her community (World Health Organization 2001).

The stories in Chapter 3 illustrated that machines can strongly influence the health and strength of people. They do so directly and indirectly (see Figure 7.1), in a short-term and long-term manner. They affect our bodies, our minds, and our social well-being. By "direct" influence, I mean that using machines has a causal relationship with our health and strength. "Indirect" influence includes phenomena that mediate or moderate the use of information technology (IT) and its effect on health. For example, an addiction to online games may reduce the social ties of an individual, and that loss of social ties may negatively impact mental health. In the following I report on studies that have been conducted on Internet use. I use the term *Internet* here as representative for various machine services, like online news services, social network platforms, gaming, and so forth.

7.1 Direct Effect of Machines on Physical Health and Strength

The direct impact of machines on physical health has been studied in the field of ergonomics, which considers factors such as safety, comfort, and performance in man–machine interaction. The journal *Ergonomics* defines the field as follows: "Drawing upon human biology, psychology, engineering and design, ergonomics aims to develop and apply knowledge and techniques to optimize system performance, whilst protecting the health, safety and wellbeing of individuals involved."[*] Researchers have published a number of ISO[†] norms on principles of ergonomics for fields in which humans and machines interact, including ISO 26800 on the general approach, principles, and concepts of ergonomics. When engineers build machines that can influence the physical health of individuals, they must begin by learning about the ergonomics standards for their field. Studies typically focus on how to apply ergonomics in specific areas like health care, navigation systems, office environments, and aviation. International organizations such as the Institute for Ergonomics and Human Factors (http://www.ergonomics.org.uk)

---

[*] Aim and scope as defined by *Ergonomics*, http://www.tandfonline.com/action/journalInformation?show=aimsScope&journalCode=terg20.

[†] ISO stands for International Organization for Standardization.

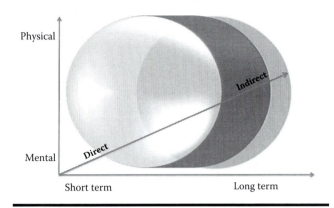

Figure 7.1 Three-dimensional influence of machines on human health and strength.

and the ISO, and relevant publications such as the *Handbook of Human Factors and Ergonomics* provide detailed guidance on how to design IT systems in the right way. Ergonomics not only explores the physical fit between humans and machines but also looks into cognitive fit and emotional reactions to machines.

But machines cannot only be designed for an optimal fit with humans. Machines can also help to *enhance* our cognitive and physical capabilities. An early example is designing IT for universal usability, giving blind and deaf people access to knowledge via IT (Shneiderman 2000). Furthermore, a whole engineering field for spare body parts is rapidly evolving. Recently, researchers developed a 3D printer for organic body parts that are customized for a particular person (IEEE 2014). Such body parts and extensions cannot only serve health purposes but also be used to extend human capabilities and strength. For example, augmented reality add-ons can be embedded in contact lenses to enhance a subject's vision (Parviz 2009). Where such sensors are not directly inserted into humans' biological system, they can be given to us in the form of physical add-on tools such as digital glasses or electronic textiles. As described in the gaming scenario, such tools may enable us to see the natural environment with an extra layer of information, for example, as an infrared or heat-map overlay. Artificial body parts and smart textiles may transcend their typical market of the elderly and those with disabilities to be used by the general population to increase physical strength (see, for example, the Tactical Assault Light Operator Suit [TALOS]*). In the retail scenario, Roger purchases a Talos suit that greatly increases his bodily strength. Sophia buys a smart glove that adds extra strength to her hand. However, in the scenario Roger's friends misjudge their physical limitations and exhaust themselves. The interplay

between digital human enhancements and humans' physical and psychological condition is largely unknown today.

The augmentation of humans' physical capability raises ethical questions: Is it good or bad to artificially enhance one's bodily condition? To what extent should users of digital body parts be granted access and manipulation possibilities to their own body devices? Must the software used in the body be open and manipulable in order to avoid, for instance, outdated proprietary systems remaining in human bodies? Societies have developed rules about the use of drugs and poisons that influence our bodies in a chemical way. Similar rules might be needed when IT is used for body enhancement.

7.2 Long-Term Effect of Machines on Physical Health and Strength

Machines can directly impact physical health, immediately and over time. The previous examples of ergonomic design, universal usability, renewable body parts, and new augmentation devices or implants are immediately observable and effective in the short term. But IT can also directly influence physical health in a way that is not immediately perceivable but has a long-term effect. Examples include radiation or gaming in unnatural body positions.

Let's start with the issue of radiation, which causes serious health concerns in some people, whereas others believe such fears to be esoteric phantasies. Health research distinguishes between thermic and athermic effects of radiation. Thermic effects (an increase in body temperature) can result from the absorption of energy by our biological body tissue, for example, absorption of radiation by the head when we hold a mobile phone to our ear. The official measure for such absorbed radiation is the specific absorption rate (SAR), which captures the relationship of watt per kilogram of body mass (W/kg). The body reacts to the level of SAR to which it is exposed. Experimental animal research has shown that radiation beyond 4 W/kg can damage the biological system (Autonome Provinz Südtirol 2002), so 4 W/kg is a threshold level for the design of IT devices such as mobile phones. Companies use accredited labs that perform SAR tests to investigate the radiation of their devices according to standards that are published by standardization bodies such as IEEE and ISO. As of 2015, a typical smartphone like the Apple iPhone has an SAR value of around 0.95 W/kg.†

---

* Denise Chow, "Military's 'Iron Man Suit' May Be Ready to Test This Summer," livescience, accessed April 14, 2014, http://www.livescience.com/43406-iron-man-suit-prototypes.html.

† Victor H., "SAR Explained: Here Is the Radiation Level of Top Smartphones (iPhone 6 Compared with Competition)," October 5, 2014, http://www.phonearena.com/news/SAR-explained-here-is-the-radiation-level-of-top-smartphones-iPhone-6-compared-with-competition_id61381.

Besides the known thermic effects of radiation, athermic biological effects may occur in response to the accumulation of smaller SAR values. So far, little scientific knowledge exists on this issue. An Austrian public report lists potential negative effects like a change in the enzyme activity ornithin decarboxylase (which is associated with tumor growth) and an impact on cells' calcium system and ionic transport (Autonome Provinz Südtirol 2002). Here a potential relationship is mentioned between high- and low-frequency fields and tumors; reproductive disorders; epilepsy; headaches; neurophysiological disorders (such as depression and disturbances of memory); disturbance of the immune system; damage of the eye tissue; and risks specific to pregnant women, children, and the elderly. The Assembly of the Council of Europe therefore strongly recommended applying the "ALARA" principle to SARs. ALARA means that the specific absorption rate for IT devices should be "as low as reasonably achievable."*

Another long-term health effect of using IT is how our bodies react to long periods of regular digital immersion, such as when we play games, sit in front of a computer screen for work, or use mobile phones. Eyestrain and problems with the back and the tendons of the hand are well known. People's posture is also influenced by their sitting position in front of screens. A core challenge is that we enjoy immersion and flow when we play digital games or do knowledge work we like. This very positive immersion causes us to forget about our bodies. As a result, some gaming companies now think about mechanisms and even business models to encourage individuals to take more breaks.† However, business models promoting health can negatively affect companies' bottom line. In the gaming scenario, Stern reflects on the business case of Playing the World, which makes players pay after a certain amount of time so that they are incentivized to not continue playing too long. In that story I also offer a potential route for game design that may be healthier: bring games back into the real world and interact through voice commands and digital glasses.

7.3 Direct Effect of Machines on Mental Health and Strength

Like physical health, machines directly influence mental health. Myriad studies have looked into how Internet use influences depression, loneliness, self-esteem, life satisfaction, and well-being. A metastudy of 40 empirical investigations with over 20,000 participants found that high Internet use is directly associated with slightly reduced well-being (Huang 2010). Yet, this overall tendency covers only part of the picture. People use the Internet for very different purposes, and each purpose affects our mental health in different ways. Using the Internet for general entertainment purposes, escape, and acquiring information does not seem to have discernible consequences for well-being.

Two forms of communication must be distinguished: using the Internet to strengthen our ties with existing friends (strengthening strong ties) or using it to find friends (creating weak ties). Studies show that people who use the Internet to communicate with existing friends and family are less likely to be depressed over time. They use the medium in a positive way to foster communication with those they feel close to. In contrast, people who use the Internet to overcome loneliness and meet new people are actually more likely to be depressed over time (Bessière et al. 2008). This latter finding resonates in another observation, which relates problematic Internet use (PIU) to mental health. A mix of behaviors characterizes PIU: a salient intensive use of digital media, mood modifications and irritation when one is not able to access the web, conflict with family and friends when access to the web is impaired, and a failure to stay away from using it even if this abstinence is desired (Ko et al. 2005). Caplan, Williams, and Yee (2009) summarize studies that report significant correlations between PIU and loneliness, depression, anxiety, shyness, aggression, introversion, and social skill deficits.

Researchers often question whether PIU causes such negative effects or whether existing individual traits such as loneliness and shyness lead to PIU in the first place.‡ A cognitive behavioral model of PIU proposed by Davis (2001) suggests that individuals who suffer from psychosocial problems are more likely to develop PIU. But research has also found that applications that are particularly social, such as online multiplayer games, foster PIU. Morahan-Martin explains "there is a growing consensus that the unique social interactions made possible by the Internet play a major role in the development of Internet abuse" (2007, p. 335). Finally, researchers have found that women, poorer people, and younger people are more likely to get depressed from using the Internet over time than others (Bessière et al. 2008). Taken together, mental health and social well-being are clearly related to the use of IT systems, but how this relationship plays out depends on who uses it, for what purposes, and how.

On the positive side: Although IT systems can add to mental problems and reduce social well-being, they are also used to relieve people in these very areas. A multitude of online services has been developed to ease dementia, phobia,

---

* Council of Europe, Parliamentary Assembly, Resolution 1815 (2011), http://www.assembly.coe.int/Mainf.asp?link=/Documents/Adopted Text/ta11/ERES1815.htm.
† "Gameplay: Balancing Enjoyment with Safety," Institute of Ergonomics & Human Factors, April 16, 2014, http://www.ergonomics.org.uk /sport-leisure/gameplay-balancing-enjoyment-with-safety/.
‡ Adam Gabbatt, "Excessive Internet Use Linked to Depression, Research Shows," *The Guardian*, February 3, 2010, http://www.theguardian .com/technology/2010/feb/03/excessive-internet-use-depression.

anxiety, insomnia, and addiction. In particular, mobile apps support people directly by giving them advice on their problems, tracking their behavior, putting them in touch with others, running them through relief games, or providing reminders about things that would otherwise be forgotten.* Early studies, including one on student stress, found that a stress management app could influence its users' weekly physical activity, engage in specific stress management methods, and exhibit decreased anxiety and family problems (Chiauzzi et al. 2008). Such findings are promising in that they suggest that IT systems can lead to some mental relief for those who need it. Perhaps they could be used to heal the same problems they may contribute to, such as PIU?

7.4 Indirect Effect of Machines on Mental Health

I have outlined that Internet use has a small direct impact on mental problems or social well-being. However, considerable research has been conducted to understand the potentially more powerful indirect influence that Internet use has.

One line of research looks at stress effects of PIU. Intensive use of the Internet can absorb our time to such an extent that we become stressed when we try to meet other life obligations. This stress, which we perceive in everyday activities, can lead to mental health problems. For example, according to Ming, students with heavy Internet use report that the Internet jeopardizes their academic performance. Poor academic performance is accompanied by high academic stress, which again impacts mental health negatively (Ming 2012).

A second line of research investigates a *social displacement* hypothesis, which suggests that computer and Internet use reduces the time we spend to maintain social resources. A lack of social ties undermines our mental health (Ming 2012) because we then need to find new friends online (Bessière et al. 2008). However, if we find true friends online and develop strong ties with them, our health can be strengthened again. Scholars refer to such developments as *social augmentation* or *social compensation* (McKenna and Bargh 2000).

A third line of research is based on the mood enhancement hypothesis, which posits that we selectively expose ourselves to media content based on our mood and can thereby disrupt a bad mood or negative ruminations on our life. The use of digital media for such purposes relieves stress, which again is good for mental health. Ming (2012) shows that mental health problems experienced by students with high academic stress can decrease as a result of exposure to mood-enhancing media.

Finally, a well-known negative indirect effect of machines on health has been observed in the field of online gaming, especially when games are addictive. Addiction to online games has physical and mental consequences such as headaches, sleep disturbance, backaches, eating irregularities, carpal tunnel syndrome, agoraphobia, and poor personal hygiene. Because online gaming addiction is such a prominent problem today, self-help organizations have been founded that provide people with information and advice.† Massive multiplayer online games (MMOGs) can be particularly addictive because they can create a feeling of social community among players. In particular, players who have a greater sense of friendship online are more likely to become addicted. As with PIU, social variables, offline friends and social ties or feelings of loneliness play a predictive role for games. Game immersion and the use of voice in games were found to encourage addiction (Caplan, Williams, and Yee 2009).

Figure 7.2 summarizes at a generic level the indirect relationships observed between computer use and mental health.

7.4.1 Mental Health Challenges in Response to Computer Use on the Job

Another indirect influence of IT on mental health involves employee burnout. Burnout is a phenomenon that expresses itself in feelings of physical exhaustion and cynicism (Green, Walkey, and Taylor 1991, p. 463). It is caused by a perceived misalignment between a job's demand and control over the job (see the job demand–control [JD–C] model; Karasek 1979, 1990). Employees get stressed and develop burnout symptoms when job control is low and job demands are high. On the contrary, more job control can attenuate the negative effects of job demands on strain. Scholars found that perceived computer self-efficacy is a key factor moderating this relationship between job control and job demand as depicted in Figure 7.3 (Salanova, Peiró, and Schaufeli 2010). Computer self-efficacy can reinforce or appease the burnout symptoms of exhaustion and cynicism. Self-efficacy refers to an individual's belief in his or her ability to perform a specific task (Bandura 1977). Computer self-efficacy is "an individual's perception of efficacy in performing specific computer-related tasks within the domain of general computing" (Marakas, Yi, and Johnson 1998, p. 128). It is driven to some extent by prior experience and by individual traits and personality variables such as age or professional orientation. However, it is also affected by specific characteristics of work on a computer, including the complexity, novelty, and difficulty of a task as well as situational support (Marakas, Yi, and Johnson 1998). When computer systems are not predictable, show incomprehensible numbers, or are not well documented and therefore difficult to understand, employees can perceive a lack of self-efficacy. This perception, combined with management's high demand for documentation and number-driven

---

* http://apps.nhs.uk/.

† "The Most Addictive Video Games," Video Game Addiction, http://www.video-game-addiction.org/most-addictive-video-games.html.

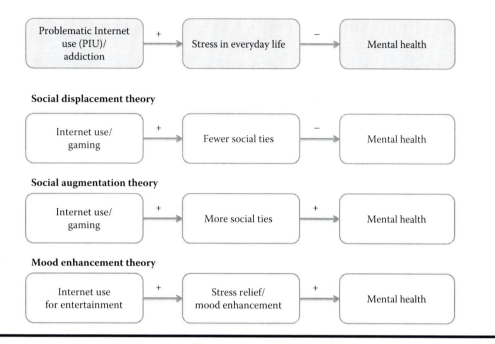

Figure 7.2   Selected indirect paths of IT influence on mental health.

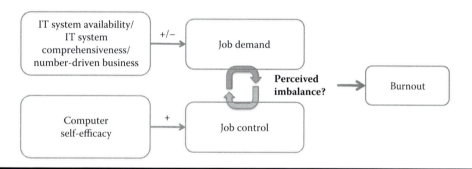

Figure 7.3   Schematic relationship between corporate IT systems, computer self-efficacy, and burnout. (Adapted from Salanova, Marisa, José M. Peiró, and Wilmar B. Schaufeli, 2010, "Self-Efficacy Specificity and Burnout Among Information Technology Workers: An Extension of the Job Demand–Control Model," *European Journal of Work and Organizational Psychology* **11(1):1–25.)**

decisions, can lead to perceptions of loss of control and then burnout. Section 5.4.1 contained a detailed description of today's data quality problems, which may be related to the burnout described in scientific burnout theories.

7.5 Indirect Effect of Machines on Physical Health

In the context of the work by the Organisation for Economic Co-operation and Development (OECD) on a "better life index,"* researchers have recognized that chronic diseases

such as cancer, cardiovascular diseases, respiratory conditions, and diabetes now cause around three-quarters of all deaths in OECD countries. Many of these diseases could be prevented if people modified their lifestyle. People who drink alcohol in moderate quantities, are physically active, eat a balanced diet, do not smoke, and are not overweight or obese have a much lower risk of early death than those who have such unhealthy habits. Against this background, we must ask whether machines can help us to be more self-aware and self-disciplined, and mentally support and coach us to live healthier lives. In the scenarios of Chapter 3, I describe the level of intelligence that machines can advance to. Taking a form like Sophia's Arthur, they might be able to continuously collect our health data and activity levels; machines could pull data from life-logging bracelets, smart textiles, or smartphones. On today's market, these types

* "Health," OECD Better Life Index, http://www.oecdbetterlifeindex .org/topics/health/.

of applications are called "life-logging" devices (European Network and Information Security Agency [ENISA] 2011) or "quantified-self" services (Swan 2012). Step counters like Fitbit, diet support apps like MyNetDiet, quit-smoking apps like Smoke Free, anti-alcoholism trackers like Change4Life Drinks Tracker, or sleep trackers like MotionX-24/7, are designed to make people aware of their behavior and support them in changing it to the positive.* At least such a positive change might be the case and more research is needed on the real effects.

On the negative side, I describe in the retail scenario how Roger worries about friends who feel pressure to meet fitness norms that may not be right for everyone. Here, a fitness-enhancing suit (Talos) creates a trade-off with one's perceived freedom. I characterize life-logging services as having an indirect effect on physical health only because the service needs to be able to motivate people first to then change their behavior both in the short and long term. Not all services will live up to this expectation.

As health-tracking applications advance, more health data will be available for analysis and predictive modeling (Manyika et al. 2011). If it was possible to track, store, and analyze individuals' health data (in a way that preserves privacy) and increase the amount of objective data on medications success, therapy effectiveness, doctor and hospital quality, and so on, we could gain a better feeling for what to do and where to go. More informed healthcare decisions may become possible in comparison to today, when differences in cost and quality are largely opaque (Manyika et al. 2011). Healthcare data also supports comparative effectiveness research (CER), which explores what medical treatments and medications work best, under what conditions, and for what kinds of people. Such data may also be used to study rare adverse drug reactions and mutual drug intolerances. Finally, machines can be used as support tools in diagnosis, helping doctors to analyze X-ray, CT, and MRI output.

However, the use of electronic health records in these ways raises many ethical questions: Should we ever be forced to share personal health data with care providers, researchers, or third parties such as insurers and health data brokers? To what extent and under what legal and technical conditions can a market for health data be legitimate? If health data such as genetic data or body measures reveal that we have a high risk for becoming sick, how and to what extent may this data be used by entities such as employers or insurers? Who is liable if the information is wrong or predictions do not play out as anticipated? To what extent could people be directly or indirectly forced to use health-monitoring applications to reduce their health risks? And who, if anyone, is allowed to exert such pressure? To what extent should in-body monitoring devices, such as chips that are inserted into the body or blood to monitor bodily conditions, be marketed to the general public? Should price discrimination on the basis of health data be allowed?

EXERCISES

1. Identify the health effects that may be created through the applications described in the scenarios.
2. Consider health data gained from a fitness application (e.g., a fitness bracelet). Think through the four stages of ethical knowledge creation and management described in Chapter 5. What requirements would a health application need to fulfill to ethically create knowledge about our health?
3. Figure 7.2 summarizes indirect effects of IT on mental health. Using research literature, the news or both, find at least three real-life examples of these effects.
4. Describe a situation in which you (or a friend or family member) experienced stress on the job that was related to IT. Explain whether and how the case relates to Figure 7.3.

References

Autonome Provinz Südtirol. 2002. "Elektromagnetische Strahlung und Gesundheit." Landesagentur für Umwelt und Arbeitsschutz. Bozen: Autonome Provinz Südtirol.

Bandura, A. 1977. "Self-Efficacy: Toward a Unified Theory of Behavioral Change." *Psychological Review* 84:191–215.

Bessière, K., S. Kiesler, R. Kraut, and B. S. Boneva. 2008. "Effects of Internet Use and Social Resources on Changes in Depression." *Information, Communication & Society* 11(1):47–70.

Caplan, S., D. Williams, and N. Yee. 2009. "Problematic Internet Use and Psychosocial Well-Being Among MMO Players." *Computers in Human Behavior* 25(6):1312–1319.

Chiauzzi, E., J. Brevard, C. Thum, S. Decembrele, and S. Lord. 2008. "MyStudentBody-Stress: An Online Stress Management Intervention for College Students." *Journal of Health Communication* 13(8):827.

Davis, R. A. 2001. "A Cognitive-Behavioral Model of Pathological Internet Use." *Computers in Human Behavior* 17(2):187–195.

European Network and Information Security Agency (ENISA). 2011. "To Log Or Not To Log? Risks and Benefits of Emerging Life-Logging Applications." Athens: European Network and Information Security Agency (ENISA).

Green, D. E., F. H. Walkey, and A. J. W. Taylor. 1991. "The Three-Factor Structure of the Maslach Burnout Inventory." *Journal of Social Behavior and Personality* 6:453–472.

Herzberg, F. 1968. "One More Time: How Do You Motivate Employees?" *Harvard Business Review* 46(1):53–62.

---

* Fitbit, http://www.fitbit.com/uk/story; MyNetDiet, http://apps.nhs .uk/app/calorie-counter-pro-by-mynetdiary/; Smoke Free, http://apps .nhs.uk/app/smoke-free/; Change4Life Drinks Tracker, http://apps .nhs.uk/app/change4life-drinks-tracker/; MotionX-24/7, http://sleep .motionx.com/.

Huang, C. 2010. "Internet Use and Psychological Well-Being: A Meta-Analysis." *Journal of Cyberpsychology, Behavior and Social Networking* 13(3):241–249.

IEEE. 2014. "Printing Body Parts: A Sampling of Progress in Biological 3D Printing." Accessed May 14, 2014. http://lifesciences.ieee.org/articles/feature-articles/332-printing-body-parts-a-sampling-of-progress-in-biological-3d-printing.

Karasek, R. 1979. "Job Demands, Job Decision Latitude, and Mental Strain: Implications for Job Redesign." *Administrative Science Quarterly* 24(2):285–308.

Karasek, R. 1990. "Lower Health Risk with Increased Job Control Among White Collar Workers." *Journal of Organizational Behavior* 11(3):171–185.

Ko, C.-H., J.-Y. Yen, C.-C. Chen, S.-H. Chen, and Y. Cheng-Fang. 2005. "Proposed Diagnostic Criteria of Internet Addiction for Adolescents." *Journal of Nervous and Mental Disease* 193(11):728–733.

Manyika, J., M. Chui, B. Brown, J. Bughin, R. Dobbs, C. Roxburgh, and A. Byers. 2011. "Big data: The Next Frontier for Innovation, Competition, and Productivity." McKinsey Global Institute (MGI).

Marakas, G. M., M. Yi, and R. Johnson. 1998. "The Multilevel and Multifaceted Character of Computer Self-Efficacy: Toward Classification of the Construct and an Integrative Framework for Research." *Information Systems Research* 9(2): 126–163.

McKenna, K. and J. A. Bargh. 2000. "Plan 9 from Cyberspace: The Implications of the Internet for Personality and Social Psychology." *Personality and Social Psychology Review* 4(1): 57–75.

Ming, A. 2012. "How Computer and Internet Use Influences Mental Health: A Five-Wave Latent Growth Model." *Asian Journal of Communication* 23(2):175–190.

Morahan-Martin, J. 2007. "Internet Use and Abuse and Psychological Problems." In *Oxford Handbook of Internet Psychology*, edited by Adam N. Joinson, Katelyn Y. A. McKenna, Tom Postmes, and Ulf-Dietrich Reips, 331–345. Oxford, UK: Oxford University Press.

Parviz, B. A. 2009. "Augmented Reality in a Contact Lens." *IEEE Spectrum*, September 1. Accessed May 14, 2014. http://spectrum.ieee.org/biomedical/bionics/augmented-reality-in-a-contact-lens/0.

Salanova, M., J. M. Peiró, and W. B. Schaufeli. 2010. "Self-Efficacy Specificity and Burnout Among Information Technology Workers: An Extension of the Job Demand–Control Model." *European Journal of Work and Organizational Psychology* 11(1):1–25.

Shneiderman, B. 2000. "Universal Usability." *Communications of the ACM* 43(5):85–91.

Swan, M. 2012. "Health 2050: The Realization of Personalized Medicine through Crowdsourcing, the Quantified Self, and the Participatory Biocitizen." *Journal of Personalized Medicine* 2:93–118.

Wiener, N. 1954. *The Human Use of Human Beings: Cybernetics and Society.* 2nd ed. Da Capo Series of Science. Boston: Da Capo Press.

World Health Organization. 1946. Preamble to the Constitution of the World Health Organization.

World Health Organization. 2001. Strengthening Mental Health Promotion.

Chapter 8

Safety, Security, and Privacy in Future IT Environments

"Those who surrender freedom for security will not have, nor do they deserve, either one."

Benjamin Franklin (1706–1790)

"Everyone has the right to life, liberty and security of person." With these words, Article 3 of the Universal Declaration of Human Rights stresses the value of security for human beings (United Nations General Assembly 1948). At the same time, Benjamin Franklin's famous words indicate that freedom might be even more important than security. Franklin's view is echoed by many liberal thinkers, who oppose what they call "Orwellian surveillance states"—states that monitor their citizens and limit liberty in the name of security. But are liberty and security mutually exclusive? Would the authors of the Universal Declaration of Human Rights have wanted to suggest that security was more important than liberty; that there is a value hierarchy between the two constructs?

8.1 Safety versus Security

The imprecise use of the term *security* creates confusion and suggests a conflict with freedom that may not exist to the extent that some believe. One reason for the confusion is that public authorities and the media use the term *security* in an inflated manner when they really mean "safety." When citizens are asked, for instance, whether freedom or security is more important for them, respondents often favor security at first.* It seems as if they positively embraced and legitimized governments' surveillance programs in a dire need for secu-

---

* "Deutsche Telekom asks citizens what is more important: Freedom or Security," Heise Online, accessed August 2, 2014, http://www.heise.de/newsticker/meldung/Telekom-fragt-Was-ist-Ihnen-wichtiger-Sicherheit-oder-Freiheit-1980972.html.

rity and at the expense of their freedom. So just the opposite of what Franklin would have advised them to say. However, it may be that they just misunderstood the question. What they really meant when answering the question was that their *safety* is more important to them than their freedom.

Why do I make this claim about confusion of terms? Unfortunately, most languages do not differentiate between safety and security. For example, security and safety are often collapsed into one term: *Sicherheit* in German, *seguridad* in Spanish, *seguranca* in Portuguese, *säkerhet* in Swedish, and so on. The commonly agreed on and current *Wikipedia* definitions of security and safety overlap as well: "Security is the degree of resistance to, or protection from, harm"; "Safety is the state of being 'safe,' the condition of being protected against … harm." So what is the difference?

The core difference can be understood when looking into the full details of *Wikipedia's* security definition: "Security is the degree of resistance to, or protection from, harm. It applies to any vulnerable and valuable asset, such as a person, dwelling, community, nation, or organization." The term *security* hence encompasses several levels of analysis. National security, organizational security, and individual security are all different things. Only when we speak about *individual* security do we approach the meaning of the word *safety*, because in its full definition, according to *Wikipedia*, safety is concerned with individual human beings: "Safety is the state of being 'safe' (from French *sauf*), the condition of being protected against physical, social, spiritual, financial, political, emotional, occupational, psychological, educational or other types or consequences of failure, damage, error, accidents, harm or any other event which could be considered non-desirable." Consequently, the first thing we should note is that when the media or market agencies ask citizens or consumers about their "security," then they should first distinguish the level of security they are actually talking about: national, organizational, or individual?

To illustrate the importance of this distinction consider the following: If people were asked whether the security of the organization they work for or their personal freedom was more important, I presume that most would choose their personal freedom. If they were asked whether the security of their country or their personal freedom was more important, the answer would probably depend on the context; in particular the degree to which a country's security seemed threatened. For example, right after September 11, 2001, many people were worried about their national security. They thought their country to be vulnerable to potential terrorist attacks. As a result, they accepted to give up some personal freedoms, such as some of their digital privacy rights, to ensure that the state would protect them through higher levels of national security. A few years later, after the context had changed, the perspective changed as well: When U.S. citizens learned how much of their personal privacy freedoms had been infringed in the name of national security, that is, through the Patriot Act, they started to debate whether surveillance at the current level is unconstitutional.*

September 11 is a good showcase to dig into the specific tension field of national security versus personal: At that time, people were evidently concerned about national security. Many feared that they or their families could be personally impacted by the developments. But would they have traded their liberty for national security? I would argue, on the contrary! In the aftermath of the September 11 terrorist attacks, I personally observed that in private many people acted in ways that were deep expressions of personal liberty. In Europe, but also in the United States, we all started to vividly discuss our personal protective measures in case of war; some considered moving to the countryside, others bought gold.† The personal freedom to plan for physical and financial actions probably strongly stabilized people's emotions and helped them feel secure again. As Abraham Maslow (1970) would argue, personal freedom and liberty of the people was a precondition for them to feel secure again at an individual level. At an individual level, liberty is a precondition for one's perceived security. The two cannot be traded off. When people are not free to personally react to a threat, they do not feel secure. And this observation brings us back to what Benjamin Franklin said (who obviously recognized the impossible trade-off): "Those who surrender freedom for security will not have … either one."

Let's come back to the definition of security again and how it relates to safety. A major linguistic pitfall is that security is often confounded with safety. Take the timely issue of airport "security": Often we are being asked whether we are willing to have our movements restricted at airports to increase the security of those airports. But do we really directly care about the security of airports? Probably not! What we are concerned about is not the level of security at the airport but the perceived safety that a respective level of security creates for us. At the organizational and national level, security is a precondition for our personal safety: If a system is secured in that it cannot be tampered with by malicious attackers, then the likelihood that it will damage people is reduced. Security is hence only indirectly important for us as individuals.

The distinction of security and safety becomes visible when comparing their academic definitions beyond *Wikipedia*. Line et al. (2006) define security as "the inability of the environment to affect the system in an undesirable way." In contrast, Line et al. define safety as "the inability of the system to affect its environment in an undesirable way." "Security is concerned with the risks originating from the environment and potentially impacting systems, whereas safety deals with the risks arising from the system and potentially impacting the environment" (Piétre-Cambacédès and Chaudet 2010, p. 59).

Although typical security threats can be identified across industries, safety standards are typically specific to every industry and type of machinery. Industry experts define safety standards for their respective (highly specialized) domain, aggregating years of knowledge and experience in how to build and maintain machines to avoid accidents. Safety is mainly linked to the avoidance of accidental risk while security is more thought of in terms of malicious attacks. An example for a safety standard is ISO 10218, the standard for industrial robots that aims to prevent accidents and harm by specifying how robots on the shop floor should operate, how they can be stopped, how fast they are allowed to move, what radius they are allowed to span, how their electric plugs are connected, and so on (ISO 2011).

Figure 8.1 illustrates this distinction between security and safety and clarifies that people's concerns (depicted as a stick figure) are triggered directly only by their desire for safety rather than security.

8.2 Safety, Cyberwar, and Cybercrime

How can information technology (IT) security impact the safety of human beings? And why does IT security become more important for safety than it used to be? As more machines are digitized, networked, or both, and as more machines receive instructions and upgrades from central

---

* For a debate on whether the Patriot Act is constitutional, see "Is the Patriot Act Unconstitutional?" Debate.org, accessed March 15, 2015, http://www.debate.org/opinions/is-the-patriot-act-unconstitutional.
† "Economic Effects Arising from the September 11 Attacks," *Wikipedia*, http://en.wikipedia.org/wiki/Economic_effects_arising_from_the_September_11_attacks.

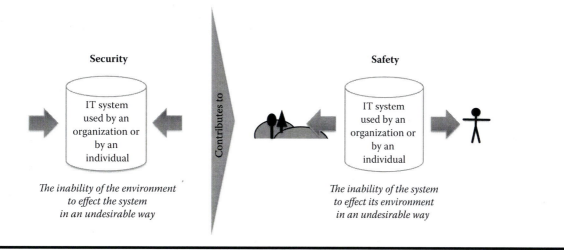

Figure 8.1 Simple illustration of the nature, difference, and normal relationship between security and safety.

computers, security is becoming an increasingly relevant factor for safety. Modern attackers can compromise the central computer systems that handle safety-critical infrastructure by infiltrating a system remotely and gaining control over it. Malicious attackers can also infect subcomponents of a system that are then built into safety-critical machinery. In any case, risk is created for system abuse. For example, cyber attackers could take over power grids and switch off the electricity supply in the area of a nuclear power plant. They could, as with the Stuxnet worm, implement a virus in parts of a critical infrastructure. Discovered in 2010, the Stuxnet worm attacked industrial programmable logic controllers, which control systems such as industrial assembly lines, amusement park rides, and centrifuges that separate nuclear material. It has been reported that Stuxnet destroyed one-fifth of Iran's nuclear centrifuges.* Scenarios like this one now populate common threat models for "cyberwar." States are worried that other states, terrorists, or even ordinary criminals could compromise the security of their critical infrastructure and gain a position in which they could undermine the safety as citizens.

Such cyberwar threats must be distinguished from "cybercrime." Cybercrime is "any crime that involves a computer and a network. The computer may have been used in the commission of a crime, or it may be the target."† Some national authorities stress that digital data processing must be essential for carrying out the crime. This means that,

for them, a crime is not a cybercrime unless digital data processing was important to commit the crime. Since so many crimes today use some kind of computer though, the term *cybercrime* got very broad in the amount and kind of activities it covers. It currently even includes activities such as computer-related copyright or trademark offenses that already happen when someone uses an illegal video-streaming platform. Figure 8.2 summarizes acts that are considered to constitute a cybercrime according to the United Nations Office on Drugs and Crime (UNODC).

Acts against confidentiality, integrity, and availability of computer data or systems

- Illegal access to a computer system
- Illegal access, interception, or acquisition of computer data
- Illegal interference with computer system or computer data
- Production, distribution, or possession of computer misuse tools
- Breach of privacy or data protection

Computer-related acts for personal or financial gain or harm

- Computer-related fraud or forgery
- Computer-related identity offenses
- Computer-related copyright or trademark offenses
- Sending or controlling sending of spam
- Computer-related acts causing personal harm
- Computer-related solicitation or "grooming" of children

Computer content-related acts

- Computer-related acts involving hate speech
- Computer-related production, distribution, or possession of child pornography
- Computer-related acts in support of terrorism offenses

Figure 8.2 Acts constituting cybercrime according to United Nations Office on Drugs and Crime. (From United Nations Office on Drugs and Crime (UNODC). 2013. "Comprehensive Study on Cybercrime.")

---

* "Stuxnet," *Wikipedia*, accessed August 2, 2014, http://en.wikipedia .org/wiki/Stuxnet; Michael B. Kelley, "The Stuxnet Attack On Iran's Nuclear Plant Was 'Far More Dangerous' Than Previously Thought," *Business Insider*, November 20, accessed August 2, 2014, http://www.businessinsider.com/stuxnet-was-far -more-dangerous-than-previous-thought-2013-11.

† "Cybercrime," *Wikipedia*, http://en.wikipedia.org/wiki/Computer _crime.

8.3 Security and Privacy Principles in Machine Engineering

The first category of cybercrimes defined by the UNODC involves the illegal access and use of computer systems resulting in a loss of confidentiality, integrity, and availability of data. Confidentiality, integrity, and availability of data are sometimes abbreviated as the "CIA principles" of data security (NIST 2013; ISO 2014a).

For companies, CIA-related acts of cybercrime are the most important threats (UNODC 2013). Companies worry that malicious attackers might access and steal intellectual property, customer data, trade secrets, and so on (cyber espionage); penetrate point-of-sale payment systems to steal and misuse customer payment cards (POS intrusions); tamper with or steal corporate data (including customer data); paralyze operations through denial-of-service attacks; or install malware that damages operations (immediately or later). Such fears are not unfounded. In 2013 alone, Verizon reported 63,437 security incidents that compromised the integrity, confidentiality, or availability of information assets. In 1367 of these cases, the data was breached, which means that the incident resulted in the disclosure or potential exposure of data (Verizon 2014). And this is only one moderate statistic on the problem.

Private individuals are often involved in this kind of crime, for example, when they respond to phishing e-mails, reveal their credentials, or have their credit card data stolen. In fact, the victimization rate for cybercrimes is significantly higher than for conventional crime forms, potentially because criminals do not need to be physically near their victims. The UNODC (2013) reports that victimization rates for online credit card fraud, identity theft, responding to a phishing attempt, and experiencing unauthorized access to an email range from 1% to 17% of the online population. This compares with typical burglary, robbery, and car theft rates of under 5% for the same 21 countries studied.

People become increasingly aware of online security threats even if they are not physically harmed by them. In 2010, 86% of consumers around the globe said that they are becoming more security conscious about their data, and 88% worried about who might have access to their personal data (Fujitsu 2010). People's concerns are relevant for companies because customers' perceived security influences the extent to which they intend to do business with a company online. The *perceived* level of data security at a company significantly influences the level of trust consumers place in that company (Chellappa and Pavlou 2002), especially in banking (Yousafzai, Foxall, and Pallister 2010). To maintain people's long-term trust in systems and avoid cybercrime damage, companies now systematically embrace relevant security goals when they build and maintain their systems. They analyze the extent to which security goals are at risk and put appropriate countermeasures in place to mitigate those risks.

Chapter 18 outlines how organizations can systematically run through risk analysis.

8.3.1 Information Security Goals

Traditionally recognized information security goals that aim to protect the information qualities threatened by cybercrime are confidentiality and integrity of data and availability of services. Well-elaborated textbooks outline the goals in detail and show how to reach them (i.e., Anderson 2008). I only give a short overview here.

To guard the *confidentiality* of information, data must be encrypted, and access and use must be confined to authorized purposes by authorized people and systems. To achieve this, information is ideally classified in terms of sensitivity, criticality, and value to the organization. Information is labeled accordingly, and access rights are set (ISO 2008, 2014a). Employees are asked to authenticate to access systems, especially those that hold sensitive, critical, or valuable information. Employees must be authorized to conduct certain operations on the information or to use the information. Furthermore, organizations should create general security awareness. Passwords need to be strong, so that they are not easily hacked, and employees must not share system access credentials.

Information has *integrity* when it is whole, complete, and uncorrupted. The integrity of information is threatened when it is exposed to corruption, damage, destruction, or other disruption of its authentic state. Corruption can occur not only while information is stored but also when it is transmitted. Many computer viruses and worms are designed explicitly to corrupt data. A key method for detecting a virus or worm is to look for changes in file integrity. Integrity can be checked, for example, by monitoring file size. A stronger method for ensuring information integrity is file hashing. Here, a special hash algorithm reads a file and uses its bits to compute a "hash value," a single number based on the individual data points of the file to be protected. This hash value is stored for each file. To ensure that a file is trustworthy, a computer system later repeats the same hashing algorithm before accessing the content of the file. If the algorithm returns a different hash value than the one stored for the file, then the file has been corrupted and the integrity of the information is lost.

Availability enables authorized users—people or computer systems—to access information without interference or obstruction and to receive it in the required format. Availability can be compromised when the service falls victim for instance to a denial-of-service attack* or has been altered so that authorized clients cannot access it any more. In a denial-of-service attack or distributed denial-of-service attack, one

---

* For more information on DoS attacks, see "Denial-of-Service attack," *Wikipedia*, http://en.wikipedia.org/wiki/Denial-of-service_attack (accessed March 15, 2015).

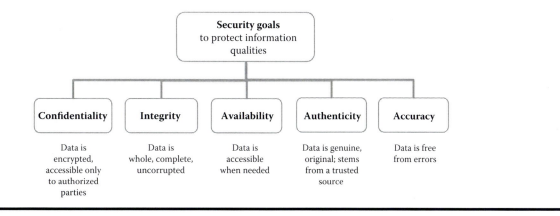

Figure 8.3 Conceptual analysis of computer security.

or several parties attack a service (or host) via the Internet; for example, they simultaneously send so many communication requests to the target machine that the target machine gets slow or even becomes unavailable for regular traffic.

Beyond these three CIA criteria, scholars and practitioners have suggested considering further security goals such as data authenticity, data accuracy, system auditability, and nonrepudiation (Cherdantseva and Hilton 2013). I will concentrate here on information qualities that must be protected instead of system characteristics that serve to protect these qualities. Therefore, I include only authenticity and accuracy, in Figure 8.3, which summarizes security protection goals.

Authenticity of information is the quality or state of being genuine or original. Information is authentic, when it is in the same state as when it was created, placed, stored, or transferred. Note the difference between integrity and authenticity: Integrity focuses on information or data not being falsified, whereas authenticity focuses on where the information originated. Suppose that attackers infiltrate an organization by using malware. They could send you information that claims to come from a trustworthy source but actually send something like a virus or phishing email. Finally, non-repudiation (that is not included in Figure 8.3) provides a system with the ability to determine whether a trusted individual took a particular action (such as creating information, sending a message, approving information or receiving a message). A system that supports nonrepudiation prevents individuals from falsely denying that they performed a particular action (NIST 2013).

Accuracy is a goal that I previously mentioned in Chapter 5, where I outlined that data quality must be maintained. Data accuracy or quality, which requires that data be free from mistakes or errors, is not only a goal to ensure the quality of knowledge but also a goal that supports the security of an organization. Imagine that, due to an error in a system database, an alarm threshold is triggered that causes damaging actions. The accuracy of data could be altered without necessarily changing its integrity. Similarly, it is possible to add erroneous information to databases based on false assumptions or to add false (even malicious) information to accounts.

8.3.2 Auditability

Auditability is recognized as a relevant system trait that can be used to ensure compliance with security goals (ISO 2012). Cherdantseva and Hilton (2013) define auditability as a system's ability "to conduct persistent, nonbypassable monitoring of all actions performed by humans or machines within the system." Auditability of systems gained significant importance after the 2002 Sarbanes–Oxley Act (SOX), which introduced new transparency and audit standards for many organizations worldwide.* It is well known how the auditing company Arthur Andersen failed to properly assess the financial risks associated with the practices of the company Enron. Enron filed for bankruptcy after financial fraud was uncovered that Arthur Andersen's auditing practices had not detected. In response to the fall of Enron and Arthur Andersen, corporations that are listed on the U.S. stock exchange must now report to the Securities and Exchange Commission (SEC) that they comply with SOX. SOX forces companies to monitor and evaluate all relevant processes that could influence their accounts. "While the topic of information security is not specifically discussed within the text of the act, the reality is that modern financial reporting systems are heavily dependent on technology and associated controls. Any review of internal controls would not be complete without addressing the information security controls around these systems. An insecure system would not be considered a source of reliable financial information because of the possibility of unauthorized transactions or manipulation of numbers" (SANS Institute 2004). SOX has hence indirectly enforced the scrutiny of information security controls. One of the main methodological standards associated with this form of scrutiny are the Common Criteria for Information Technology Security Evaluation, which are elaborated by the ISO in ISO/IEC 15408 (ISO 2012). *The Common Criteria* are used as a framework to

---

* "Sarbanes–Oxley Act," *Wikipedia*, accessed March 19, 2015, http://en.wikipedia.org/wiki/Sarbanes%E2%80%93Oxley_Act.

formulate security requirements for systems. The security requirements are then used to analyze and configure systems' security levels. One of these criteria is the use of audit trails for all security-relevant activities.

8.3.2.1 Security Audits versus Privacy Efforts

In earlier chapters I regularly commented on people's privacy concerns. Privacy often comes to mind when we talk about security engineering. For ordinary people, security-relevant activities seem to be those that equally help to protect their personal information. However, end-user data protection or privacy is not really in the spotlight when systems are audited for security goals. Only those personal information databases that are also financially relevant to the company are part of a security auditing effort. For example, a bank will take great care to protect its customers' bank account details. The financial information of the customer is, in this case, a crucial part of the bank's assets. Here privacy and security go hand in hand. In contrast, a car manufacturer may hold some information about the buyers of its cars, but a security audit would not prioritize the protection of this buyer data. Instead, the audit would focus more on protecting those systems that handle manufacturing and logistics processes because these processes are financially more relevant to the car manufacturer.

It is important to recognize that security efforts and privacy efforts are not necessarily the same because the terms *security* and *privacy* are often confounded. Laymen easily equate the two and believe that their personal data is automatically better protected when companies talk about improving their security. That said, security audits can embrace people's privacy concerns. In this case, "privacy targets" become an integral part of security-relevant activities.

Companies can pursue two strategies to meet privacy targets beyond data encryption and typical CIA measures. One is data minimization (as recognized in the Common Criteria's privacy section). The other is to control data flows by using policies and audit trails.

8.3.2.2 Data Minimization for Security and Privacy Reasons

Data minimization means that a company keeps only personal data records that are needed for its business. All of the rest of the personal data, customer transaction histories, and so on are deleted or anonymized so that they can no longer be attributed to a unique individual (ISO 2012). Data minimization through anonymization has been a very efficient strategy for companies to avoid privacy and security problems with customers. Companies presume that if data is anonymized, customers cannot be harmed and data protection law does not apply. At the same time, the anonymized data can be used for research purposes. Ohm (2010) commented that "nearly every

information privacy law or regulation grants a get-out-of-jail-free card to those who anonymize their data" (p. 1704).

Box 5.1 in Chapter 5 describes anonymization and pseudonymization techniques. In addition, unlinkability and unobservability are data minimization strategies which protect personal data. *Unlinkability* means "a user may make multiple uses of resources or services without others being able to link these uses together" (ISO 2012, p. 122). For example, a website operator might not log and link multiple visits of its customers to form an interaction profile over time or track how users move through sites. *Unobservability* means that "a user may use a resource or service without others, especially third parties, being able to observe that the resource or service is being used" (ISO 2012, p. 123). For example, a website operator might not log the IP addresses of those who read the content he provides.

8.3.2.3 Audit Trails for Privacy Management

Some security experts argue that data minimization efforts (as well as the pursuit of CIA principles) fully address people's digital privacy. They equate their security efforts with the creation of privacy. From their perspective, privacy is part of the overall security effort in a company and nothing more. And certainly, many of the privacy harms described in Section 5.6, such as unauthorized secondary data uses, breach of confidentiality, public disclosure, and exposure, could be avoided if personal data records were systematically minimized as part of a companies' security efforts. That said, other scholars, in particular those from the legal studies, nongovernmental organizations (NGOs) or social sciences, counter that security efforts do not suffice to create privacy. They see security as just one piece of the puzzle to ensure people's privacy in terms of information self-determination. From these thinkers' perspective, which is strongly adopted in Europe, privacy is not only a security issue. Privacy is not only something that can be passively harmed. Privacy is also an active right that allows people to freely determine who can use their data, when, and for what purposes. From this perspective, privacy is defined in terms of control over access to the self: "Privacy, as a whole or in part, represents control over transactions between person(s) and other(s), the ultimate aim of which is to enhance autonomy and/or minimize vulnerability" (Margulis 2003, p. 245). This autonomy-embracing definition of privacy is mirrored in European privacy legislation. Here, citizens need to opt-in to the use of their data before a data collector can use it (note that, in the United States, citizens can only opt out of the use of their data). European citizens need to be informed up front on the purposes of data use, they can access their data after they reveal it, withdraw their consent to its use, and so on (European Parliament and the Council of Europe 1995). From this informational self-determination perspective on privacy, security is just one part of a larger human rights endeavor.

Privacy scholars who share the broader European perspective on "privacy as informational self-determination" advise companies to ensure people's participation and/or potential control over data exchange and data use (see, for example, the verdict on the census of the German Federal Constitutional Court [BVerfG 1983]). They argue that personal data should be treated as a shared asset in a way that is negotiated with customers in policies (ISO/IEC 29101). Privacy policies can specify data usage rights and restrictions. They may for instance be negotiated prior to the data exchange with the help of a protocol such as P3P (Cranor et al. 2006).

A web protocol such as HTTPA (HTTP with Accountability) can equally be used to transmit usage restriction policies between web servers and clients (Seneviratne and Kagal 2014). Companies create a detailed and transparent history of access requests for personal data, and they track the transfer, processing, and disclosure of privacy-critical information. For example, the HTTPA protocol creates audit logs every time a party wants to access and use personal data, and people can check what happened to their data (Seneviratne and Kagal 2014). The logs can later serve customers, auditors, and those "accountable" in organizations to check whether personal data usage was compliant with privacy policies.

8.3.3 Accountability

A major lever for privacy as well as security is that companies become more accountable for the way in which they handle the personal data of their customers. In fact, according to various standards, guidelines and laws, companies are accountable already for the security and privacy of their customers' data (Organisation for Economic Co-operation and Development [OECD] 1980; Alhadeff, Van Alsenoy, and Dumortier 2011). Yet, unfortunately, the term *accountability* is imprecise. It can refer to different forms of responsibility at various levels. In computer science, the term *accountability* is used mainly in reference to auditability, the use of nonrepudiation mechanisms, or both. At a higher organizational level, accountability is situated with an individual in an organization who must safeguard and control equipment and information. This individual is also responsible to the proper authorities for the loss or misuse of that equipment or information (CNSS 2010). At a still higher level, accountability simply denotes that a company must be held responsible for its actions. No matter at what level accountability is defined, individuals (and their organizations) can meaningfully bear responsibility for transactions only if their IT systems provide them with the necessary information. Weitzner et al. (2008) clarify: "Information accountability means the use of information should be *transparent* so it is possible to determine whether a particular use is appropriate under a given set of rules" (p. 84). (Recall Section 5.4.3 for more detail on transparency.)

BOX 8.1 ACCOUNTABILITY FOR PERSONAL DATA ACCORDING TO ISO/IEC 29100:2011 (E), SEC. 5.10: INFORMATION TECHNOLOGY—SECURITY TECHNIQUES—PRIVACY FRAMEWORK

DEFINITIONS

PII = personally identifiable information: any information that (a) can be used to identify the PII principal to whom such information relates, or (b) is or might be directly or indirectly linked to a PII principal

PII controller = privacy stakeholder (or privacy stakeholders) that determines the purposes and means for processing personally identifiable information (PII) other than natural persons who use data for personal purposes

PII principal = natural person to whom the personally identifiable information (PII) relates

"The processing of PII entails a duty of care and the adoption of concrete and practical measures for its protection. Adhering to the accountability principle means:

■ Documenting and communicating as appropriate all privacy-related *policies,* procedures and practices;

■ Assigning to a specified individual within the organization (who might in turn delegate to others in the organization as appropriate) the task of implementing the privacy-related *policies,* procedures and practices;

■ When transferring PII to third parties, ensuring that the third party recipient will be bound to provide an equivalent level of privacy protection through contractual or other means such as mandatory internal policies (applicable law can contain additional requirements regarding international data transfers);

■ Providing suitable training for the personnel of the PII controller who will have access to PII;

■ Setting up efficient internal complaint handling and redress procedures for use by PII principals;

■ Informing PII principals about privacy breaches that can lead to substantial damage to them (unless prohibited, e.g., while working with law enforcement) as well as the measures taken for resolution;

■ Notifying all relevant privacy stakeholders about privacy breaches as required in some jurisdictions (e.g., the data protection authorities) and depending on the level of risk;

■ Allowing an aggrieved PII principal access to appropriate and effective sanctions and/or remedies, such as rectification, expungement or restitution if a privacy breach has occurred; and

■ Considering procedures for compensation for situations in which it will be difficult or impossible to bring the natural person's privacy status back to a position as if nothing had occurred.

Measures to remediate a privacy breach should be proportionate to the risks associated with the breach but they should be implemented as quickly as possible (unless otherwise prohibited, e.g., interference with a lawful investigation)."

To prove that they are legally compliant and accountable to users and lawmakers, organizations can take concrete accountability measures for personal data, such as those standardized in the ISO/IEC 29100 standard called **"Information technology—Security techniques—Privacy framework"** (ISO 2014b). Box 8.1 cites these accountability measures for personally identifiable information (PII).

8.4 Privacy and Surveillance

Let's assume a perfect world: Companies have done their best to optimize the security and privacy levels for personal data. They encrypt personal data where possible and verify information quality according to the CIA criteria as well as the authenticity and accuracy of the data. They also act accountably in line with ISO/IEC 29100, ensuring that people's consent-based privacy policies are respected within and beyond corporate boundaries. Information self-determination is respected by the corporate world. People can participate as sovereigns in personal data markets; they can share their data for research purposes or marketing campaigns for appropriate returns or choose to keep their data private without fearing any social sorting or discrimination.

If all of these measures were taken diligently, organizations would minimize the risk of causing privacy harms. People may be less concerned about their security and feel more in control of their data than they do today. Their identities would be stolen or misrepresented less often. When future robotic systems physically interact with humans, the risk that such systems could be controlled remotely or misrepresent users would be reduced. Also, software agents

that act on people's behalf (based on data exchange policies) would not betray their owners by disclosing unauthorized personal data to data collectors without the owner's consent. Social sorting and systematic discrimination (as we often see it today) would be prohibited. In such a perfect world, people could feel sufficiently safe and secure vis-à-vis their machines, at least at a basic level (in Maslow's sense). People could trust their machines.

But one major ethical challenge remains untouched here immediately and springs to mind when we talk about "security": What degree of surveillance is acceptable in the name of security to guard people's liberty?

8.4.1 Surveillance and Dataveillance

Analyzing surveillance torts over the past 100 years and more Solove wrote: *"Surveillance* is the watching, listening to, or recording of an individual's activities" (Solove 2006, p. 490). In the late 80s when IT was already widely deployed, Roger Clarke (1988) refined this baseline definition of surveillance and distinguished what he calls "dataveillance." Dataveillance refers to the systematic use of personal data systems in the investigation or monitoring of the actions or communications of one or more persons. Clarke distinguishes between "personal dataveillance" of previously identified individuals and the "mass dataveillance" of groups of people. With this distinction, Clarke hints at qualitative differences between the classical forms of surveillance torts that Daniel Solove refers to and the kind of dataveillance modern machines enable. The distinction is important. Unlike classical surveillance in the analog world, dataveillance is marked by invisibility, remoteness, networked pervasiveness, impartiality, and new forms of consent to data collection (Figure 8.4). What does this mean?

Traditional surveillance achieved a certain degree of secrecy that could still be uncovered. For instance when the secret service spied on people by their houses. Todays dataveillance, technologies like ubiquitous sensor technology, cameras, and mobile phones record without the knowledge of the people being observed. Secrecy aggravates to invisibility. Some efforts are made to put up warning signs for video cameras or radio-frequency identification (RFID). Therefore, it is sometimes still possible to spot some of the recording devices used. But most people today probably do not know about the vast amount of records collected about them through their digital devices. The old "hunter–hunted game," where the hunted could detect the attacker, flee, and hide, or even play around with his attacker has gone. A surveying entity today is like a ghost, and it cannot be evaded.

Traditional surveillance was naturally physically close. In contrast, today's surveillance has removed the physical aspect. Surveillance only materializes remotely now, potentially in some security monitoring room where unknown security folks stare at their displays. This *remoteness* constitutes an

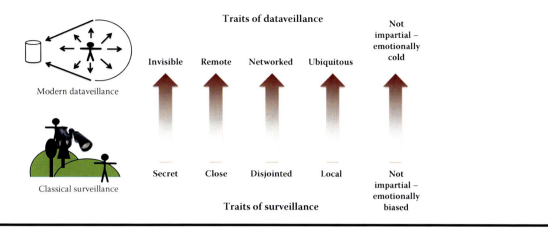

Figure 8.4   How the traits of dataveillance aggravate the traits of classical surveillance.

essential difference between the original kind of surveillance, which people emotionally rejected, and our modern form of surveillance. Humans can perceive the penetrating stare of a close by human being. But this natural instinct is lost in digital surveillance environments. As we do not feel the observer, we cannot really perceive a threat. We are not built for this, just as a lion does not recognize the presence of a huntsman approaching downwind.

The third difference between traditional surveillance and dataveillance is its *pervasiveness*. In states like the former East Germany, where the government tried to spy on huge parts of its population, they could do so only by using human spies and their analog equipment. Observation was limited and imperfect, and activities could occur in unobserved niches. An individual could argue that the observer had missed many aspects of his or her true life and convictions. Thoughts remained unobserved. Current and future forms of surveillance are different. Pervasive computing does not only comprise unlimited geographic coverage for all those places on earth where there is an Internet connection and pervades all of our objects, but it also penetrates and encloses our bodies. Talos suits, smart bracelets, or smart glasses as described in the scenarios in Chapter 3 measure every bodily process, and every blink of attention or disinterest that our pupillary dilation reveals. This pervasiveness would not be too threatening if the technologies were isolated, serving only local purposes. But these technologies can be networked, and their information can be integrated to create a holistic view of individuals' lives. Figure 8.5 depicts an idea of the pervasive surveillance infrastructure and possible connections between different data collection entities. Figure 8.5 is not a complete nor exact representation of the surveillance infrastructure as it is today or will be, but the image depicts the core devices and services involved of modern surveillance.

Finally, a major difference between classical surveillance and dataveillance is the kind of *impartiality* the observing entity practices. An Eastern German, a spy who was reporting

on his neighbor was probably not always impartial. If the spy disliked his neighbor, he might have selectively reported every little detail that suggested disobedience. Victims could be badly misrepresented. But, of course, also the contrary happened, where spies liked their neighbors. In contrast, machines do not care whether the person they observe is a spouse or a neighbor. They are emotionally cold. They are not necessarily more objective, because the way they classify and interpret data may contain the bias of the programmer.

8.4.2 Pros and Cons of Surveillance

Surveillance tends to be discussed as a threat to democratic society. While I personally share the opinion that surveillance is a great threat to societies, I also believe that ethically designed surveillance systems can impede some of these threats while creating local technological benefits. Take the example from my future stories in Chapter 3 of how a company could monitor its employees not only for unidirectional security reasons (as we have it today) but for the mutual benefit of companies and employees:

> Encrypted work activity logging was part of United Games' work terms and conditions for employment. In fact, the integration of activity logging into work contracts in many companies was celebrated years ago as a major achievement of the labor unions. The encrypted activity logging process came as a response to a steep rise in burnout and workplace bullying, which seriously impacted companies' productivity and damaged people's health, mental stability, and well-being. A compromise on the mode of surveillance was struck between unions and employers. Prior to these negotiations, employers had conducted video surveillance in a unidirectional way that undermined employees' privacy while providing

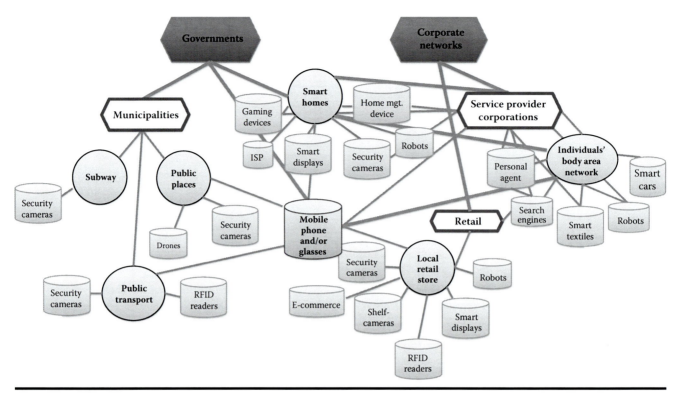

Figure 8.5 Illustrative excerpt of existing and potential future networks of data collecting services and devices (cylinders), hubs integrating these devices/services (circles), and governmental or corporate recipients of the data (hexagons). Sources and networks could eventually be used for systematic electronic monitoring/surveillance (thick lines).

no benefits to them. As part of the new process, employee activities and conversations would be logged in all rooms as well as VR facilities and stored in an encrypted way under the full control of employees (in their personal data clouds). With this system no one, not even the CEO of the company, could view the original data. However, when a security incident happened, employees were informed and asked to share their data. In serious cases of burnout or bullying, employees themselves could initiate a process of data analysis, handing over their secret key so that a designated representative could recover their data, text, and voice streams, and perform a conflict analysis. Data-mining technology would then look for patterns of behavior typical for mobbing or burnout as well as cognitive and emotional states. The streams could also be used to replay specific situations in which conflict had occurred. However, these replays would occur only in the presence of a trained coach or mediator. This practice had not only reduced bullying in recent years but also helped employees to better understand their own communication patterns and behavior. Finally, the encrypted data was also used to extract aggregated heat maps of

the company's general emotional state. This practice helped upper management to better grasp the true emotional "state of their corporate nation."

Of course, many readers of this scenario may perceive this scenario as chilling. Should all work activities really be logged and potentially analyzed? Still, the scenario has benefits. The threatening traits of dataveillance are maintained. But thanks to the ethical data governance and humane use of the technology, at least some of the threats are mitigated (see Figure 8.6). In the context of the scenario, dataveillance is *open and transparent* to employees. Personal data is used only if employees consent and give their private keys to decrypt the data, self-initiating the use of the data at the individual level. *Human judgment* is integrated into the use of the data, which supports self-understanding. Video sequences are analyzed in the presence of a coach. Because the dataveillance infrastructure is ethically designed, it cannot only provide extra security for the company but also reduce employees' misbehavior at work. When things go wrong, as in the bullying case, people can learn about themselves and enter states of self-observation supported by technology. Technologies' cold impartiality can be positive in that people are less offended by the feedback than if this feedback came from a human. Soberly, they can see beyond what they think happened to what really happened. They can review their past behavior

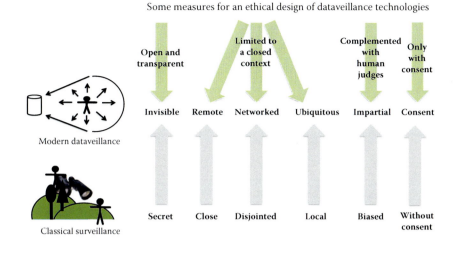

Figure 8.6 Some measures can help to ease the negative effects of modern surveillance technologies.

and try to use their insight to resolve conflicts. Machines could also use the vast amounts of behavioral data to detect patterns of behavior that help us to better understand behavior. Great learning at the individual, organizational, and societal level may become possible, underpinning centuries of philosophical reasoning with more complete facts.

Note also that, in the work scenario, dataveillance is *technically limited to a closed context*. The decentralized isolation of dataveillance facilities, which does not network them at a higher level, reflects what Helen Nissenbaum (2004) called the "contextual integrity" of data collection and use (see again Section 5.6.3). Such contextual integrity through closed-context monitoring can balance people's desire for safety with the potential abuses of networked pervasive surveillance. Of course, this model requires that the people who are observed trust key players in the surveillance network, in particular, the data hubs that collect and pool personal data (denoted as circles in Figure 8.5). In Chapter 9, I explain what trustworthiness means and how technology companies can build trust.

The learning benefits we can derive from "Big Data" are complemented by another argument that is often brought forward, which is that people seem to appreciate surveillance. When surveillance infrastructure is in place, some people feel safer in places that are traditionally unsafe but cannot be avoided, such as parking lots, underground stations, or parks at night. Solove reports that Britain's closed circuit television (CCTV)—a network of 1.5 million to 2 million public surveillance cameras—is widely perceived as "a friendly eye in the sky" (2006, p. 494). Jeffrey Rosen (2005) reports on students' reactions to body scanners at airports. He observes that quite a few welcome being naked for various reasons, ranging from security fears to wanting to demonstrate their "purity."

As described earlier, large-scale public polls also tend to suggest that people prefer security to liberty. In his book *The*

Naked Crowd, Rosen explains why people embrace surveillance: On a higher sociological level he argues, the "crowd's unrealistic demand for a zero risk society is related to our anxieties about identity. Because we can no longer rely on traditional markers of status to decide whom to trust (i.e., clothes, family, religion, face-to-face meetings…), the crowd demands that individuals in the crowd prove their trustworthiness by exposing as much personal information as possible" (2004, Prologue) through the technologies that are set up. Rosen questions whether politics should be driven by such feel-good investments into surveillance technologies that appease ordinary people's sentiments. Crowds are vulnerable he says to systematic errors and biases in judgments; they are driven by "pseudo-events" in the press that make them misjudge the true risks in their daily lives. "Why should we care about the emotionalism of the Naked Crowd?" Rosen asks provocatively.

There are several reasons why many intellectuals criticize the build-up of a technological surveillance infrastructure. One is the potential for abuse. Aristotle (1998) warned us that democracies have historically lacked stability. We know from our own history how the precise recordings of Jews' whereabouts in some European countries led to their systematic persecution during the Holocaust (see Chapter 5, Figure 5.5). What would happen if a networked dataveillance infrastructure similar to the one depicted in Figure 8.5 fell into the wrong hands? Would citizens in future societies constantly need to fear being watched and lose the right to free speech, as in George Orwell's *1984* or Dave Eggers 2013 novel *The Circle*? Both Orwell and Eggers powerfully illustrate how data-power asymmetries translate into totalitarianism; only that in Orwell's *1984* totalitarianism is driven by the state, whereas in Eggers's *The Circle*, unregulated corporate power leads to economic totalitarianism.

Totalitarianism and power asymmetries are complemented by a perverse system of self-censorship even if data is not

Figure 8.7 The Presidio Modelo was a "model prison" of the Panopticon design, built on the Island de la Juvental in Cuba. (© CC BY-SA 3.0, Friman 2005.)

necessarily available and used. Jeremy Bentham powerfully demonstrated how self-censorship works once a surveillance architecture is set up. Bentham designed and described a prison architecture, which he called the "Panopticon" (Figure 8.7): The Panopticon design allows guards to observe all inmates of a prison without the inmates being able to tell whether they are being watched. The inmates are in cells around a central circular tower structure. Although it is physically impossible for the guard to observe all cells at once, inmates cannot know when they are being watched, so they act as though they are being watched at all times. As a result, they constantly self-censor their behavior. Figure 8.7 shows a prison in Cuba that was built along the concept of the Panopticon.

Why is self-censorship problematic? After all, some people argue that citizens and prisoners should behave well, that some self-censorship is therefore good for society, and that those who have nothing to hide do not feel followed either. However, remember the definition of positive liberty: People need to be able to make decisions of their own free will. Their behavior should be driven by their own desires and not by some external manipulative force. As we become more conscious of being watched, the motivation of our behavior may no longer come from ourselves. We may act well just because we are being watched. As a result, we degrade ourselves to the state of slaves in a surveillance machine. Edward Snowden* (2014) brought up this point when he wrote: "When we know we're being watched, we impose restraints on our behavior—even clearly innocent activities—just as surely as if we were ordered to do so. The mass surveillance systems of today, systems that pre-emptively automate the indiscriminate seizure of private records, constitute a sort of surveillance time-machine—a machine that simply cannot

---

* Edward Snowden is a whistleblower who in 2013 uncovered massive surveillance activities undertaken by the U.S. National Security Agency and international secret service partners.

operate without violating our liberty on the broadest scale. And it permits governments to go back and scrutinize every decision you've ever made, every friend you've ever spoken to, and derive suspicion from an innocent life. Even a well-intentioned mistake can turn a life upside down."

8.4.3 Reaching Golden Means in Surveillance?

Figure 8.5 shows that dataveillance is a bottom-up phenomenon: Each device or service operates more or less individually depending on the level of decentralization embedded in the technological design. Devices and services then connect to data hubs to the extent required by the technical architecture. Devices, services and hubs can then be integrated into pervasive data sharing networks. The overall surveillance system operates like a hierarchal network in that it works only if the original data sources supply personal data. The sources of dataveillance can fuel mass surveillance if they are accessed and provide even more information when they are connected. This architecture puts tremendous responsibility on the design of each data source.

Each decentralized technical data source should therefore be built with privacy controls inside. Such efforts are recognized today on a political level and are called "privacy by design" (Cavoukian 2011; Spiekermann 2012). Chapters 17 and 18 outline in detail what engineers can do to build systems with privacy inside. This practice can significantly influence the extent of dataveillance. But that said, governments and industry buyers of technology influence how technology is built. Their demand for technical features drives the type of technology that is supplied. The extent of their demand also determines the number of firms and the scale of production of surveillance technologies. Companies and governments therefore constantly need to determine the extent of data collection through their systems. For example, a gaming service might collect, store, analyze, and share fine-grained emotional player data, as outlined in the gaming scenario. A university or corporation might monitor students or employees. Or governments might use surveillance cameras, drones, or robots to monitor citizens. In all these cases, a small group of people makes an initial decision that then affects many others. How can this small group make surveillance decisions reasonably and wisely?

Rosen (2005, Prologue) shows that "technologies and laws demanded by a fearful public often have no connection to the practical realities of the threats that we face." Leaders cannot respond numbly to public polls or the hungry sales efforts of security equipment firms. Instead, they need to make wise judgments on the extent of surveillance they want to support in their organization, consciously balancing security fears, privacy concerns, and threats to liberty. There is no absolute answer on how to resolve this trade-off. All technological deployment decisions are unique. But the question of

BOX 8.2 A GOLDEN MEAN PROCESS FOR INDIVIDUAL DECISION MAKERS WHO DECIDE ON THE EXTENT OF NECESSARY SURVEILLANCE

How can we find a golden mean in our surveillance practices? One way is for leaders to be courageous enough to publicize and transparently share their initial judgment on what the golden mean is. They should share how they arrived at their judgment based on reason and facts. The general public can then publically react to the initial judgment, potentially challenging it. This kind of public request for comments is abbreviated as RFC. Requests for comments are very common in the technical world today. *Wikipedia's* dispute resolution system works on an RFC basis. The Internet Engineering Task Force (IETF), a principal technical development and standards-setting body for the Internet, also uses RFCs. The result of such a stakeholder feedback and collaboration process can be an adjustment or rebalancing of the initial judgment based on a more widely shared agreement of what constitutes a "golden mean." Simplistically speaking, transparency establishes checks, which then allow for balances of an initial judgment. Kant argued that every legislator should give "his laws in such a way that they *could* have arisen from the united will of a whole people and to regard each subject, insofar as he wants to be a citizen, as if he has joined in voting for such a will."[1] This process does not necessarily mean that the fears of the general public should determine the judgment. Experience has shown that transparent and identified commenting systems are frequented by experts more often than by the general public. As reasons and facts could be shared between decision makers and a self-elected polis, a good middle ground could be found and argued. Figure 8.8 illustrates what I call the "golden mean process."

The golden mean process depends strongly on the wisdom inherent in the initial judgment. One source of good judgment is good data and a focus on outcomes. As Michael Porter put it: "Successful collaboration will be data driven, clearly linked to defined outcomes, well connected to the goals of the stakeholders, and tracked with clear metrics."[2] If the initial judgment is too extreme, then the polis responding to it will equally fall into extremes. This polarization of the problem space can easily create conflict where really compromise is required. For a wise initial proposal we need wise leaders that ideally possess what Aristotle would call the virtues of sophrosyne (temperance), philotimia (the right level of ambition) as well as the courage to publish and defend their opinion (andreia).[3] The same virtues are important in adjusting the initial judgment.

NOTES

1. Immanuel Kant, 1793/1999, "On the Common Saying: That May Be Correct in Theory, But Is of No Use in Practice," in *Practical Philosophy,* edited by M. J. Gregor and A. W. Wood, New York: Cambridge University Press, 297.
2. Michael Porter and Mark R. Kramer, 2011, "Creating Shared Value," *Harvard Business Review* 89(1).
3. Aristotle, 2000, *Nicomachean Ethics,* translated by Robert Crisp, Cambridge Texts in the History of Philosophy, Cambridge: Cambridge University Press.

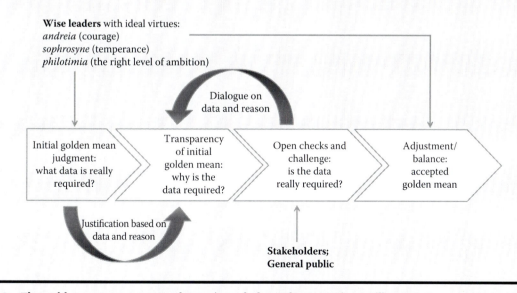

Figure 8.8   The golden mean process to determine a balanced extent of surveillance.

extent has recurred in ethical practice for millennia. It is asking for the "golden mean" (Aristotle) in our practices, "the middle way" (Buddhism), or what the Chinese consider their "doctrine of the mean" (Chinese: 中庸; pinyin: zhōng yōng). Box 8.2 makes a suggestion of what a golden mean process for surveillance could look like.

EXERCISES

1. Describe a situation in which you felt that your security was threatened. What did you do to feel secure again? Did any of these measures relate to your liberty? How did your positive and negative liberty help to restore your feeling of security?

2. Reflect on the security operations of a bank that you use. What does the bank need to do to secure your financial information?

3. Think of the Alpha1 robots described in Chapter 3. What security measures are vital for the robot to embed so that it does not undermine user safety?

4. List and reflect on the strategies that companies use to secure the privacy of their customer data. What do you think makes more sense for a company: to anonymize/pseudonymize its data (see Chapter 5, Box 5.1) or to keep it identified and pursue strict policy management (see the auditability section in this chapter)? Apply your thinking to the emotional profiles that United Games collects from its virtual reality players.

5. Think about the Panopticon design illustrated in Figure 8.7. In your own life, are there situations where a Panopticon effect causes you to change your behavior?

6. Identify examples where a Panopticon is established beyond security reasons. Try to describe how our societies have built Panopticons that might influence our behavior, leading us to control ourselves or act in ways that are unnatural for us.

7. Using Figure 8.5, fill in the IT systems (names of devices, service providers, manufacturers, etc.) that you directly or indirectly use yourself. After you complete the picture, identify who possesses the greatest dataveillance power over yourself.

8. Split the class into three teams. The task of team 1 is to choose a case that is relevant in the immediate environment and where a video surveillance investment was made for security reasons. Team 1 should find out how many and where the cameras were installed, who uses them, for what purpose, and so on. Once this information is accumulated it should not be revealed to the other classmates outside of team 1. The other classmates are just informed about the surveillance case that has been researched (i.e., video surveillance on the university campus). Split the rest of the classmates into two teams. One team represents the IT investors who are the "guards" that want to implement video surveillance infrastructure. The other team represents citizens, the "polis." The guards start a negotiation process with the citizens. Based on reason and numbers, they propose how much video surveillance they think is needed. The other team, the polis, then scrutinizes and challenges this proposal. The two teams need to agree on the amount of and conditions of surveillance that is appropriate. After the teams agree, compare their decision to what really happened (based on the information accumulated by team 1).

9. Investigate a video surveillance infrastructure from your daily work life or private life. Try to find out how many cameras there are, where they are placed, and for what purposes. Also get crime statistics for places similar to the one you are investigating. Then run through the golden mean process. Judge whether the surveillance infrastructure around you balances the desire for protection against crime with your privacy. Does it achieve this balance? Were the builders of the infrastructure wise in the Aristotelian sense?

References

Alhadeff, J., B. Van Alsenoy, and J. Dumortier. 2011. "The Accountability Principle in Data Protection Regulation: Origin, Development and Future Directions." Paper presented at Privacy and Accountability, Berlin, Germany, April 5–6.

Anderson, R. 2008. *Security Engineering: A Guide to Building Dependable Distributed Systems.* 2nd ed. Indianapolis: Wiley Publishing.

Aristotle. 1998. *Politics.* Translated by C.D.C Reeve. Indianapolis, IN: Hacket Publishing.

BVerfG. 1983. "Verdict on the Census." BvR 209, 269, 362, 420, 440, 484/83. German Federal Constitutional Court, Bundesverfasungsgericht (BVerfG).

Cavoukian, A. 2011. *Privacy by Design … Take the Challenge.* Information and Privacy Commissioner of Ontario, Canada.

Chellappa, R. K. and P. A. Pavlou. 2002. "Perceived Information Security, Financial Liability and Consumer Trust in Electronic Commerce Transactions." *Logistics Information Management* 15(5):358–368. doi: 10.1108/09576050210447046.

Cherdantseva, Y. and J. Hilton. 2013. "A Reference Model of Information Assurance & Security." Paper presented at 8th International Conference on Availability, Reliability and Security (ARES), Regensburg, Germany, September 2–6.

Clarke, R. 1988. "Information Technology and Dataveillance." *Communications of the ACM* 31(5):498–512.

CNSS. 2010. National Information Assurance (IA) Glossary. Fort Meade, MD: Committee on National Security Systems.

Cranor, L. F., B. Dobbs, S. Egelman, G. Hogben, J. Humphrey, and M. Schunter. 2006. *The Platform for Privacy Preferences 1.1 (P3P1.1) Specification: W3C Working Group Note 13 November 2006.* World Wide Web Consortium (W3C)—P3P Working Group 2006. http://www.w3.org/TR/P3P11/, accessed July 17, 2007.

European Parliament and the Council of Europe. 1995. Directive 95/46/EC of the European Parliament and of the Council of 24 October 1995 on the protection of Individuals with regard to the processing of personal data and on the free movement of such data. L 281/31. Official Journal of the European Communities.

Fujitsu. 2010. "Personal Data in the Cloud: A Global Survey of Consumer Attitudes." Tokyo, Japan.

ISO. 2008. ISO/IEC 27005 Information Technology—Security Techniques—Information Security Risk Management. International Organization for Standardization.

ISO. 2011. EN ISO 10218-1: Robots for Industrial Environments—Safety Requirements (Part 1). ISO.

ISO. 2012. ISO/IEC 15408: Common Criteria for Information Technology Security Evaluation.

ISO. 2014a. ISO/IEC 27000 Information Technology—Security Techniques—Information Security Management Systems—Overview and Vocabulary. International Organization for Standardization.

ISO. 2014b. ISO/IEC 29100: Information Technology—Security Techniques—Privacy Architecture Framework. DIN Deutsches Institut für Normung e.V.

Line, M. B., O. Nordland, L. Rostad, and I. A. Tondel. 2006. "Safety vs. Security?" Paper presented at 8th International Conference on Probabilistic Safety Assessment and Management (PSAM), New Orleans, Louisiana, May 14–18.

Margulis, S. 2003. "Privacy as a Social Issue and Behavioral Concept." *Journal of Social Issues* 59(2):243–261.

Maslow, A. 1970. *Motivation and Personality*. 2nd ed. New York: Harper & Row Publishers.

National Institute of Standards and Technology (NIST). 2013. NIST 800-53: Security and Privacy Controls for Federal Information Systems and Organizations. Gaithersburg, MD: U.S. Department of Commerce.

Nissenbaum, H. 2004. "Privacy as Contextual Integrity." *Washington Law Review* 79(1).

Ohm, P. 2010. "Broken Promises of Privacy: Responding to the Surprising Failure of Anonymization." *UCLA Law Review* 57:1701–1777.

Organisation for Economic Co-operation and Development (OECD). 1980. OECD Guidelines on the Protection of Privacy and Transborder Flows of Personal Data.

Piétre-Cambacédès, L. and C. Chaudet. 2010. "The SEMA Referential Framework: Avoiding Ambiguities in the Terms 'Security' and 'Safety.'" *International Journal of Critical Infrastructure Protection* 3(2):55–66.

Rosen, J. 2004. *The Naked Crowd: Reclaiming Security and Freedom in an Anxious Age*. New York: Random House.

SANS Institute. 2004. "An Overview of Sarbanes–Oxley for the Information Security Professional." SANS Institute InfoSec Reading Room. Swansea, UK.

Seneviratne, O. and L. Kagal. 2014. "Enabling Privacy Through Transparency." Paper presented at IEEE Conference on Privacy, Security and Trust, Toronto, Canada, July 23–24.

Snowden, E. 2014. "On Liberty: Edward Snowden and Top Writers on What Freedom Means to Them." *The Guardian*, February 21. http://www.theguardian.com/books/2014/feb/21/on-liberty-edward-snowden-freedom, accessed August 9, 2014.

Solove, D. J. 2006. "A Taxonomy of Privacy." *University of Pennsylvania Law Review* 154 (3):477–560.

Spiekermann, S. 2012. "The Challenges of Privacy by Design." *Communications of the ACM* 55(7).

United Nations General Assembly. 1948. Universal Declaration of Human Rights.

United Nations Office on Drugs and Crime (UNODC). 2013. "Comprehensive Study on Cybercrime."

Verizon. 2014. "2013 Data Breach Investigations Report." Verizon Trademark Services LLC.

Weitzner, D., H. Abelson, T. Berners-Lee, J. Feigenbaum, J. Hendler, and G. J. Sussman. 2008. "Information Accountability." *Communications of the ACM* 51(6):82–87.

Yousafzai, S. Y., G. R. Foxall, and J. G. Pallister. 2010. "Explaining Internet Banking Behavior: Theory of Reasoned Action, Theory of Planned Behavior, or Technology Acceptance Model?" *Journal of Applied Psychology* 40(5):1172–1202.

Chapter 9

Trust in Future IT Environments

"Whatever matters to human beings, trust is the atmosphere in which it thrives."

Sissela Bok (1978)

In her book *Moral Repair: Reconstructing Moral Relations after Wrongdoing*, Margaret Urban Walker (2006) observes that humans interact with the help of "default trust." We perceive "zones of default trust," spaces and circumstances in which we can use trust as a shortcut when we decide to cooperate with others: "Sometimes when people refer to their 'communities,' either as networks of people or as geographical locations or both, they capture this sense of the place where one feels relatively safe. This is not because one believes one is utterly protected, but because one believes one knows what to expect and from whom to expect it, and one knows what is normal and what is out of place. One knows, in a word, what to expect and whom to trust. This practical outlook of ease, comfort, or complacency that relies on the good or tolerable behavior of others is the form of trust I call 'default trust'" (Urban Walker 2006, p. 85).

Trust has long been an integral part of functioning social systems (Luhmann 1979). Many people have a disposition to trust by character (Rotter 1971). These preconditions are a valuable starting point for the machine age. Yet, trust is dangerous and can be betrayed. As machines move from engineers' playgrounds and media analysts' imaginations into the real world, we might encounter machines that do not warrant our trust. Machines often do not work the way we expect them to, nor do they necessarily work in our best interest and respect our expressed or implicit preferences. They may even turn against us at some point, as described in the robot scenario in Chapter 3. Some scholars, reflecting on our developing technological environment, talk about an emerging "credibility crisis" (Cohen 2012, p. 1924). Of course, some believers in technology point to the extent to which we are willing to rely on technology on a daily basis.

However, as I will show, there is a difference between relying on machines because we have confidence in using them and really "trusting" them. Engineers must understand this difference and delve deeply into the concept of trust so that they can build machines that can be deemed trustworthy.

9.1 What Is Trust?

In the Chapter 3 scenarios about the future, many machines require trust in order to be embraced. Just think of the enormous trust that would be required for people to allow governments to have humanoid Alpha1 robots patrol the streets. People would need to trust that the robots would not wrongly hurt anyone and would be benevolent rather than dangerous. Enormous trust would also need to be placed in workplace systems. Agent Hal manages almost all operations at the robot manufacturer Future Lab. It drives manufacturing as well as part of the sales operations. It decides about what and when to inform company employees and top management about company operations. Finally, both United Games and Halloville Mall promise to be reliable when it comes to data handling practices. Both economic entities collect vast amounts of data about employees and customers. But United Games promises to analyze the data only with the consent of employees. And the Halloville Mall promises to respect some customers' desire to stay anonymous during their shopping trips. In all of these cases, people trust their computer systems and, indirectly, the service providers of the system. People trust that the Alpha1 systems are competent and that the robots will act benevolently in the citizens' interest. People trust in Hal's competence to judge Future Lab's operations and expect it to act predictably. Finally, they trust in the moral integrity or honesty of United Games and the Halloville Mall to not abuse the systems. These examples show that the four most prevalent trusting beliefs in the trust literature—*benevolence, competence, predictability*, and *honesty*—are just as relevant in machine environments as in human environments (McKnight and Chervany 1996). We psychologically transfer our trusting beliefs to machines (robots, agents), expecting humanlike characteristics of trustworthiness from them (Reeves and Nass 1996). We also place our trust in the providers of these machines, trusting them to not misuse or abuse the power of the machinery against our interests (Figure 9.1). Finally, in their work on trusting beliefs in e-commerce contexts, David Gefen and his colleagues identified a fifth trust belief that is important for online business environments in particular: the *absence of opportunism* (Gefen, Elena, and Straub 2003).

Given these five trusting beliefs, we can understand the expectations that are inherent in the definition of trust. Niklas Luhmann (1979) saw trust as a willingness to behave based on expectations about the behavior of others when considering the risk involved. With respect to the machines and their providers, these expectations are *competence, benevolence, honesty, predictability*, and *nonopportunism*.

Luhmann's definition hints at another important dimension of trust: the presence of risk. Trust and risk are in an unconditional positive relationship with each other. The more risk there is, the more we need to trust that things will work out well. Trust is required only when there is no further risk reduction possible and when the trustor is hence vulnerable. Vulnerability can be appeased if the machines or operators signal that they are competent enough to handle the risk. But beyond being competent, the trusted party must also *commit* to act well and in line with the expectations of the trustor. Here, we must therefore also consider commitment and motive on the side of the trusted party.

9.2 Distinguishing Trust from Confidence and Reliance

Philosophers differ on how they treat the question of commitment needed from the trusted (McLeod 2011): For some philosophers, it is just important that a trusted party signals his or her commitment. For others, the origins of such a commitment are equally important. Commitment can be "calculative" when it is motivated by selfish interests or when people are engaged in a kind of "social contract," based, for instance, on a public declaration or legal act. In contrast, commitment can also be based on the goodwill or moral integrity of the trusted entity. Commitment comes from the notion of *care* that the trusted party has for the trustor. McLeod (2011) argues that we can really only trust when commitment comes from care and moral integrity. Otherwise, she argues, we cannot trust but merrily *rely* on the other party for the time being: "The particular reason why care is central is that it allows us to distinguish between trust and mere reliance" (McLeod 2011, p. 5).

Differentiating trust from reliance is only one distinction philosophers have made to carve out the true nature of trust. Another distinction is between trust and confidence. Trust is an active *decision* by a trustor to delegate to a trustee some aspect of importance to achieve a goal (Grodzinsky, Miller, and Wolf 2011). In fact, many instances where we rely on machines today do not involve active decisions and are therefore not really expressions of trust. Wolter Pieters (2011) explains the difference between confidence and trust: "Confidence means self-assurance of the safety or security of a system without knowing the risks or considering alternatives. Trust [in contrast] means self-assurance by assessment of risks and alternatives … We have confidence in electricity supply, in people obeying traffic rules, etc. When there are different options possible, such as in choosing a bank for one's savings, a comparison needs to be made, and trust takes the place of confidence" (p. 56). It is important to know about the difference between reliance, confidence, and trust, because many would argue that people trust machines already and will continue to do so in the future. However, what they really observe is not trust but reliance, which may be supported by more or less confidence.

Finally, new forms of trust relationships will emerge in the machine age. Scholars talk about "e-trust," which is "specifically developed in digital contexts and/or involving artificial agents" (Taddeo and Floridi 2011, p. 1). E-trust is subject to the same underlying dynamics of trust that I described for our physical human world, but we interact with different entities or at least perceive them differently. Figure 9.1 is a simplified representation of e-trust depicting the three entities people need to trust online: (1) machines, (2) other people whom they encounter through machines, and (3) providers of machines. Indirectly, people also need

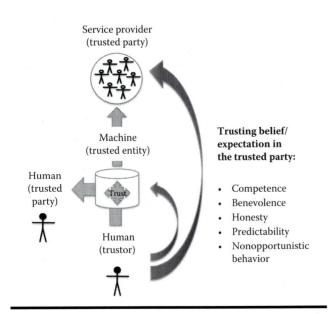

Figure 9.1   A conceptual analysis of e-trust.

	Human–Human	Human–Agent	Agent–Human	Agent–Agent
Physical Encounter	HHP	HAP	AHP	AAP
Virtual Encounter	HHV	HAV	AHV	AAV

- HHP-trust: traditional notion of human "face-to-face" trust
- HHV-trust: humans trust each other, but mediated by electronic means
- HAP-trust: human trusts a physically present agent, e.g., a robot
- HAV-trust: human trusts a virtual agent, e.g., embodied interface agent
- AHP-trust: an artificial entity (i.e., a robot) trusts a human who is physically present
- AHV-trust: an artificial entity (i.e., a software program) trusts a human who is virtually present
- AAP-trust: an artificial agent trusts another artificial agent in a physical encounter (i.e., two robots interacting)
- AAV-trust: an artificial agent trusts another artificial agent in a virtual encounter (i.e., two web bots)

Figure 9.2 Forms of e-trust derived from Grodzinsky, Miller, and Wolf.

to trust the engineers who built the machines, because engineers are responsible for how the machine works. Grodzinsky, Miller, and Wolf (2011) define the forms of e-trust that are shown in Figure 9.2. Their summary of e-trust relationships is more complete than other summaries because it considers machine-to-machine communication as well. However, it does not contain a vital trustee: the providers of machines, who determine how machines are ultimately used and hence how trustworthy the machines are in terms of competence (determined to some extent by the financial investment made into them) and commitment (determined to some extent by the moral attitude of the operator).

9.3 Trust Mechanisms in Machines

Trust is one of those things in life that we cannot want or demand. We need to earn it and provide evidence that we are trustworthy. This evidence can be created in various ways. Pettit (2004) distinguishes between evidence of face, evidence of file, and evidence of frame. *Evidence of face* is really important when trust is built between humans in physical encounters; how someone says something, his or her body language, and the (often involuntary) emotions expressed in people's faces are key to building trust. At the moment, this kind of trust building is still rare in human–machine interaction. But as machines become physical in the form of robots, this kind of evidence may gain importance. Engineers already work on integrating facial expression into robots that aim to create trust. An almost historic example for this kind of work is MIT's (Massachusetts Institute of Technology)

robot Kismet, which reproduces emotions in the form of various facial expressions.[*]

Evidence of file is the interaction we have with another person or entity over time. We consciously or unconsciously track the dynamics of being with others and either build up trust or become cautious. Unlike the evidence of face, evidence of file is harder to fake because the trustee needs to show consistent behavior over time to be trusted. Some scholars refer to this kind of trust as "knowledge-based trust" or "familiarity" (Gefen, Elena, and Straub 2003). I will discuss in Section 9.4 how this kind of evidence can be created by using reputation systems.

A third way of building trust is to provide *evidence of frame*. Evidence of frame is created by observing how a person or entity treats others, or how others testify to the trustworthiness of the trustee. Again, reputation systems are a very valuable way to provide this kind of evidence. But when it comes to machines and their trustworthy functioning, seals and symbols can also confirm that the machine complies with certain standards of behavior and construction quality.

Thomas Simpson (2011) adds two further types of evidence to Pettit's list: evidence of context and evidence of identity. *Evidence of context* means that aspects in a situation can push the trusted party to behave in a good way. Pieters (2011) outlines how "explanations for confidence" are particularly suited to provide evidence of context. For instance, public statements of guarantees, long warranties, and strong regulation of a technology, accompanied by sanctions for misconduct, help to build people's calculative trust in machines. The goal of these explanations of confidence (in a context) is not to show people how an IT system functions in detail but to make them comfortable enough to use it. Going a step further, Pieters argues, we can identify "explanations for trust." These lay open how a system works. The goal here is to create transparency around a system. This transparency creates trust because it supports an active choice for using a particular system over another.

Simpson (2011) points to the importance of *evidence of identity*. A person's occupation, religious creed, or way of living may be evidence of trust in respective situations where such characteristics become important. For example, people trust that a doctor can help when an accident happens. Transferring this form of evidence to a company context, some machine service providers have built up strong reputations for the performance of their machines. At the beginning, evidence of file is necessary for brand building. But after an initial period of performance proof, the brand alone often inspires trust. This is then called evidence of identity.

---

[*] "Kismet, the Robot," accessed August 15, 2014, http://www.ai.mit .edu/projects/sociable/baby-bits.html.

Finally, Gefen, Elena, and Straub (2003) have identified "situation normality" and "ease of use" as factors that are particularly important in e-commerce contexts to build trust. Situation normality seems to correspond to the predictability belief. In this view, "people tend to extend greater trust when the nature of the interaction is in accordance with what they consider to be typical and, thus, anticipated … In contrast with familiarity, however, situational normality does not deal with knowledge about the actual vendor; rather, it deals with the extent that the interaction with that vendor is normal compared with similar sites" (Gefen, Elena, and Straub 2003, p. 64). For machine design, this means that machines can build trust by using typical steps and forms of interaction that users recognize from comparable systems as well as information requests comparable to other systems. This observation hints at the importance of common design standards for multiple future systems. Common standards for system use are already widely known. For example, many diverse systems use the same symbols to signal on/off functionality. Reference architectures are used for application design. Both of these foster *situation normality*, which again greatly increases the perceived ease of use of a system and fosters trust in it.

9.4 How Computer Scientists Understand Trust

Note that computer science students learn about trust in a different manner. In computer science textbooks, trustworthiness of systems is often discussed under the alternative term of "system dependability." System dependability is seen as a nonfunctional requirement. Ian Sommerville summarizes this construct as follows: "The dependability of a computer system is a property of the system that reflects its trustworthiness. Trustworthiness here essentially means the degree of confidence a user has that the system will operate as they expect, and that the system will not *fail* in normal use" (2011, p. 291).

Summerville then specifies what dependability means, outlining that dependability (trustworthiness) depends on the security of a system, the safety of a system, and its reliability. Both security and safety were defined earlier. The system trait of *reliability* is similar to what scholars in the information systems (IS) literature call "situation normality." It means "the probability, over a given period of time, that the system will correctly deliver services as expected from the user" (Sommerville 2011, p. 292). Note though that the definition of reliability for computer science readers is more precise than situation normality is for IS readers. Reliability is defined in terms of a "probability over time" and hence viewed as a measurable system variable that the system can be tested for.

From this discrepancy of the trust value's definition we learn two things: The most important one is that the computer science perspective is at this moment much narrower than the general social perspective. Although computer scientists learn to think of trust in terms of dependability, Figure 9.3 makes plain that this is a very limited view of what makes a system trustworthy from a user's perspective (or from the perspective of society at large). Dependability is just one form of system evidence. And even if it is the most important one (recognized by the bolted line around dependability in Figure 9.3), engineers are just as demanded when it comes to create emotional user interaction, to ensure transparency of the system and to continuously improve its ease of use. Engineers will also be involved in creating evidence of frame through certification of the system or ramping it up for a quality seal.

That said, the second learning here is that engineers cannot be made responsible for creating trust in systems by themselves. In general, managers, such as product managers, need to work on providing all the other forms of evidence required for trust. They need to think about warranties and guarantees they can give, nourish trust in the service brand as a whole, and ensure appropriate media voice around the service. This again implies that managers need to be close to engineers and understand the system well enough to create the right buzz around it. Too often systems cannot deliver on false promises made by managers. The result is not only a loss of face of the people involved but also a general damage to the brand.

Figure 9.3 summarizes the main trust-building mechanisms that are identified in the literature; each of them requires considerable investment in system design, certification, and marketing.

9.5 Reputation Systems

One of the most powerful ways to signal trustworthiness is to score well in reputation systems. A reputation system collects, distributes, and aggregates feedback about participants' past behavior or about goods and services (Resnick 2000). Typically, it does so within a community or domain. A well-known example is TripAdvisor's history of comments for hotels or Amazon's star system for books. Many travel decisions today are made with the support of reputation systems. Reputation systems allow for evidence of file because they often contain a history of transactions with the trusted party; they also allow for evidence of frame because other customers or independent assessors have also investigated the object.

Based on a critical discussion of existing reputation systems and the forms of trust evidence described previously, some measures can help to optimize the value of online reputation (see Figure 9.4 as well as Simpson 2011). The most

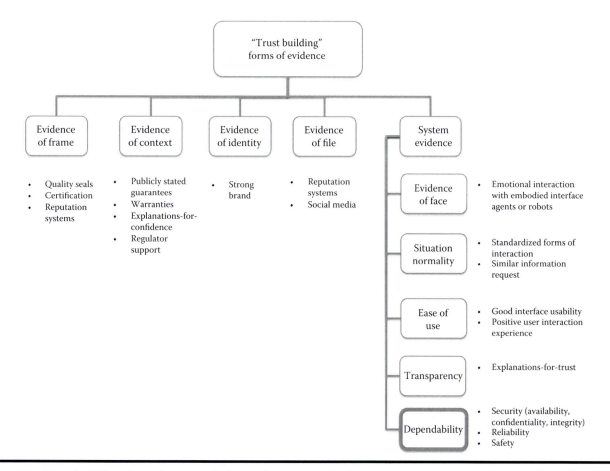

Figure 9.3 **Trust building through system design and system marketing.**

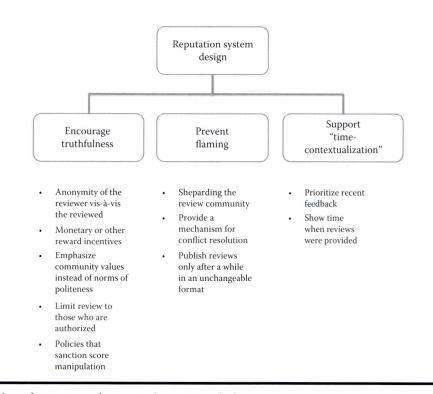

Figure 9.4 **Mechanisms for trustworthy reputation system design.**

important characteristic of a reputation system is to ensure a high level of *truthfulness*. The entities that are rated in a reputation system have an incentive to look good. Therefore, manipulation of system results or even fraud is more likely to occur than in other systems. At the same time, social norms of politeness often impede people from leaving negative comments online. For example, 99.1% of eBay's reputation system comments are positive (Simpson 2011). These behaviors can easily undermine the value of a reputation system.

Operators of reputation systems can encourage truthfulness in various ways: They can offer monetary or other reward incentives to motivate high quality reviews that are balanced. Reviewers can be assured that their identity will not be shared with those that are being rated. Loyalty to the community could be explicitly encouraged over politeness. Operators might also restrict reviewing to those who prove that they really used a service. Such an entitlement measure is problematic because it limits the number of reviewers. However, controlling who reviews also helps to prevent the entities that are being reviewed from writing reviews themselves. If reviews are not controlled, ratings can be inflated. Nonlegal sanctions, such as a complete delisting of a person, good, or service, are powerful ways of thwarting manipulation.

Reputation systems must also address flaming or "shitstorm" behavior that can sometimes be observed online. This involves overly negative comments and scores on entities without true justification. Sometimes there is also tit-for-tat negative reviewing. Simpson (2011) reports for instance that sellers leave negative or neutral feedback on a buyer 61.7% of the time that that buyer leaves negative or neutral feedback on them. However, such tit-for-tat behavior does not support truth building. Service providers can counter this kind of behavior by actively resolving conflicts through a mediator. Or, feedback can be published only after both parties have submitted it, without providing the ability to change comments afterward.

Respecting the *time context* of a reputation score mirrors how humans judge trustworthiness and has produced more reliable reputation scores (Novotny and Spiekermann 2014). A challenge for digital reputation is that those with very strong reputations can afford to be untrustworthy on occasion, relying on the system to view sudden negative feedback as an outlier. A way to avoid this is to give recent or timely feedback greater weight. Prioritization of recent feedback over old feedback also allows people to rebuild a reputation if they had a negative score in the past.

EXERCISES

1. See how many trust relationships you can identify in the robot scenario from Chapter 3. Use Figure 9.2 as a structural baseline for your analysis.

2. Reflect on the difference between reliance, trust, and confidence. From the technologies that you use, find examples where you rely, where you need to trust, and where you are confident that things will work out.

3. Analyze the explanations that are given by the operators of a system you have confidence in and another system that you need to trust. Compare the explanations that are given by the operators to gain your trust.

4. Choose a reputation system that you regularly use and analyze its trust-building mechanisms alongside Figure 9.4.

References

Cohen, J. E. 2012. "What Privacy Is For." *Harvard Law Review* 126:1904–1933.

Gefen, D., K. Elena, and D. W. Straub. 2003. "Trust and TAM in Online Shopping: An Integrated Model." *MIS Quarterly* 27(1):51–90.

Grodzinsky, F. S., K. W. Miller, and M. J. Wolf. 2011. "Developing Artificial Agents Worthy of Trust: 'Would you Buy a Used Car from This Artificial Agent'?" *Ethics and Information Technology* 13(1):17–27.

Luhmann, N. 1979. *Trust and Power*. Chichester, UK: Johan Wiley & Sons.

McKnight, H. D. and N. L. Chervany. 1996. *The Meaning of Trust*. Minneapolis, MN: University of Minnesota.

McLeod, C. 2011. "Trust." In *The Stanford Encyclopedia of Philosophy*, edited by Edward N. Zalta. Stanford, CA: The Metaphysics Research Lab.

Novotny, A. and S. Spiekermann. 2014. "Oblivion on the Web: An Inquiry of User Needs and Technologies." Paper presented at European Conference on Information Systems (ECIS 2014), Tel Aviv, Israel, June 9–11.

Pettit, P. 2004. "Trust, Reliance and the Internet." *Analyse und Kritik* 26(1):108–121.

Pieters, W. 2011. "Explanation and Trust: What to Tell the User in Security and AI?" *Ethics and Information Technology* 13(1): 53–64.

Reeves, B. and C. Nass. 1996. *The Media Equation: How People Treat Computers, Television, and New Media Like Real People and Places*. New York: Cambridge University Press.

Resnick, P. 2000. "Reputation Systems." *Communications of the ACM* 43(12):45–48.

Rotter, J. B. 1971. "Expectancies for Interpersonal Trust." *American Psychologist* 35:1–7.

Simpson, T. W. 2011. "e-Trust and Reputation." *Ethics and Information Technology* 13(1):29–38.

Sommerville, I. 2011. *Software Engineering*. 9th ed. Boston: Pearson.

Taddeo, M. and L. Floridi. 2011. "The Case of E-Trust." *Ethics and Information Technology* 13(1):1–3.

Urban Walker, M. 2006. *Moral Repair: Reconstructing Moral Relations after Wrongdoing*. Cambridge, UK: Cambridge University Press.

Chapter 10

Friendship in Future IT Environments

"What is a friend? A single soul dwelling in two bodies."

Aristotle (384 BC–322 BC)

The quality of our relationships is core to our well-being. As a result, it is not surprising that we have seen significant debate about whether our new IT services destroy true friendship or, in contrast, enrich friendship by providing new means of communication. In 2010, 86% of respondents to a global study by Fujitsu agreed or strongly agreed that we are becoming socially isolated because all communication will be done with computers and not with people (Fujitsu 2010). I describe this trend in my stories in Chapter 3 where Agent Arthur becomes Sophia's friend, Jeremy loves to be accompanied by a robot in the Halloville Mall, and elderly people use robots (instead of family members) in their households to help them. In all of these scenarios original and authentic human encounters are replaced by the smooth and superficial surface of machines. Intellectuals who observe and study this trend are worried: "The idea of the 'original' is in crisis," writes Sherry Turkle (2011, p. 74), a leading scholar in the field of human-robot interaction. She bewails that we are developing a "culture of simulation" in which "authenticity is for us what sex was to the Victorians: taboo and fascination, threat and preoccupation" (Turkle 2011, p. 74).

We must take this criticism seriously. But we must also recognize that while authentic face-to-face communication is reduced, new digital media have created many new forms of constant connection between people (Roberts and Koliska 2014). Families and friends constantly update each other on location, news, arts, and personal moments through presence apps and social media. Video telephony helps remote friends stay close. Some people meet in virtual worlds. Scholars wonder whether such short-term, feel-good connections come at the cost of true friendship. The *social augmentation hypothesis* states that this is not the case. The use of new media augments people's total social resources, in particular, existing strong ties.

From this perspective, digital media provides an additional avenue to be social with each other. People can coordinate their personal networks more easily via e-mail or messaging, and they can stay in tune with what happens to their friends.

In contrast, the *social displacement hypothesis* states that people who are more active online are less available for real-world engagements. And if they are, what quality do these offline relationships have when people constantly interrupt their face-to-face communication through the use of their smartphones? Does a person's network size tell us anything about what is really happening within those friendships? Compared to those who do not use the Internet, American Internet users are 42% more likely to visit a public park or plaza, and 45% are more likely to visit a coffee shop or café (Hampton et al. 2009). U.S. bloggers are even 61% more likely to visit a public park than people who do not maintain a blog. But do Internet users and bloggers speak to anyone in these public hangout places? Or are they "alone together," as Turkle has critically posed?

To better understand how machines influence human relationships, and to potentially build machines that support friendship, we need to better understand what the social construct of friendship really means.

10.1 What Is Philia (Friendship)?

Philosophers of all ages have identified three kinds of love: agape (ἀγάπη, dilectio, caritas), philia (φιλία, amicitia), and eros (ἔρως, amor) (Helm 2013; Hoff 2013). Agape is unconditional love of the kind people can have for God or for humankind in general. The word is often translated as "charity." This kind of love does not depend on any particular traits of the beloved. In contrast, eros and philia are both triggered by our responsiveness to others. Eros is a desire for someone, often sexual in nature. Philia is what is most associated with our term *friendship*. It expresses itself as an affectionate regard or positive feeling toward another. At the same time, philia is not necessarily restricted to the term *friend* as we use it today. It also embraces people like family members or close colleagues; it is the broad kind of friendship we find on social network platforms. In this chapter, I concentrate on philia in terms of the strong and weak ties we may have with others (and not eros or charity).

Philia takes a central role in the creation of happiness and a fulfilled life. Aristotle regarded a good life as inherently social and believed that a social life was the soil in which people's virtues and good character root, receive nourishment, and grow. Terrell Bynum wrote: "Aristotle clearly saw [that] autonomy is not sufficient for flourishing, because human beings are fundamentally social and they cannot flourish on their own ... Knowledge and science, wisdom and ethics, justice and law are all social achievements" (2006, p. 160). Aristotle explicitly distinguished between two kinds of friendship: the imperfect friendships of utility and pleasure on one side, and the perfect friendship of virtue on the other. In the latter form of friendship, each participant altruistically wishes well for the other without considering their own personal utility or pleasure (Munn 2012). As part of this form of friendship, a participant might criticize their friend to help that friend understand his or her weaknesses. Virtuous friendship was important for Aristotle because he promoted "virtue ethics," a stream of philosophy that sees "the good" as something arising from people's habitual virtuous character rather than from a utilitarian calculus or mutual pleasure only.

Shannon Vallor (2010, 2012) outlines how Aristotelian thinking is relevant for the analysis of friendship in online social media. She criticizes the narrow focus of traditional studies of social network platforms on feelings of happiness in terms of personal pleasure and utility only. Variables such as life satisfaction, self-esteem, and social capital have been at the forefront of investigation. And it seems like the feel-good strategy of a social network platform like Facebook, which offers Like buttons but no Dislike buttons, caters to the more superficial dimensions of pleasurable community. However, while these "psychosocial goods" are important, they are not enough for friendship or a good life—at least not in the view of Aristotle. For him, friendship requires virtue. And to develop such a quality of character, we must have true friends with whom we can go beyond the feel-good factor (Vallor 2010, 2012).

But what are the characteristics of true and virtuous friendships? And how do we develop them? Across multiple works, Aristotle and other philosophers (Aristoteles 1969; MacIntyre 1984; Vallor 2012; Helm 2013) have identified various characteristics of friendship, in particular reciprocity, shared activity, the development of self-knowledge as well as empathy, and care and intimacy for and with the other (Figure 10.1).

For Aristotle, *reciprocity* or "the reciprocal sharing of good [...] is the glue of all friendship" (Vallor 2012). It is the ability of people to give and to take. Giving and taking pleasure and utility is one way to cultivate reciprocity. For example, friends might exchange presents, provide help, or support each other professionally. Most important, however, complete (teleia) friendship goes further than just creating pleasure and utility. It also involves the exchange of respect, love, knowledge, and virtue. A good friend can give us an honest opinion or help us to understand something. It is such mutual feedback that helps us to correct ourselves and grow over time.

Reciprocity implies that friends spend time together. A shared life or *shared activities* are therefore an important driver of friendship. However, it is not only the time spent together that counts. Colleagues in a company also share time together but are not always friends. Friendship

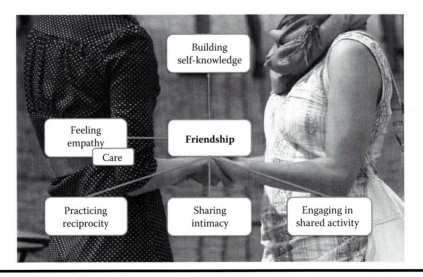

Figure 10.1 Characteristics of a true and virtuous friendship. (©cc by 2.0 Wrote 2008.)

manifests itself in those shared activities, where both friends enjoy the thing they do together and particularly enjoy doing this thing in the company of their friend (Munn 2012; Helm 2013).

While spending time together and giving and taking from each other, friends can grow together and learn from each other. *Self-knowledge* is the result of such continuous learning. We gain knowledge about our own being in the world, a well-rounded and realistic understanding of the social world around us, and about how we relate to the world and fit into it. Unlike most of today's interpretations of self-knowledge—which involve digging deeply into our own selves, our childhood, and so forth—Aristotle understood self-knowledge more as a matter of understanding our role in the world. He wrote, "we are not able to see who we are from ourselves" (Aristotle 1915, p. 1213a15–16). He even wrote "if a human being surveys himself, we censure him as stupid" (Aristotle 1915, p. 1213a5–6). Instead, the "self-sufficing man will require friendship in order to know himself" (Aristotle 1915, p. 1213a26–27). Friends mirror each other's behavior and thereby help each other to develop. Some authors even argue that friends become each other's "procreators" (Millgram 1987).

Just observing a friend's behavior can create some self-knowledge. But *intimate* exchanges are equally important: with friends, we can share very private concerns and hope to get advice. Some scholars therefore view intimacy as a major pillar of friendship: a mutual self-disclosure or sharing of secrets that goes beyond the kind of conversations we would have with a colleague at work or some acquaintance (Helm 2013).

Finally, the genuine feeling of sympathy is highly important for friendship. Philosophers often distinguish between empathy and care when they write about the feelings underlying friendship. *Empathy* is a spontaneous emotive or perceptual capacity to feel with another person, to coexperience the joys and sufferings of the other person. "Someone who shares in the sorrows and joys of his friend," Aristotle wrote (2000, p. 1166a5–10). But empathy is not a given. It depends on many tiny gestures and observations that people either appreciate or reject in each other. The "non-voluntary self-disclosures" that become apparent when people spend time together (Cocking and Matthews 2000) can breed empathy or separation.

Separate from empathy is the concept of *care*. Whereas empathy is triggered by a friend's situation, which we may pity or take joy in, care is unidirectional. We can care for a friend without him or her doing anything. Helm (2013) discusses how care is similar to the unconditional love of agape described above. Care bestows value on a friend without any contemplation.

I will now use these dimensions of friendship to explain how machines are able to influence friendship. First, machines can influence how existing offline friendships are conducted. People use the communication functions of machines to stay in touch, plan activities, share ideas, and develop new procedures more easily and hence more frequently. Second, machines can be used to form new bonds of friendship through shared avatar activity online. In virtual worlds strangers meet and spend time together. Sometimes these virtual friendships lead to offline relationships. And third, people may form friendly ties with artificial beings such as robots or the kind of virtual personal agent called Arthur in the scenarios. Figure 10.2 summarizes the three areas of artificial relations.

How do these forms of human–machine interaction alter humans' perceptions of belonging and friendship? Can machines be built to strengthen our friendships along the lines of the relevant dimensions depicted in Figure 10.1? Let's start with classic human-to-human friendships that are mediated via the Internet, e-mail, chat, social networks, and so on.

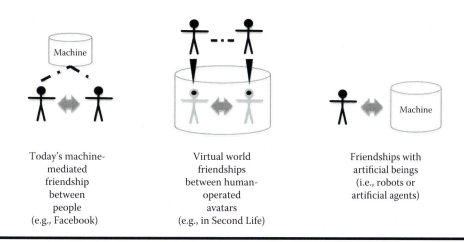

Today's machine-mediated friendship between people (e.g., Facebook)

Virtual world friendships between human-operated avatars (e.g., in Second Life)

Friendships with artificial beings (i.e., robots or artificial agents)

Figure 10.2   Three ways in which machines influence friendship in the machine age.

10.2 Machines as Mediators of Friendship

In 2010, about 700 students from 10 different countries in North and South America, Europe, Asia, and Africa participated in a study in which they were asked to spend one entire day offline, without digital media. After that day, they were asked to report on their experience. "What does this unplugging reveal about being plugged in?" asked Jessica Roberts and Michael Koliska (2014), the authors of the study. About half of the students were unable to complete the day and dropped out of the experiment. The major experience that all of the students reported about the day—including those who dropped out—was a perception of dependence and addiction. The majority of students who did not drop out still had a hard time staying offline. The feeling of dependence was accompanied by anxiety and distress. However, the third most common feeling was relief about being offline.

The findings from Roberts and Koliska's study show how important digital connectedness has become for today's relationships, including friendships. A core reason for the dependence and distress that was felt by the participants was "a sense of having left an existing 'world', in which they feel everyone else lives, and that being outside this environment was challenging and difficult" (Roberts and Koliska 2014). Comments on the offline day were: "It was not an easy experience because I felt I was in kind of another world—left out" (student from Uganda); "…all I wanted to do was pick up my phone and become a part of the human race again" (student from the United Kingdom); "I felt isolated, without information and limited to the people around me" (student from Slovakia).

As IT devices have become so vital in new relationships it is important that they are designed such that they optimally cater to and strengthen ties.

What functions can we use to support the utility, pleasure, and virtue of friendships? What services foster reciprocity, shared activity, accumulation of self-knowledge, or the sharing of empathy and intimacy? Figure 10.3 presents a selection of current technical features and the dimensions of friendship they support.

Online *reciprocity* can support the creation of utility, pleasure, and virtue. Today's first-generation machines not only enable us to exchange pleasures by sending each other messages and emoticons, retweeting the other person, spreading a joint work, and so forth. Utility is also enhanced by the reciprocity inherent in the online medium. The Internet facilitates the sharing of resources, physical goods, services, and information. Of course, virtuous friendship cannot be built exclusively online. But it can be supported by digital media: Ideas, thoughts, or concerns can be exchanged in an in-depth form such as e-mail. Friends can exchange ideas by offering thoughtful feedback to a blog post or jointly working on an online project such as a forum or wiki. Online reciprocity has found new forms of symbolic language. Emoticons and small symbols are regularly used in playful digital exchanges. The smiley face symbol has reached such ubiquity that it triggers the same brain impulses as a real smiling face (Churchesa et al. 2013).

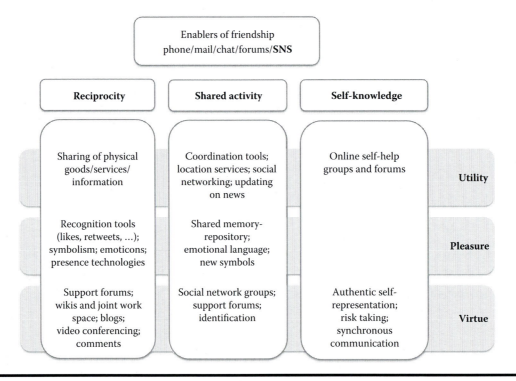

Figure 10.3   How first-generation machines enable various dimensions of friendship.

Machines do not merely support reciprocity but also shared activity. Mobile phones, chat, and location services allow us to more easily coordinate and schedule offline activities. Social networks help to find new comrades-in-arms for offline matters, to update each other on developments, and to easily share memories of shared activities. Videoconferences can maintain long-distance friendships and facilitate time together that would physically be impossible. Forums for special interest group forums, such as programming platforms, are a venue where users can give each other extensive help. In short, people can meet around and through online media.

Although reciprocity and shared activity flourish in online groups and forums, these forums are often anonymous. Friendship, in contrast, is built between identified individuals. When virtuous friendships and strong ties are the goal, then anonymity may not be the best path to take. Anonymity is widely heralded as an important ethical trait of online environments because it protects people's privacy, benefits free speech, and enhances people's deliberation. I have described how personal data can be anonymized, and I believe that anonymization is an effective measure to thwart mass surveillance. But when it comes to web forums, news-portal comments, and social networks, which by their nature refer to real-world activities, identities, or both, I question the benefit of anonymity or pseudonymity. In this context, anonymity undermines accountability and responsibility in communication.

Current digital services and social networks seem to support reciprocity and shared activity, but their ability to foster self-knowledge is questionable. Self-knowledge develops in our friendships through mutual observation, the exchange of honest feedback and joint experiences that we can learn from. But such immediate exchange is not available online. Many online media are asynchronous. And social networks, though built to foster friendship, primarily encourage one-to-many or many-to-many kinds of communication. This kind of "splintered mirror" communication (Vallor 2012) often dominates the richer one-to-one communication styles typical for friendships. If social networks wanted to support true friendships, they would need to support more synchronous communication for strong tie building. Features like messaging and video conferencing are a good start. But it would be fruitful to think about even richer channels for one-on-one exchanges such as unique friendship spaces with joint digital goods like books or music files, video messaging, shared gaming resources, shared photos and experiences, and private message repositories.

Another hurdle to building self-knowledge on social networks is that many people on these platforms engage in some kind of impression management. They reveal only selective content and build alternate identities that embellish or trivialize their real life. In a recent study on European Facebook, 60% of respondents said that they do not believe people present themselves how they really are. As a result, "friends" receive feedback only on the shallow information objects

they actually publicize. A holistic exchange is not possible. Nietzsche once famously pointed out that personal harvesting, which could be interpreted as building self-knowledge, is a matter of exposing oneself to risky endeavors: "For—believe me—the secret of harvesting from existence the greatest fruitfulness and the greatest enjoyment is—*to live dangerously!*" (Nietzsche 1882/2001, sec. 283). People on social networks normally do not take that risk. In contrast, they only show a polished façade.

10.3 Shared Life and Learning in Virtual Worlds

More risk—at least in fictitious form—is taken in virtual reality worlds. As of 2014, over 1 billion users were registered within virtual worlds at least 500,000 of them being active inhabitants of the virtual world Second Life.[*] Top games such as *World of Warcraft* (WOW) or *League of Legends* attract around 1.9 million hours of play per year.[†] On average, players spend between 10 to 30 hours per week playing these games, a time amount that hints to problematic Internet use (PIU; see Chapter 7 on health).[‡] Considering these numbers is important, even from an academic perspective, because they document the extent to which friendships are now built and lived in digital worlds instead of the real world. People of all ages have joined these places to play, meet, and socialize with existing offline friends, their families, or to meet new people.[§]

This rise of virtual worlds and their role in human relationships is highly controversial. Intellectuals often see virtual worlds as a threat to true human bonding, a dangerous

---

[*] Audrey Watters, "Number of Virtual World Users Breaks 1 Billion, Roughly Half Under Age 15," ReadWrite, October 1, 2010, http://readwrite.com/2010/10/01/number_of_virtual_world_users _breaks_the_1_billion; Wagner James Wu, "Second Life Turns 10: What It Did Wrong, and Why It May Have Its Own Second Life," June 23, 2013, accessed September 1, 2014, http://gigaom .com/2013/06/23/second-life-turns-10-what-it-did-wrong-and-why -it-will-have-its-own-second-life/.

[†] Sigmund Leominster, "Why Have Virtual Worlds Declined," *The Metaverse Tribune*, May 15, 2013, accessed September 1, 2014, http://metaversetribune.com/2013/05/15/why-have-virtual-worlds -declined/.

[‡] The estimate of 10 to 20 hours is based on Yee (2014) and a 2012 statistic from *The Metaverse Tribune* (http://metaversetribune. com/2013/05/15/why-have-virtual-worlds-declined/). The average player takes 372 hours (2 full months of work) to reach the maximum level in the game *World of Warcraft* (Yee 2014).

[§] It is important to recognize that the widely held stereotype of virtual world players being mostly young teenage boys does not seem to be correct. According to Yee (2014), the average age of players in virtual worlds is 30. Only 20% of online gamers are teenage boys. Boys and girls equally enjoy playing in virtual worlds. Only immersive war-focused games, such as *World of Warcraft*, have 80% male players.

displacement of the real by the virtual. In his book *On the Internet*, Hubert Dreyfus writes, "The temptation is to live in a world of stimulating images and simulated commitments and thus to lead a simulated life … [T]he present age … transforms the task itself into an unreal feat of artifice, and reality into a theatre" (2009, p. 88). In contrast to such powerful critics, younger scholars who have spent a lot of time in virtual worlds themselves and have closely observed how people use these worlds are much more balanced about the developments. "Gaming can be beneficial when it's part of a healthy palette of social interactions," writes Nick Yee, a virtual reality (VR) specialist. "Family members who play online games together report more family communication time and better communication quality … 41% of online gamers felt that their game friendships—with people who they first met in online games—were comparable to or better than those with their real-life friends" (Yee 2014, p. 36).

So who is right? An important source of academic research for understanding the true social dynamics of virtual worlds has been the U.S.-based Daedalus Project.* In the past 15 years Yee, the initiator of the project, surveyed more than 35,000 players of massive multiplayer online games (MMOGs). In his book *The Proteus Paradox* (2014), Yee summarizes his findings. The data reported hereafter on virtual worlds as well as the player comments cited are taken from this source.

Looking into the design and current use of virtual worlds, it becomes clear that they can help to build friendships along the dimensions introduced earlier. First, virtual worlds create a stimulating virtual place where people can meet for all kinds of adventurous and fantastic endeavors. We all know how important places such as the dinner table are for personal bonding. Virtual worlds can be a modern form of a dinner table (except that only virtual food is served). In fact, 19% of virtual world visitors play with at least one family member. Also, 80% know the people they meet there from the offline world. One-fourth of the players regularly visit virtual worlds with a romantic partner.

Virtual hangout places are created around the idea of shared activity in its truest sense. In many games, players need to form large persistent social groups, sometimes known as guilds, that help them to kill monsters, jointly survive, and advance in the game. People in virtual worlds are therefore hardly ever idle (Soraker 2012). That said, the technical design of virtual worlds directly influences how much cooperation and reciprocity occurs in those worlds. In games like the original *EverQuest*, avatars could advance in the game only if they helped each other. Yee illustrates how *EverQuest* players experienced death in the game, relying on other players to help them revive their avatars. When *EverQuest* avatars were killed (for example, in a monster raid), they were stripped naked and had to recover their body and equipment in a limited time frame. "'To succeed in *Everquest* you need to form relationships with people you can trust. The game does a wonderful job of forcing people in this situation. Real life (RL) rarely offers this opportunity as technological advances mean we have little reliance on others' (*EverQuest* player, male, 29) … [T]he willingness to spend an hour to help a friend to retrieve a corpse isn't something that can be faked" (Yee 2014, p. 183).

The design around defeat and advancement in *EverQuest* forced people to bond and practice reciprocity. However, this aspect of design is not the only way to encourage friendship in these games. The need to share game resources can also make people join forces. In *EverQuest*, for example, players regularly needed to share spells for instance in order to advance. Yee further notes the importance of idle time and access to information about game function as ways to encourage shared activity and reciprocity. In older games, players regularly had to deal with downtime (mainly for technical reasons). Although this was annoying, it also presented an opportunity for players to chat and get to know each other. In addition, because many games are complex to play, people spend considerable time working to understand how commands work and how to achieve specific ends in the game. This exchange of information about game function could be an inherent part of the in-game experience. A game could force players to ask each other for help rather than outsourcing this activity to a separate information interface (for example, the *World of Warcraft* Thottbot† application). If gaming companies wanted to, the concept of RTFM ("read the fucking manual") could be replaced by a rich in-game reciprocity.

An important dimension of friendship is the ability to learn from each other. As I have noted, social networks like Facebook limit self-knowledge because people often share only the good parts of their lives, manipulating their platform image. In contrast, virtual worlds force people to be more real. Despite their artificial interface and use of avatars for representation, engaging in games in virtual worlds brings forth people's true character. As they game intensively, people forget about their masks and get to know each other largely as they really are. Research shows that character traits (such as the Big Five personality dimensions,‡ namely, openness to experience, conscientiousness, extraversion, agreeableness, neuroticism) are carried into the virtual world (Yee et al. 2011). Ten to 20 hours of intense play per week in a highly complex environment simply undermines people's

---

* The Daedalus Project, accessed September 1, 2014, http://www.nickyee.com/daedalus/.

† "Thottbot," http://wow.gamepedia.com/Thottbot.

‡ "Big Five personality traits," *Wikipedia*, http://en.wikipedia.org/wiki/Big_Five_personality_traits.

ability to maintain a role.* Consequently, virtual worlds see frequent nonvoluntary self-disclosures, a quality that is recognized as vital for friendship formation. This aspect of gaming allows people to receive feedback on their behaviors. Finally, when people's real voices replace text-based communication or when people's real faces are morphed into an avatar (see Figure 10.5), the experience of being with another *real* person is even more vivid, regardless of whether the person appears in a virtual body.

The notion that people reveal many dimensions of their true personality in virtual worlds becomes evident in stories of how people fall in love in these worlds. About 10% of online gamers have dated someone they first met in a virtual world. Obviously, lovers do not report the experience of falling in love at first sight. But playing together and observing how the other person reacts over time creates a nonsuperficial way of meeting. "Virtual worlds can negate some of the superficial aspects of face-to-face relationships," writes Yee (2014, p. 134). He goes on to cite one of the players who fell in love: "On the outside we seem totally opposite. But we work so well on the inside. I guess that is what comes of meeting inside out :p" (*World of Warcraft*, female, 25). "Inside out" is the term that online gamers use to refer to this reversed model of forming relationships.

The roles people cast in virtual world groups or in guilds can also contribute to their self-knowledge. For example, leading a guild can make people collect management experiences that prepare them for real-life situations. People from all age groups, continents, and backgrounds play together, and even though everyone turns up as their avatar, their different cultures and learning experiences are still present. "Slaying a dragon is actually quite straightforward once you've figured out how to manage a team of two dozen people to help you. And this is the crucial management problem that every successful guild leader must solve … Being a guild leader has taught me about personality types and how to manage people more than any job I've ever worked on" (*World of Warcraft*, female, 27).

An indirect way of building self-knowledge in virtual worlds is interaction with others in different roles and sexes. People can choose any gender and select avatars from a range of different races (e.g., elves, trolls, humans) and classes (e.g., wizard, mage, cleric). While the in-depth refinement of fictitious virtual personalities is a niche in virtual world games today, simple gender bending or the maintenance of multiple avatars is already common. Over half of all male players, for instance, have at least one female avatar. Gender bending allows people to directly gain the experience of how it feels to be in the skin of the other sex. This experience can breed

empathy and understanding. One male player confessed: "I'm amazed how thoughtless some people can be, how amazingly inept men are at flirting and starting a conversation with a female, and how it really does take more effort to be taken seriously as a female versus a male" (*EverQuest*, male, 24). Gender bending is not the only way to learn: Avatar appearance and size influence how people play their roles online, and people also transfer some of this virtual experience into the offline world (Yee, Bailenson, and Ducheneaut 2009). For example, tall and good-looking avatars keep less physical distance in virtual worlds. Being more confident encourages this behavior. If, let's say, a rather unattractive real person plays a beautiful avatar, he or she might learn how it feels to be confident and transfer this feeling into real life.

Finally, virtual worlds allow us to observe ourselves, to look over our own shoulder in how we interact with others. In an extreme form, this kind of self-observation is regularly practiced by some male players. They create and maintain a second female character in the game whose sole role is to watch the main male avatar play. In fact, the Daedalus project found that "by far the most widely adopted male explanation [for having a female character] is that the third-person perspective in these games means that players spend a great deal of time looking at the back of their character" (Yee 2014, p. 111).

Our current knowledge of virtual world games suggests that shared activity, reciprocity and self-knowledge development are to quite some extent present in virtual world environments. The characteristics of friendship can be supported by certain game designs, stories, dependencies and functionality. Figure 10.4 summarizes the enablers of

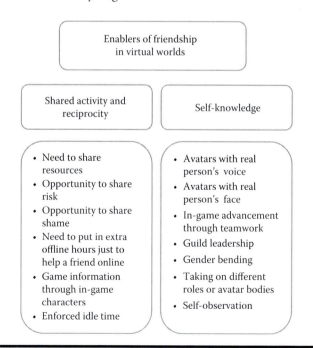

Figure 10.4 A selection of enablers for friendship building in virtual worlds.

---

* Of course, the games are called "role-playing games." Players can select from a range of races (e.g., elves, trolls, humans) and classes (e.g., wizard, mage, cleric). But continued role-playing is actually a niche in those games. For example, only a handful of the hundreds of available servers on *World of Warcraft* are explicitly reserved for role-playing.

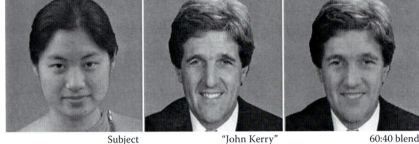

Figure 10.5 An example of digital morphing of faces. (From Wikipedia, http://en.wikipedia.org/wiki/File:Candidate_morphs.jpg, accessed August 30, 2014. Public domain.)

virtual friendship creation. Another important dimension of friendship building and maintenance is the emotional part of friendships, which includes characteristics such as sympathy, empathy, and intimacy. The next section will delve in into this question.

10.4 Empathy in Virtual Worlds

At the core of every friendship is an emotional attraction that expresses itself in sympathy and empathy. But can we have genuine sympathy for someone we meet in a virtual world in an avatar body? Social science research shows that, to some extent, human mating choices and judgments about attractiveness are artificially creatable and predictable. They are a function of how much someone looks like us (Penton-Voak, Perrett, and Peirce 1999) or our family members (Bereczkei, Hegedus, and Hajnal 2009). This human psychology can be used to artificially create sympathy in virtual worlds. It seems that if we are to like an avatar, it only has to adapt its artificial face to our own facial features or that of our parents. Studies in 2004 and 2006 showed how morphing of faces can influence election results: The faces of election candidates were morphed with voter's faces. Figure 10.5 shows what a male's and female's morphed faces look like when they are blended with George Bush or John Kerry. In the top row 40% of George Bush's facial features are morphed with another man's face. The bottom row illustrates the same effect for a 40% blend between a female and John Kerry. Voters without strong political preferences were swayed by the influence of

the morphed politician's face with their own. They did not recognize the manipulation and voted for the figure that looked like them (Bailenson et al. 2008).

Although attractiveness can be artificially manipulated through morphing techniques, creating true empathy in virtual worlds seems to be more challenging. Empathy involves feeling with another person and sharing that person's happiness or grief. Empathy research shows that humans (and animals) perceive what is happening to peers, but as they do so they physically share in the experience of the peer. The peer's experience resonates in the body of the observer and initiates action in the observer, such as a desire to help (Preston and de Waal 2002). A part of this behavioral phenomenon of empathy seems to be related to humans' system of mirror neurons. Mirror neurons mirror in our body the behavior of someone we observe as though we were ourselves acting. Mirror neurons alone apparently do not produce empathy, but they provide cellular evidence for a shared representation of perception and action (Preston and de Waal 2002; Jabbi, Swart, and Keysersa 2007). An open question is how mirror neurons function when we observe others in virtual worlds versus observing them in the real world. Experts currently think that mirror neurons work best in real life, when people are physically close. Virtual reality and videos are imperfect substitutes.* As a result, virtual reality would provide us with

---

* Sandra Blakeslee, "Cells That Read Minds," *New York Times,* January 10, 2006, http://www.nytimes.com/2006/01/10/science/10mirr.html?pagewanted=all&_r=0.

fewer experiences of empathy. A vital ingredient for friendship and important constituent of our inner emotional landscape suffers. More research on this one is certainly needed.

Mapping the bodily feelings of others onto our own internal body states is one of several indications for the importance of full bodily presence in high-quality relationships. But this finding is not the only one that encourages full bodily presence. A study in the field of sociology observed how physical proximity affects the spread of happiness in groups. In a longitudinal study of physical social networks over 20 years, James Fowler and Nicholas Christakis (2008) found that happy friends who live closer to one's house promote personal happiness more than friends who live farther away. In fact, a friend who lives within a mile and who becomes happy increases one's probability to be equally happy by 25% (Fowler and Christakis 2008).

Although these studies indicate the importance of physical presence, some scholars do not believe in it. They argue for the supreme importance of humans' mental states and doubt the importance of bodily presence. An illustration of their argument is how we feel when we read a good novel or watch an emotional movie. Even though we are not physically with the characters, we feel passionate solidarity with them. We cry and laugh while we read or watch. Good books and movies prove a deep connection between imagination and empathy as well. So who is right? For over 300 years, philosophers have debated the importance of our bodily presence compared to a pure mental presence. Ever since René Descartes formulated his famous sentence "I think, therefore I am," modernists have argued that our human essence resides in our brains. But this belief is not universally shared. Box 19.2 in Chapter 19 gives a short overview of the philosophical tension over the importance of body and mind. Our understanding of body–mind unity or independence will ultimately determine our conclusions on the relative value of virtual friendship.

10.5 Intimacy and Disinhibition in Online Environments

Many observations have been made on how intimate people get when their communication is digitally mediated. Walther (1996) called this phenomenon "hyperpersonal interaction." The "boundary regulation process" (Altman 1975) that people use to manage their privacy vis-à-vis others therefore seems to follow different dynamics online than for face-to-face encounters. Generally, people open up more when their communication is digitally mediated, a phenomenon that has been called the "online disinhibition effect" (Suler 2004). However, we must distinguish between two kinds of digital encounters: One is communication with other people online at various degrees of anonymity or identification, for example, on web forums or on social network platforms. The other is talking to a machine, be it a virtual interface agent, such as Agent Arthur, or a robot.

Let's start with human-to-human communication that is digitally mediated. On many news portals and forums where commentators' identities are protected by anonymity, toxic disinhibition is a common phenomenon. Many people—often called "trolls"—use aggressive and demeaning language to express negative feelings. Conversely, there can be benign disinhibition: people being exceptionally kind, generous and enthusiastic toward others. Overall, people are more frank online.

One reason people open up is that many online platforms promise anonymity. Anonymity in relation to communication partners makes the senders of information less vulnerable, particularly when the senders trust that their online actions are totally separate from their real offline selves. As Johan Suler writes on the effects of dissociative anonymity: "the person can avert responsibility for those behaviors, almost as if superego restrictions and moral cognitive processes have been temporarily suspended from the online psyche" (2004, p. 322). Anonymity is, of course, not a basis for real friendship, which requires that people reveal their identities at some point. So general-information forums on the Internet are not the place to be intimate with one another. But in virtual worlds, people often transition from initial anonymity to identification. They reveal who they are to a select group of others. What has been said anonymously then becomes important.

Even when identities are revealed, online mediation seems to encourage more open communication. In the Daedalus project, 24% of virtual world players said that they told personal issues or secrets to their online friends that they had never told their offline friends (Yee 2014). Scholars believe that players reveal more in virtual worlds because they are invisible. When avatars speak to each other or people chat, nonverbal cues are filtered out. People do not have to worry about how they look or sound and, most important, they also do not see the other person's reaction to what they say. In traditional psychoanalytic theory, the therapist sits behind the patient for the same reason: people open up more when they do not see to whom they are speaking. One could argue that online friendship benefits from invisibility in the same way as a psychoanalytic treatment is beneficial. Secrets may be shared more openly. Yet, visible reactions and the sound of a friend's voice additionally create reciprocity and support the building of self-knowledge. By seeing how our friends react to what we tell them, we learn about ourselves. So, taken together, the technical reality of cue-free communication creates a contradictory effect: It increases intimacy, but reduces reciprocity and learning.

Another reason for online communication being more genuine is that people are initially equalized and less prejudiced. All

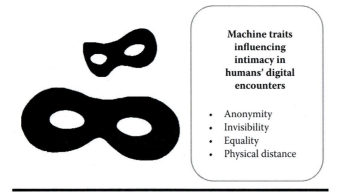

Machine traits influencing intimacy in humans' digital encounters

- Anonymity
- Invisibility
- Equality
- Physical distance

Figure 10.6 Machine traits that foster online intimacy.

real-world signs of status, wealth, race, and so on are largely leveled out. In virtual worlds, everyone can look however they want to look. On social networks, where people know each other from offline encounters, they can and do post rather favorable images of themselves and their lives. In a way, the nature of the digital medium itself makes the world flat: The enforced two-dimensionality of the screen brings everyone symbolically to the same level. This representational and positive equality lowers communication barriers that exist in the real world. Furthermore, it encourages people to recognize skills that are often suppressed by inequality biases in the real world; such skills include writing skills, humor, the quality of one's ideas, and technical know-how. This shifting of relevant skills makes some people open up more than they would offline. For example, shy people and people with physical handicaps participate more online than they do offline (Kowert, Domahidi, and Quandt 2014). In the Daedalus project, people said that they open up better in the virtual world and therefore learn more about their lovers there than in the real world.

Some authors have argued that it is not equality that makes people open up in virtual worlds but rather the *distance* of mediated friendships (Briggle 2008). Less courage is required to be candid in an environment where you can execute an "emotional hit-and-run" by just closing the computer, knowing that the other person is probably far away (Suler 2004). This argument is anecdotally supported by a phenomenon we observe when strangers on a plane reveal intimate information to each other, expecting that the information they share will not catch up with them later. Could interfaces foster honesty and intimacy by indicating the real physical distance between parties?

Figure 10.6 summarizes the machine traits that foster human-to-human intimacy online.

10.6 Intimacy with Artificial Beings

A special form of becoming intimate in the machine age is when we share information directly with an artificial entity instead of another human being. Conversations with artificial agents (such as Agent Arthur in my scenarios), with

preprogrammed figures in virtual worlds, or with robots are examples of this kind of interaction.

Over and over again, research has shown that people get exceptionally intimate with artificial beings. This finding was demonstrated for the first time in Robert Weizenbaum's ELIZA experiments at the Massachusetts Institute of Technology in the 1960s (Weizenbaum 1977). ELIZA was a computer program that employed an early form of natural language processing and simple pattern matching. People would type in sentences like "My head hurts" and ELIZA would respond with something like "Why do you say your head hurts?" Many people who used the computer system started to really like it and extensively share personal information with it. In fact, Weizenbaum described how his secretary got hold of the program and got so fond of it that she started to talk to it on a regular basis, sharing extensive parts of her private life.

Scholars have investigated why people disclose so much to computers and how this disclosure can be manipulated (Reeves and Nass 1996; Moon 2000). Based on such research, the *theory of social response* postulates that humans treat machines in the same way as other human beings even when they know that the machines do not possess feelings or "selves" (Reeves and Nass 1996). Scholars explain this behavior by noting that humans evolved as social beings and apply their learned heuristics to machines. They think that people are "mindless" in a way, failing to reflect on the difference between other humans and machines (Nass and Moon 2000).

Against the background of social response theory, Youngme Moon investigated how conversational interface strategies can be used to make people self-disclose. One of these strategies is to mimic the reciprocity of self-disclosure. From human interactions, it is known that disclosure begets disclosure. People who receive information from others feel obliged to share something about themselves (Derlega et al. 1993). Disclosure is also much more likely to occur if requests for information gradually escalate. Relationships proceed from casual exchanges to increasingly intimate ones over time (Altman and Taylor 1973). Moon found that when these strategies are implemented in machines, such as conversational agents, people disclose increasingly more intimate information. Figure 10.7 shows one of the manipulations that Moon used in her experiments. This experimental example shows how easily humans' intimate disclosure can be manipulated through machine interaction strategies. Turkle et al. (2006) comment that, faced with relational agents, people fantasize about a "mutual connection."

In her studies on human–robot interaction Turkle found another dimension that seems important. She quotes an elderly man, Jonathan (74), who lives in a nursing home and has used the "My Real Baby Robot" for a while. "The robot wouldn't criticize me," says the old man. From a reaction

Nondisclosing machine	Disclosing machine
Machine says: "What do you dislike about your physical appearance?"	**Machine says:** "You may have noticed that this computer looks just like most other PCs on campus. In fact, 90% of all computers are beige, so this computer is not very distinctive in its appearance. What do you dislike about your physical appearance?"
Participant answers: "I could lose some pudginess and gain more tone, which requires effort."	**Participant answers:** "I hate my big hips. I'm a sugar freak, and all that sugar sits on my hips. I also don't like that I have relatively small breasts, but that is nothing compared to the way the size of my hips bothers me. No amount of running or lifting or anything else seems to slim them."

Figure 10.7 Varying machine strategies to make people disclose. (Based on Moon, Youngme, 2000, "Intimate Exchanges: Using Computers to Elicit Self-Disclosure from Consumers," *Journal of Consumer Research* **26(4):323–339, Figure 1, p. 330.)**

like this one, we might speculate whether interaction with artificial agents is simply easier, less chaotic, or less entropic for people.

Humans like to be inert to a certain extent, and so sharing intimate information with a lifeless artifact seems easiest. Unlike a human, an artificial agent is unlikely to object or cause turbulence. It seems to me as if people like to keep moving in a straight line and at constant velocity, just as Newton observed for all physical bodies in his first law in *Philosophiae Naturalis Principia Mathematica*: "The vis insita, or innate force of matter, is a power of resisting by which every body, as much as in it lies, endeavors to preserve its present state, whether it be of rest or of moving uniformly forward in a straight line" (Newton 1687/1848, p. 73). "Do not disturb my circles!" said Archimedes (287–212 BC), introducing a saying that we often use to express this very desire to be undisturbed. More research is needed to investigate this relationship between human inertia and the pleasure people take in exchanging information with machines.

Finally, authors have speculated that intimacy or online disinhibition could be a result of "solipsistic introjection" (Suler 2004). The term *solipsism* stems from the Latin word *solus*, meaning "alone," and *ipse*, meaning "self." Solipsism is the philosophical idea that only our own mind is sure to exist,

a thinking that ties up with modernism's mantra "I think, therefore I am." Solipsistic introjection means that digital companions such as ELIZA become real characters within our intrapsychic world (just as other people could simply be representations in that world). Suler (2004) describes how we unconsciously experience conversations with digital companions as if we were talking to ourselves: "People fantasize about flirting, arguing with a boss, or honestly confronting a friend about what they feel. In their imagination, where it's safe, people feel free to say and do things they would not in reality. At that moment, reality is one's imagination. Online text communication can evolve into an introjected psychological tapestry in which a person's mind weaves these fantasy role-plays, usually unconsciously and with considerable disinhibition. Cyberspace may become a stage, and we are merely players" (p. 232).

Similarly, Turkle (2011, p. 70) points to research by Heinz Kohut, who described how people shore up their sense of self by turning other persons or objects into "self-objects" that complete them. In this role, the other—in our case, the machine—is experienced as part of the self. By addressing this other self, people can balance their inner states. Turkle recounts a rather sad example: "In a nursing home study on robots and the elderly, Ruth, 72, is comforted by a robot Paro after her son has broken off contact with her. Ruth, depressed about her son's abandonment, comes to regard the robot as being equally depressed. She turns to Paro, strokes him and says, 'Yes you're sad, aren't you. It's tough out there. Yet, it's hard.' Ruth strokes the robot once again, attempting to comfort it, and in so doing, comforts herself" (2011, p. 71).

Both the philosophical idea of solipsism and the psychological research on self-objects argue that talking to machines, digital agents, robots, or virtual characters is a kind of narcissistic experience. We open up to machines because we like to mirror ourselves without objection (Figure 10.8). This activity can have beneficial therapeutic effects in that it may help people overcome some of their isolation, but the question is to what extent it benefits the development of humans' social character. "The question raised by relational artifacts are not so much about the machines' capabilities, but our vulnerabilities—not about whether the objects really have emotion or intelligence but about what they evoke in us" (Turkle 2011, p. 68).

In sum, we can build machines that foster narcissistic tendencies by mimicking behavior, mirroring moods (Shibata 2004; Turkle 2011), or flattering the user (Reeves and Nass 1996). Although these manipulations positively foster intimacy and disinhibition, we must carefully balance the benefits with potentially negative effects on character formation. Instead of being neutral or offering enough praise to foster narcissism, machines could become our better selves or coaches. Susan Leigh Anderson, one of the pioneers of ethical

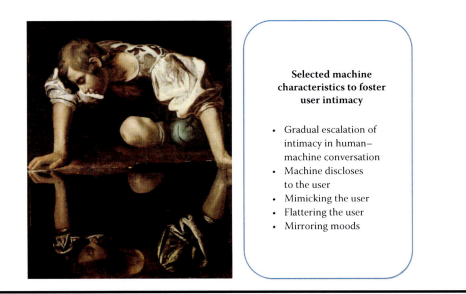

Selected machine characteristics to foster user intimacy

- Gradual escalation of intimacy in human–machine conversation
- Machine discloses to the user
- Mimicking the user
- Flattering the user
- Mirroring moods

Figure 10.8 *Narcissus* by Caravaggio (between 1594 and 1596). Talking to machines is a kind of narcissistic experience, reflecting and mirroring ourselves, and being happy therein.

machines, envisions this path when she writes, "I believe … that interacting with 'ethical' machines might inspire us to behave more ethically ourselves" (2011, p. 524). Her idea is to grant artificial agents access to ethical theory and reasoning, and make this knowledge accessible to human beings through interaction. Instead of agreeing with users or accepting our behavior without objection, machines could give us honest and frank feedback. Of course, this behavior must be carefully designed as well. Machines should not become paternalistic, prescribing actions and nudging us too often—so often that they infringe on our liberties in the name of ethics. I have envisioned the possibility of balanced ethical feedback when describing how Sophia interacts with her Agent Arthur:

> Sophia almost cannot live without Arthur's judgment anymore. She really loves him like a friend even though he recently started to criticize her sometimes; for example, when she was lazy or unfair to a friend.

10.7 Final Thoughts on Friendship in Future IT Environments

The topic of building friendships in the machine age presents a unique challenge for this book. Up to this point, I could think about values, decompose them, and then argue that respecting their various conceptual dimensions in IT design would make the world a better place. It is ethical to cater machine design to a respective value, to build trust, transparency, and security into machines. But writing about IT design and friendship is different.

Designing a machine to foster or mimic friendship could negatively impact real-world friendships as we know them today.

First, take the example of virtual worlds. If we further strengthen friendship mechanisms in virtual worlds (i.e., with the help of this chapter), might people spend even more time there than their current 20 hours per week? What time is then left for real friends and family? As we strengthen social mechanisms in virtual worlds, we risk weakening offline ties and thereby risk weakening perhaps our ability to empathize. Unlike phones or social networks that bring people together in the real world, virtual worlds make people mentally go away and wander in virtual fantasies. Can this be good in an ethical sense? Can this be morally right?

The bonds formed online are strong and vivid in people's minds. If we believe that humans should live in the real world, then we must carefully balance how many hours they spend immersed in fantasies and mirror worlds, and how many hours they spend with their physical peers, families, and real friends. This decision is a matter of individual human judgment or a family decision, but it could also be a political decision. In terms of technical design, the implementation is extremely simple: A timer or an off switch does the trick.

But there is also a third way for virtual worlds: bringing virtual representations into the real world. In the gaming scenario, I describe the potential of augmented reality technology to introduce digital play and professional communication into the real world. A "virtual overlay" on top of the real world would allow us to stay physically connected to other human beings while still having exciting games to play. This technology could potentially strengthen our relationships beyond even what we have today:

> Children have started to meet outdoors again in the woods to fight virtual characters. This trend was covered in the press because most modern children had rarely left home lately, instead staying in VR tubes to play virtual games. Now children suddenly spent hours in fresh air, and the first medical studies show that the physical activity and emotional stability of players is significantly increased.

The second area of machine research where friendship plays a role is robot design. Again, building robots to mimic humans or to incorporate qualities that increase our attachment to them may be unethical. For example, Turkle hinted at how powerful it is to build robots so that humans have to care for them and nurture them. She alludes to the famous Tamagotchi devices that were popular in the late 1990s. Tamagotchis were sold as creatures from another planet that needed human nurturance, both physical and emotional. As Tamagotchis grew from childhood and became adults, they needed to be cleaned when dirty, nursed when sick, amused when bored, and fed when hungry. If its needs were not met, the Tamagotchi would expire. Parents had to care for Tamagotchis while kids were at school, even during business meetings. Turkle concludes that "when it comes to bonding with computers, nurturance is the 'killer app'" (2011, p. 67). But can nurturing a robot, which clearly fosters bonding with the machine, be good for our development as human beings? Is it ethical to develop machines with such manipulative mechanisms in mind?

Albert Borgmann's device paradigm would clearly deny that it is good for humans to nurture lifeless machines, even though these machines may appear to us as "beings." He describes how technology turns aspects of our lives into interactions with various black boxes that we can no longer engage with or even understand (Sullins 2008). The result is a superficial "commodification" of our personal relationships. For instance, a commodification of the phenomenon of nurturance. If we get used to the mechanisms of commoditized relationships that we experience with artificial beings, we could reach a point where "we … see our family, and ultimately ourselves, as mere dysfunctional devices … and might work to replace them with our perfect robotic companions" (Sullins 2008, p. 155). Such a replacement is already a common subject for transhumanists (Kurzweil 2006). In their view, human beings are suboptimal information-processing entities and a mere intermediary stage in the evolution of information (see also Chapter 19 and Kurzweil 2006). In contrast, scholars such as Johan Sullins want to avoid replacement. He thinks

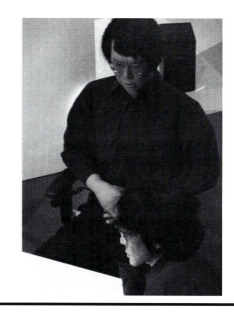

Figure 10.9 Geminoid is a humanoid robot built after his creator Hiroshi Ishiguro who kneels next to him. (By Shervinafshar, CC BY-SA 3.0.)

Figure 10.10 R2D2 as shown in the *Star Wars* films. (By Gage Skidmore, CC BY-SA 2.0.)

about technical measures that may help to create the right degree of differentiation between people and robots, avoiding functions and traits that make robots resemble people in their reactions and looks. For example, he recommends that robot emotions should remain iconic or cartoonish so that they can be easily identified as synthetic (Sullins 2008). More generally, he thinks that robots should not be built to be wholesale replacements for human interaction. However, Sullins's perspective stands in sharp contrast to current technological advances in the field, most notably the development of humanoid robots. Scholars like Hiroshi Ishiguro's explicit goal is to create robots that perfectly resemble human beings and can displace humans in environments such as service jobs. Figures 10.9 and 10.10

contrast the different perspectives of how robots should be built.*

EXERCISES

1. Apply your friendships to Figure 10.1. How does one of your good offline friendships correspond to the traits of friendship? Next, take someone you mostly know from an online environment and try to apply Figure 10.1 to that relationship. How do the two friendships compare? Where do you see most of the differences in friendship quality?

2. Take one of the virtual worlds that you know best and play often (if you play online; otherwise do not do this exercise). Consider the enablers of friendship in virtual worlds as they are shown in Figure 10.4. How does the virtual world you play in enable friendship? What could be further improved to foster friendship?

3. From a friendship perspective, discuss the pros and cons of using humanoid robots in nursing homes to care for elderly people.

4. In-class activity: Work with a fellow student to write a dialogue between a disclosing machine and a human (similar to Figure 10.7). One of you will play the role of the machine, and the other will play the role of the human. Build the conversation such that it fosters intimacy by considering the characteristics outlined in Figure 10.8.

5. Should virtual worlds be allowed to morph faces to influence people's choices for a good cause? Give reasons for your view.

6. Is it possible to build friendships in virtual worlds that are just as good as those in the real world? Give reasons for your opinion.

References

Altman, I. 1975. *The Environment and Social Behavior: Privacy, Personal Space, Territory, Crowding.* Monterey, CA: Brooks/Cole.

Altman, I. and D. Taylor. 1973. *Social Penetration: The Development of Interpersonal Relationships.* New York: Holt, Rinehart & Winston.

Anderson, S. L. 2011. "How Machines Might Help Us Achieve Breakthroughs in Ethical Theory and Inspire Us to Behave Better." In *Machine Ethics*, edited by Michael Anderson and Susan Leigh Anderson. New York: Cambridge University Press.

Aristoteles. 1969. *Nikomachische Ethik.* Stuttgart, Germany: Reclam Verlag.

Aristotle. 1915. *Magna Moralia.* Edited by W. D. Ross. Translated by St. George Stock. Oxford: Clarendon Press.

Aristotle. 2000. *Nicomachean Ethics.* Translated by Robert Crisp. Cambridge Texts in the History of Philosophy. Cambridge: Cambridge University Press.

Bailenson, J. N., S. Iyengar, N. Yee, and N. A. Collins. 2008. "Facial Similarity between Voters and Candidates Causes Influence." *Public Opinion Quarterly* 72(5):935–961.

Bereczkei, T., G. Hegedus, and G. Hajnal. 2009. "Facialmetric Similarities Mediate Mate Choice: Sexual Imprinting on Opposite-Sex Parents." *Proceedings of the Royal Society* 276 (1654):91–98.

Briggle, A. 2008. "Real Friends: How the Internet Can Foster Friendship." *Ethics and Information Technology* 10(1):71–79.

Bynum, T. W. 2006. "Flourishing Ethics." *Ethics and Information Technology* 8(4):157–173.

Churchesa, O., M. Nichollsa, M. Thiessenb, M. Kohlerc, and H. Keagec. 2013. "Emoticons in Mind: An Event-Related Potential Study." *Social Neuroscience* 9(2):196–202.

Cocking, D. and S. Matthews. 2000. "Unreal Friends." *Ethics and Information Technology* 2(4):223–231.

Derlega, V. J., S. Metts, S. Petronio, and S. T. Margulis. 1993. *Self-Disclosure.* Newbury Park, CA: Sage.

Dreyfus, H. L. 2009. *On the Internet.* Edited by Simon Critchley and Richard Kearney. Thinking in Action. New York: Routledge.

Fowler, J. H. and Christakis, N. A. 2008. "Dynamic Spread of Happiness in a Large Social Network: Longitudinal Analysis over 20 Years in the Framingham Heart Study." *British Medical Journal (BMJ)* 337(a2338).

Fujitsu. 2010. "Personal Data in the Cloud: A Global Survey of Consumer Attitudes." Tokyo, Japan.

Hampton, K. N., L. F. Sessions, E. J. Her, and L. Rainie. 2009. "Social Isolation and New Technology: How the Internet and Mobile Phones Impact Americans' Social Networks." Pew Internet and American Life Project, November 4.

Helm, B. 2013. "Friendship." In *The Stanford Encyclopedia of Philosophy*. Stanford, CA: The Metaphysics Research Lab.

Hoff, J. 2013. *The Analogical Turn: Rethinking Modernity with Nicholas of Cusa.* Cambridge, UK: William B. Eerdmans Publishing Company.

Jabbi, M., M. Swart, and C. Keysersa. 2007. "Empathy for Positive and Negative Emotions in the Gustatory Cortex." *NeuroImage* 34(4):1744–1753.

Kowert, R., E. Domahidi, and T. Quandt. 2014. "The Relationship between Online Video Game Involvement and Gaming-Related Friendships among Emotionally Sensitive Individuals." *Cyberpsychology, Behavior, and Social Networking* 17(7):447–453.

Kurzweil, R. 2006. *The Singularity Is Near: When Humans Transcend Biology.* London: Penguin Group.

MacIntyre, A. 1984. *After Virtue: A Study in Moral Theory.* 2nd ed. Notre Dame, IN: University of Notre Dame Press.

Millgram, E. 1987. "Aristotle on Making Other Selves." *Canadian Journal of Philosophy* 17(2):361–376.

Moon, Y. 2000. "Intimate Exchanges: Using Computers to Elicit Self-Disclosure from Consumers." *Journal of Consumer Research* 26(4):323–339.

Munn, N. J. 2012. "The Reality of Friendship within Immersive Virtual Worlds." *Ethics and Information Technology* 14(1):1–10.

---

* For an overview of state-of-the-art humanoid robots, see the website of Hiroshi Ishiguro Laboratories, http://www.geminoid.jp/en/index.html (accessed September 9, 2014).

Nass, C. and Y. Moon. 2000. "Machines and Mindlessness: Social Responses to Computers." *Journal of Social Issues* 56(1):81–103.

Newton, I. 1687/1848. *Newton's Principia: The Mathematical Principles of Natural Philosophy*. Translated by Andrew Motte. New York: Daniel Adee.

Nietzsche, F. 1882/2001. *The Gay Science*. Edited by Bernard Williams. Translated by Josefine Nauckhoff and Adrian Del Caro. Cambridge Texts in the History of Philosophy. Cambridge, New York: Cambridge University Press.

Penton-Voak, I. S., D. I. Perrett, and J. W. Peirce. 1999. "Computer Graphic Studies of the Role of Facial Similarity in Judgements of Attractiveness." *Current Psychology* 18(1):104–117.

Preston, S. D. and F. B. M. de Waal. 2002. "Empathy: Its Ultimate and Proximate Bases." *Behavioral and Brain Sciences* 25(1):1–72.

Reeves, B. and C. Nass. 1996. *The Media Equation: How People Treat Computers, Television, and New Media Like Real People and Places*. New York: Cambridge University Press.

Roberts, J. and M. Koliska. 2014. "The Effects of Ambient Media: What Unplugging Reveals about Being Plugged In." *First Monday* 19(8).

Shibata, T. 2004. "An Overview of Human Interactive Robots for Psychological Enrichment." *Proceedings of IEEE* 92(11): 1749–1758.

Soraker, J. H. 2012. "How Shall I Compare Thee? Comparing the Prudential Value of Actual and Virtual Friendship." *Ethics and Information Technology* 14:209–219.

Suler, J. 2004. "The Online Disinhibition Effect." *Cyberpsychology & Behavior* 7(3):321–326.

Sullins, J. P. 2008. "Friends by Design: A Design Philosophy for Personal Robotics Technology." In *Philosophy and Design: From Engineering to Architecture*, edited by Pieter E. Vermaas, Peter Kroes, Andrew Light, and Stevan A. Moore. Milton Keynes, UK: Springer Science.

Turkle, S. 2011. "Authenticity in the Age of Digital Companions." In *Machine Ethics*, edited by Michael Anderson and Susan Leigh Anderson, 62–76. New York: Cambridge University Press.

Turkle, S., W. Taggart, C. Kidd, and O. Dasté. 2006. "Relational Artefacts with Children and Elders: The Complexities of Cybercompanionship." *Connection Science* 18(4):347–361.

Vallor, S. 2010. "Social Networking Technology and the Virtues." *Ethics and Information Technology* 12(2):157–179.

Vallor, S. 2012. "Flourishing on Facebook: Virtue Friendship & New Social Media." *Ethics and Information Technology* 14(3):185–199.

Walther, J. B. 1996. "Computer Mediated Communications: Impersonal, Interpersonal, and Hyperpersonal Interaction." *Communication Research* 23:3–43.

Weizenbaum, J. 1977. Die Macht der Computer und die Ohnmacht der Vernunft Frankfurt: Suhrkamp Verlag.

Yee, N. 2014. *The Proteus Paradox*. New Haven, CT: Yale University Press.

Yee, N., J. N. Bailenson, and N. Ducheneaut. 2009. "The Proteus Effect: Implications of Transformed Digital Self-Representation on Online and Offline Behavior." *Communication Research* 36(2):285–312.

Yee, N., N. Ducheneaut, L. Nelson, and P. Likarish. 2011. "Introverted Elves and Conscious Gnomes." Paper presented at Computer Human Interaction (CHI), Vancouver, Canada, May 7–12.

Chapter 11

Dignity and Respect in the Machine Age

"To regard or treat someone as merely an object for aesthetic appreciation or scientific observation or technological management, or as a prey, or as a machine or tool, or as raw material or resource, or as a commodity or investment, or as obstacle, or as dirt or vermin, or as nothing is to insult and demean their dignity as persons and to violate the moral obligation to respect persons."

Robin Dillon (2010)

In his seminal work on human motivation and personality, Maslow (1970) distinguished two aspects of self-respect or self-esteem: first, self-respect, which stems from personal experiences of achievement and confidence in the face of the world; and second, reputation and prestige, which we receive from others in the form of respect. The two forms of respect may re-inforce each other. Taken together, they constitute to a large extent what philosophers call human dignity (Nussbaum 2004; Ashcroft 2005). Therefore, let us begin by looking at the concept of dignity and see how philosophers view self-respect and respect by others as integral components of human dignity.

11.1 Dignity and Respect

The respect for human dignity has been recognized by Christian and secular scholars for at least 500 years. One of the earliest works to reflect on human dignity was probably Pico della Mirandola's "Oration on the Dignity of Man," from 1486. In the sixteenth century, Dominican monk Francesco de Vitoria was the first to talk about an "intrinsic dignity" of men, referring to Native Americans who he sought to defend against Spanish colonial conquerors. Widespread adoption of the concept of human dignity is, however, associated with the enlightenment period. Kant (1784) regarded dignity as founded in three human traits. First, equality is the idea that all human beings are born equal and have the right to be respected as rational beings, not animals. This view means that no matter how a person behaves he or she has the right to be treated as a person. The second trait that constitutes human dignity is agency. Humans have the ability but also the responsibility to act autonomously. Dignity constitutes itself in us when we act responsibly and make decisions in accordance with what we perceive to be worthwhile and fitting with our convictions. And third, humans can autonomously define themselves. They are the masters of their identities. We can live a life that gives expression to ideals and pursue projects that help us to form and live out our identity.

Not all global cultures share the idea of dignity or believe in these three particular traits. But in Western cultures, this thinking has been extremely powerful and constitutes the root of our current legal and political systems. As a result, many Western national constitutions, directives, and documents, such as The Universal Declaration of Human Rights and The European Convention on Human Rights, embrace this thinking. These documents often start with an explicit expression of the idea that "all human beings are born free and equal in dignity and rights" (Article 1; United Nations General Assembly 1948).

If we believe that humans are born with dignity, we also believe that they naturally deserve a certain *recognition respect*. Robin S. Dillon, one of the leading contemporary scholars on respect, writes "recognition respect is the only fitting response to the moral worth of dignity, the response that dignity mandates" (2010, p. 22). We give recognition respect to other people when we take their wishes, attitudes, or desires into account before we act ourselves. Countering selfish wants, we recognize the dignity of others by respecting their autonomy, their choices, their privacy, their property, and their physical needs. Some scholars call recognition respect "consideration respect."

For those who construct and operate machines, recognition respect means acting based on many values addressed in this book: building and operating machines that are fair to humans and treat them without bias, collecting personal data only with people's consent, giving people control over automation, protecting people from exposure, respecting people's freedom of thought and action, and giving people the right to be let alone. Most of these aspects have been treated in Chapters 5 and 6 on ethical data collection and freedom. From an ethical perspective, engineers and managers should engage in ethical IT design to give recognition respect to the people who use the machines and to those who are exposed to them.

Besides recognition respect, philosophers recognize another form of respect, *evaluative respect* (Dillon 2010). Evaluative respect is a kind of appraisal for our achievements. Evaluative respect recognizes that we all try to live up to certain standards of worthiness by which we then tend to judge others and ourselves. Evaluative respect for oneself or for others can therefore be measured in degrees, depending on the extent to which the object of appraisal meets a standard. "It is the kind of respect which we might have a great deal of for some individuals, little of for others, or lose for those whose clay feet or dirty laundry becomes apparent" (Dillon 2010, p. 20). A wise leader who makes careful decisions for his company and employees may receive evaluative respect from them (see Chapter 15).

Members of today's capitalist societies tend to have evaluative respect for people with possessions or property, for people who have something to say (i.e., in the media) or, most important, for people in good jobs or positions. Even though these three main drivers of evaluative respect may not necessarily lead us to be a virtuous society, I will focus hereafter on these three particular drivers of respect.

I will discuss how these are influenced by our current and future machine world:

■ Personal property of machines or ownership of digital goods (e.g., software agents, digital music, and personal data) might contribute to people's evaluative self-respect. We can design information markets and digital services to foster perceptions of ownership as well as real ownership.

■ People can receive evaluative respect from others for what they say. Deliberate communication, but also brilliant software code that is shared, can earn people positive attention capital (Franck 1998) and respect from the community (Coleman 2013). A prerequisite for earning respect in this way is freedom of speech.

In the software world it may be the four software freedoms (see Section 11.5).

■ When it comes to jobs, we are stuck in a dilemma. Naturally machines replace human labor as tasks are automated. As a result, machines tend to be a threat to many individual's positive growth rather than a boon. At least this is true for the generation of employees whose work is easily replaceable by machines and who cannot easily switch to other positions or professions. I discuss this dilemma in Section 11.7.

Before I delve into these three subject domains, I first want to deconstruct and analyze the respect value as such. I want to outline how the act of respecting someone manifests itself in practice and how a practice of recognition respect can be translated into respectful and polite machines.

11.2 Respectful Machines

Respect stems from the Latin word *respicere*, which means "to look at or to look again." This verbal root indicates that respect involves paying attention to someone or something, not in the sense of staring at someone, but in terms of considering what someone has to say and taking his or her position seriously. Dillon notes that when we are attentive to someone, we have to be careful to not automatically categorize people. Instead, we must first try to understand who they really are or how they want to be seen (Dillon 2010). Can machines integrate this kind of respect?

Machines observe minute details about us. They are extremely attentive to what we do, and they collect behavioral data about us when we do things like surf the web, pay for something, move in public places, travel, or use game consoles. However, machines cannot use this data to really understand us; rather, they categorize us and deconstruct our identities to form segments for which the machine has an internal representation. For example, marketing segmentation may categorize someone as a young, middle-class, male hedonist who wants to be rich or as an old, greedy, female widow. Segments can also be much more fine-grained; predicting buying intentions, pregnancy, marriage plans, and so forth. But machines are not good at really understanding us in the full emotional individuality that respect requires. As I outlined in Chapter 10, empathy is the capability of humans to understand each other and grasp each other's experiences. Our mirror neuron system seems to support this capability.

We do not know at the moment whether machines can ever live up to human levels of emotional understanding and intelligence. Machines can interpret minimal changes in humans' facial expressions and measure some of their emotions through sensors (e.g., through skin conductance sensors, body temperature sensors, pupillary dilation). They can interpret gestures and spoken words or text. Combining this data might give machines some insight into the states we are in and also into our personal stiles of behavior. If humans wanted to expose this degree of personal feeling to machine operators, machines would probably achieve higher degrees of attentive understanding of us than they do today. The question is, however, whether people are willing to share such highly private information with machine operators in the long run and under what conditions. If machines are built to preserve privacy and give full control to people (see Chapters 5 and 12), some might be willing to expose their feelings in this way. But even then it is unclear to what extent the collected data can be combined to approximate a truly intelligent and respectful characterization of humans; including an adaptive understanding of how a person *wants* to be seen.

The second aspect of respecting another person is to respond in a respectful way. A respectful response is fitting or appropriate to the observed behavior. But what is an appropriate response? Between humans, responsiveness is certainly driven by individual spontaneity and sympathy for the vis-à-vis; a unique situational phenomenon. But some of it is typically also governed by a "judgment of groundedness" (Dillon 2010, p. 10). This means that the way we respond to others is a response to how they treat us and what character traits become apparent form their behavior. For instance, we may respect someone for his or her character of commitment and reliability, and therefore be polite to that person. The IT service world could embed this kind of respectful response to good character much more than it does today. Take, for example, the rewards offered to mobile phone subscribers when they call a call center to report a problem. Today, customers with high monthly phone bills are more likely to receive priority routing to a service assistant or be offered something like a free upgrade for a new mobile phone. In contrast, loyalty and longevity as a customer, regular bill payment (signaling reliability and trustworthiness), care for hardware such as mobile phones (signaling care), or rare complaints (frugality) are traits that are less typically rewarded by machines or machine operators. Thus, so far, machine responsiveness is driven more by the short-term financial utility of an operator. Respectful responses derived from a grounded judgment, however, should be utility independent. They could be focused more on character traits of customers.

Besides a judgment for groundedness, Dillon outlines that respect implies an interest-independent valuing component. This component ties into Kant's perspective on human dignity. From this perspective, human beings deserve response just because they are humans. So even if neither their character nor financial utility promise a fruitful exchange, machines or machine operators should still respond somehow to people requesting an exchange. Today, they often do not. For example, when people living in a poor part of town request a mail-order catalog, it is often not sent

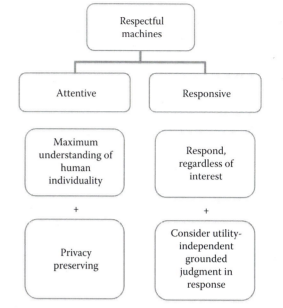

Figure 11.1 Conceptual investigation of the value of respect (for machine design).

to them because the provider machine assumes that they cannot pay for the products in the catalog anyways. Figure 11.1 summarizes the outlined dimensions of respectful behavior that could be considered in machines' design.

But machines can do more than just incorporating the rudimentary dimensions of respect. We can design machines that flatter us and cater to our self-esteem. We can partially achieve this design goal by building machines that are even polite.

11.3 Polite Machines

As we interact more with machines, it will become more important that machines interact politely with us. The perceived politeness of machines such as robots and personal agents will influence whether we embrace and appreciate this new species or avoid and detest them. A form of polite interaction is described in the retail scenario:

> [Sophia] really loves [her agent Arthur] like a friend even though he recently started to criticize her sometimes; for example, when she was lazy or unfair to a friend. … [But] Arthur is always extremely polite when advising her. His tone of voice is always soft and friendly. He relates his criticism to some history of her behavior and also garnishes his suggestions with some reference to philosophy, history, or statistics he is aware of. But most important, he is really selective of when he makes a remark.

Forms of politeness are embedded in all cultures but involve different norms of behavior or etiquette. In the Western world, a culture of politeness developed in the seventeenth century, the time of Enlightenment. The early eighteenth century philosopher Abel Boyer defined politeness as "a dext'rous management of our words and actions, whereby we make other people have a better opinion of us and themselves" (1702, p. 106).

Brian Whitworth has refined this definition, arguing that the core of politeness is the concept of choice (Whitworth and Liu 2008). When we say thank you we imply that the other party had a choice to say no. When we say please, we signal awareness that the other party does not need to comply with our wishes and has the choice not to. We do not interrupt the person we are talking to while she speaks because we want to leave her the room to choose when to finish. In polite communication, the locus of choice control passes back and forth between the parties.

Choice can be understood from two angles that allow us to distinguish between positive and negative politeness. *Positive politeness* gives the other party a perception of choice to act as they wish or creates positive room for conversation and action. An example is saying excuse me, because we give the other party room to judge our behavior. Another example is agreeing with another party and confirming their viewpoint. We give the other party the impression of acceptance. We can also engage in *negative politeness* when we politely disagree. We then take choice away from the other party in a way that is agreeable and that leaves that party room for objection. "If you don't mind …" or "If it isn't too much trouble …" are ways to make requests less infringing. Indirect speech is a common strategy in negative politeness. Finally, if choice is taken away from the other party and we cannot comply with the other party's wishes, we typically express pity. We say sorry in these cases.

How can polite behavior be transferred to machines? The most obvious measures involve implementing polite language in machines. Currently, machines often beep loudly at us or send cryptic error messages. The adjustment of acoustic levels or a "soft voice" are a primitive first step in the right direction. Also, the flow of interaction can be designed to the norms of the culture in which the machines are deployed. Robots in Japan, for example, may use different interaction process flows or gestures than those deployed in Europe. Today, little is known about the measures that would be required to take account of cultural differences in this domain.

But a polite voice and language flow does not suffice. As anyone who has watched Stanley Kubrick's film *2001: A Space Odyssey* knows, such a voice can even be vicious. In the film, the computer Hal has a polite voice and language but betrays the crew. What is not fully clear for most parts of the film is who is actually behind Hal's straying intelligence. We only realize later that a higher extraterrestrial intelligence is influencing Hal to the detriment of the crew. In a less deadly but similarly invisible form, we often do not know today who operates the digital services we use. Who surveys our mobile data traffic beyond the operator we signed up with? Who is behind the identification systems we use? As of 2015, each time we open a browser and start surfing the web, an average of 56 remote parties are likely to monitor what we do online (Angwin 2012). If we want to identify the parties that are watching us, we must meticulously download, install, and operate extra privacy tools. And even then, we receive only cryptic identity information on the data collectors such as Krux Digital, Dynamic Yield, and New Relic. Hardly any contact details or background information is given on any of the parties that have some role in our interactions online. In contrast, polite machines or services would disclose these parties in full detail and give us the possibility to contact them.

The core of machine politeness is, however, that they give and respect human choices (Dillon 2010). To meet this requirement, a machine must first offer desired choices. A user experience study can identify the choices that people want to have. For example, many people are concerned about losing their privacy and want more control over their personal data. People want to choose whether their surfing behavior is tracked, and a polite machine would offer this choice. In Chapter 5, I outlined how informed consent can work. However, such a choice is useful only if it is easy to understand and easy to exercise. Choices can therefore be accompanied by detailed and clear, easy-to-understand explanations. Visualization may be used for illustrating potential choice dependencies (to demonstrate how one choice influences other options). Background information can be made available. And easy-to-use control panels can give users the means to exercise their choice. Of course, a challenge here is to not overwhelm people. As outlined in Section 5.5 on ethical knowledge aggregation, it is not the goal of transparency to give people complete detailed information on everything that there is to know. Transparency requires only meaningful, useful, truthful, comprehensive, accessible, and appropriate information on available choices (recall Figure 5.14).

A second dimension of choice design in polite machines is to respect the *choices made by users*. You do not hold a door open for someone and then cut them off to walk through it yourself. Such behavior would be considered rude in personal interactions. However, in digital interactions, a user's choice is less obvious. If there are no audit logs or feedback signals from the operator that confirm the mutual agreement, then there is no way for a user to know whether the machine operator actually respects the choice the user made (see Chapter 8 on safety and accountability). Therefore, a polite machine will acknowledge a choice and indicate that the choice will be respected.

Another way to respect user choices is to give people room to decide by *avoiding preemptive actions*. Even if a user welcomes higher levels of automation after first use, machines should initially allow them to specify preferences instead of setting defaults (see Chapter 6 on freedom and automation). Nudging people by setting defaults is a powerful way to influence decision making. However, it is not really polite and respectful because it challenges people's liberty. A polite default choice that forces people to decide on later defaults may be a better way to ensure that machine actions are in line with people's preferences. Whitworth and Liu (2008) call this "meta-choice." Meta-choices should be used in machines especially when the usage of common resources is at stake (resources shared by the machine and the machine owner). Most users perceive and consider common resources such as their personal data, attention, desktop space, and Facebook wall to be theirs (see next section). A polite machine respects this "psychological ownership status" and does not act autonomously on the resources before providing meta-choices.

Finally, machines should have a memory of interactions so that they do not force people to constantly make the same choices. Although sufficient interaction memory is often framed only as a usability issue, it is relevant here. It is relevant, because some machine operators today deliberately prompt users to repeatedly make the same choices in order to nag and persuade them to consent to activities the operator prefers. A simple current example is the checkbox for newsletters that people can opt in or opt out to receive advertising information. Companies want users to choose to receive advertising. As a result, when forms are reloaded because a user forgot to enter required information, the newsletter checkbox is often changed back to the default choice of subscribing to the newsletter, even if the user previously actively denied the newsletter receipt. Experiments have shown that people treat machines just like other social actors and apply the same rules of politeness to them that they do to people. For example, in one experiment, people preferred to share negative information on a computer with a third computer in a separate room than with a computer in the same room (Reeves and Nass 1996). Why should machines not treat us equally politely? Figure 11.2 summarizes what constitutes a polite machine.

11.4 Psychological Ownership in Future IT Environments

Respect and politeness are ethical expressions of a valuation of others. According to Kant and many thinkers of classical modernity, human beings deserve respect and an expression thereof because they are born equal. However, people who are socialized in a Western, capitalist society make a lot of this respect and politeness dependent on the status one achieves through personal possessions.

"A man's Self is the sum total of all that he can call his, not only his body and his psychic powers, but his clothes and his house, his wife and children, his ancestors and friends, his reputation and works, his lands and yacht and banc-account. All these things give him the same emotions. If they wax and prosper, he feels triumphant; if they dwindle and die away, he feels cast down," according to William James (1890).

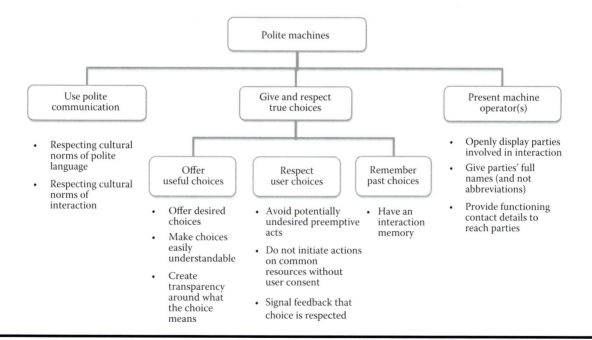

Figure 11.2 Conceptual analysis of politeness for machine design.

As James outlines, an important part of our self-identity relies on personal possessions. In his book *Being and Nothingness*, the French philosopher Jean-Paul Sartre (1992) argued that the only way we can know who we are is by observing what we have. Many authors have recognized that possessions are psychologically like "extensions of the self" (Belk 1988). Pierce, Kostova, and Dirks (2003) reviewed the motives that facilitate development of psychological ownership for objects: They found that owning something caters to our desire for efficacy and effectance. The ability to control our environment stimulates us and gives us a sense of security. In our modern times it is—whether we like that or not—highly related to the formation of our self-identity. Possessions help us to understand who we are, express our identity to others, and serve as a continuation of ourselves when we associate memories with them. Finally, people have a deep desire to have a place. Like animals, we tend to define our territory, a home that provides us with not only physical and psychic security but also satisfaction and stimulation (Porteous 1976).

Property is about more than just legal ownership; it is a mental state in which individuals feel as if the target of ownership is theirs. "I suggest that … it is most productive to examine property as a dual creation, part attitude, part object, part in mind, part in 'real,'" wrote Amitai Etzioni in his reflection on the socio-economics of property (1991, p. 466). Just think of a gardener who after a certain time feels that the garden belongs to him even though it may be public property. Pierce calls this cognitive affective mental state "psychological ownership" (Pierce, Kostova, and Dirks 2003), which is created when we use and control an object over a longer period of time, start to know it intimately and invest ourselves in it. Simone Weil once wrote, "All men have an invincible inclination to appropriate in their own minds, anything which over a long, uninterrupted period they have used for their work, pleasure, or the necessities of life" (1952, p. 33).

When it comes to digital services and devices, the concept of psychological ownership is just as powerful as in the physical world. It has been shown that people perceive psychological ownership for instance for their personal data on Facebook (Spiekermann, Korunovska, and Christine 2012). People buy digital goods to play online. And people create content online and are like artistic gardeners in their virtual space; building up ownership perceptions for it. That said, legal ownership of digital information goods, services, machines, and so on is in contrast organized through licensing schemes, which in turn are based on copyright and patent law (see Section 11.6). So when we create and use digital services, we often enter gray zones of ownership. For example, when people use a social network like Facebook and fill it with their personal data, such as their photographs, jokes, and ideas, who should be the rightful owner of that content? Legally, Facebook has secured itself a usage right to this content. But does Facebook reduce psychological ownership of its content for its users by denying them exclusive usage rights to and full control over their personal data, communication, ideas, and friends? How about mash-ups of films and music files, which people create based on their own and other people's (and companies') content? For example, take a film collage artfully created by someone and then shared on YouTube. In these collages, private individuals take existing material from copyrighted sources and meticulously cut and mix them into something new. What is the best way to assign ownership rights in such a case, given peoples' ownership psychology, the attachment to their digital creations, and the goal of companies to have people come back? My point is that it may be beneficial for companies to consider psychological ownership mechanisms in their business models and IT designs.

How can psychological ownership be theoretically maximized? Psychological ownership is deeply rooted in people's control over their objects, their in-depth knowledge of them, and their investments of self into them. Belk (1988) and Pierce, Kostova, and Dirks (2003) summarize these three complementary and probably additive causes of psychological ownership, and Figure 11.3 relates these causes to machines.

Lita Furby extensively researched the first root of psychological ownership: personal control over objects. She argues that greater amounts of control over an object promote a person's experience of the object as part of the self (Furby 1978). Controlling involves the ability to use the object, which extends beyond a legal right to actual operation. Work by David Mick and Susan Fournier (1998) has shown how people sometimes abandon technical objects that they legally own instead of taking psychological ownership of them. The two main reasons for this abandonment are typically that the systems are too complicated to use (Venkatesh et al. 2003)

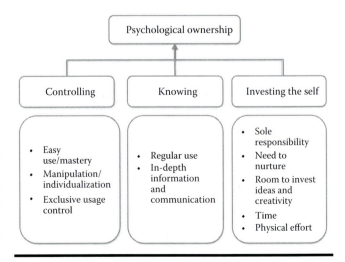

Figure 11.3   Conceptual analysis of psychological ownership.

or are not compatible enough with the way we live our lives (Rogers 2003). These findings point to the need to be able to manipulate, personalize and control how our systems work in order to build up ownership perceptions. Many desktop computers and smartphones already integrate functionality that allows us to customize features such as when they ring or notify us, and how things are organized. Still, many systems also deprive us of control and thereby undermine our psychological ownership of them. As of 2015, operating system providers, mobile phone manufacturers, and other companies tend to remotely access people's devices, upload information without legal knowledge and consent, run applications that are incomprehensible (if at all accessible), place warning messages on the screen that cannot be ignored, and so on. These practices are mostly done under the guise of security, but they are still examples of how organizations control machines that they do not own anymore (Whitworth and Liu 2008). With these practices, service providers and device sellers deprive object owners of the ability to fully control access to their possessions (Rudmin and Berry 1987). People seem to accept the practice. But it should be noted that people normally strive for the exclusive use of what they own. And when they share or admit access, they typically want to determine and choose for themselves with whom, when, and how often.

The second cause of psychological ownership is intimate knowledge of the target object. As we get acquainted with using our devices and experience them, we start to appropriate them. Regular usage can foster this appropriation process. During this process of learning about the object, it is vital that we accumulate information about it and think that we comprehend it better than other people do. IT manufacturers and service providers can support this belief by providing customers with in-depth knowledge about their products. This knowledge cannot be provided only through fancy package inserts or extensive background material on the web. Knowledge about products is also conveyed through social media in forms such as interactive services, individualized product homepages (facilitated by Internet of Things, technologies like radio-frequency identification [RFID]), artificial agents that can be asked for help, and so on.

Finally, a third cause of psychological ownership is the ability to invest oneself and one's own creativity into the target object. John Locke (1689/1988) argued that because we own our labor and ourselves, we are likely to feel that we own that which we create, shape, or produce. So any technology that gives us room to personalize the objects or services and to adapt them based on our own ideas, time, and effort will foster psychological ownership. Another key construct that can spur self-investment is when objects need to be nurtured. Just think back to the famous Tamagotchi device that people took care of.

The following extract from the robot scenario in Chapter 3 illustrates the power of user control, knowledge and self-investment:

> [Jeremy's] dream would be to own a robot himself. Some people do and walk around with them proudly, like others walk their dogs. The more fancy robots are personalized in terms of voice, hair, eyes, size, and so on, and the more they have learned from their owners (including various software upgrades), the more people get attached to them.
>
> … Future Lab's philosophy is that robots should be devoted human servants; completely owned and controlled by their owners and never replacing humans. This product and sales philosophy has gained the company wide respect and recognition from the public; a public that has increasingly become wary of remote-controlled robot devices that replaced many industry jobs.
>
> The idea of the human–robot hierarchy (with humans always on top) is deeply embedded in Future Lab's design process. … Future Lab's robots are embedded with powerful artificial intelligence (AI) technologies including voice, face, and emotional recognition. But these AI functions run independently in dedicated sandboxes contained in the device that does not need to be networked to function. The robots learn locally after their initial setup and so become pretty unique creatures depending on their owners. … Future Lab's robots are designed with a view to total user control and excellent feedback functionality. Users cannot only command Future Lab's robots through easy and direct voice control but also switch them off completely through a designated "off command." Users can easily repair and replace most of the fully recyclable plug-and-play hardware components by using 3D plotters. This possibility to deconstruct robots like Lego parts has also led to very fancy personalization efforts of the community. The related software is completely open and constantly improved by the opensource community.

Unlike the example, modern IT companies probably envision remotely controlled robots. In fact, most will need to, because early robot technology will not be advanced enough to embed powerful AI (like speech recognition) into fully decentralized and nonnetworked systems. Other requirements that call for centralized and remotely controlled robot architectures include remotely servicing robots (similar to today's operating systems), overseeing security, and harvesting personal data that is collected from users. These requirements could result in business models where largely standardized robots are leased or rented to people instead of being sold outright. From a psychological ownership perspective, however, such a business decision could limit the market success of robots and the potential of these devices to foster people's perceived ownership driven, self-respect.

Although I believe in the power of ownership psychology, I do not want to miss pointing to two critiques. The first one is that the power of ownership psychology may not be the same in all cultures. Collectivist or socialist cultures may put less emphasis on the need to exclusively own and control something. The desire to control the device, rooted in an individualistic effectance motive, may be less salient in collectivist cultures than it is in individualistic cultures (Hofstede 1980).

The second critique relates to the general philosophical perspective on the importance of having ownership. Philosopher, Erich Fromm, criticized the "radical hedonism" inherent in a strive for more "having." In his influential work To Have or to Be, Fromm (1976) suggested that the orientation of wanting to possess should be critically questioned and replaced with an emphasis on sharing, giving, and sacrificing. Philosophers like Karl Marx (1867/1978) have criticized "commodity fetishism," instead pointing to the importance of "doing." Marx believed that real happiness and human growth can be achieved only when people do meaningful and properly rewarded work. John Rawls (2001) noted that the opportunity for meaningful work is the social basis for self-respect (Moriarty 2009). How this "doing" might be challenged in the machine age is discussed in the last section of this chapter on machines' impact on work. Finally, and more recently philosophers acknowledge that "we no longer understand what it means to be!" (Hoff 2015). Seen the importance to go beyond the aspect of having I want to show in the following section how free software can foster not only a perception of "having" through psychological ownership, but also higher satisfactions by supporting programmers' doing and being.

11.5 Self-Esteem through Open and Free Software

I have outlined so far that in capitalist societies "being" becomes determined to some extent by one's possessions (having) and one's work (doing), both of which create self-respect as well as respect from others, fostering one's dignity. Although the perception of our being is dominated by what we have and do, some of that perception is also separate from these factors. It is about how we relate to others. How do we feel embedded in shared practices with others (affiliation)? What impact (power) do we have on the life over others?

How can software freedoms impact our way of having, doing, and being? I would argue that a powerful emotional mix of these individual needs of having, doing, and being expresses itself in the free software movement and the effects it has on those who program.

The desire to have is present among software programmers, not necessarily in terms of a legal property right to the code they create but in terms of the psychological ownership thereof. Software developers or engineers who code often feel like artists or poets (Black 2002). They invest themselves creatively. They deeply know the machine they work on (or the part of it that they build). And they enjoy having control over the machine through the code they master. In fact, I would argue that the pleasure of being wizards, controlling something others cannot understand, having intimate knowledge about it is a psychological mechanism that motivates many programmers.

The ability to quickly master and control machines, and to create functionality in a gratifying way is strongly dependent on the existing code base. Programmers today constantly use and expand code libraries. Code libraries contain encapsulated code in files that permit the distribution of discrete units of functionality or "code behavior." The behavior of a piece of code can be inherited by a new piece of software that a programmer composes. The programmer can also alter the existing code base. The only prerequisite for this sharing to work is that the code is written in the same or in a compatible programming language and, of course, that it is "free." A whole generation of programmers is now used to sharing free code. Consequently, it is not surprising that the mantra of the software community and hence a large part of today's programming world is dedicated to the "freedom to run, copy, distribute, study, change and improve the software."[*] Programmers' "doing" depends on this freedom. But entrepreneurs also benefit from free and open source code libraries as well as whole programs and service components, which can be combined to create value bundles at much lower cost than if everything needed to be built from scratch.

The Free Software Foundation (FSF) lists four freedoms for software users:[†]

- Freedom 0—The freedom to run the program as you wish, for any purpose.
- Freedom 1—The freedom to study how the program works and change it so it does your computing as you wish. (Access to the source code is a precondition for this freedom.)
- Freedom 2—The freedom to redistribute copies so you can help your neighbor.

---

[*] "What Is Free Software?" Free Software Foundation, accessed March 1, 2015, https://www.gnu.org/philosophy/free-sw.en.html.
[†] Ibid.

■ Freedom 3—The freedom to distribute copies of your modified versions to others. (By distributing, you can give the whole community a chance to benefit from your changes. Access to the source code is a precondition for this freedom.)

An important part of freedom 0 is that programmers must be able to tinker with technological systems and use them for anything they want. This negative liberty of being free from any external use restrictions is more important for the FSF than preventing individual moral abuses of a piece of software. For example, this freedom allows programmers to use software for purposes that the original author would morally not support, such as for military or surveillance purposes. In this line of thinking, the community is more important than the individual programmer: If a programmer wants to distribute free code under FSF's GNU General Public License (GPL), then he often has to give up his "droit morale" (moral right) to determine what the software may or may not be used for. The community comes first and dominates a programmer's perceived ownership right to control what he created. But there are good reasons for this choice, most important being that the FSF wants free software to spread and become the dominant way in which software is distributed.

To understand software, users often need to be able to run it (freedom 0). Only by running a program can software developers understand the behavior of the code, observe effects, and make changes if needed (freedom 1). So, understanding is the essence of coding freedom, because understanding code in conjunction with the right to change it gives a programmer or user control over the software (Coleman 2013). On FSF's homepage, Richard Stallman (2015) outlines why this control is so essential: "Freedom means having control over your own life. If you use a program to carry out activities in your life, your freedom depends on your having control over the program. You deserve to have control over the programs you use, and all the more so when you use them for something important in your life." I outlined in Chapter 6 how control is essential for liberty and how perceived autonomy vis-à-vis machines requires machine accessibility.

The FSF relates the control over code to power: "When users don't control the program, we call it a 'nonfree' or 'proprietary' program. The nonfree program controls the users, and the developer controls the program; this makes the program an instrument of unjust power." I described unjust power abuse in the robot scenario.

> The BeeXL drone was a bit bigger than the regular device but carried small doses of extremely powerful teargas combined with a hypnotic. When gangs attacked Alpha1s, these bees came to the robot's support and sprayed gas onto the attackers, who quickly fell and could be picked up and

arrested by the police. To optimize reaction times, BeeXL drones were set to autonomously intervene and spray gas as soon as they detected violence. Recently, though, a debate started in the press on this kind of autonomous action by bee robots. When an old lady with dementia danced violently in a public square, a bee drone had mistakenly identified here as a criminal, intoxicating her in front of a stupefied crowd of witnesses.

The FSF believes that the threat of such an abuse of power by governmental or corporate institutions (as well as software mistakes) can be avoided or mitigated if everyone—including end users—can access program code and freely change it. In this way, the community takes care of the power balance between men and machines. Of course, not everyone can program and change things. But the belief is that enough good minds and hands are available to watch out for, change, and influence negative developments.

Another very different and positive way of looking at the concept of power is to recognize that those who master code also feel *empowered* by the process of mastering it. The feeling of controlling a machine is a very positive reward for the effort that is expended to understand it. Free code that can serve as a starting point for this understanding nourishes the power motif that influences personal motivation (McClelland 2009). Figure 11.4 illustrates how the software freedoms are important to the three concepts of having, doing, and being that I have outlined.

Freedoms 2 and 3 are then essential for the further use and commercialization of free software products and hence entrepreneurship. If a company wants to use free software that is published under version 2 or version 3 of FSF's

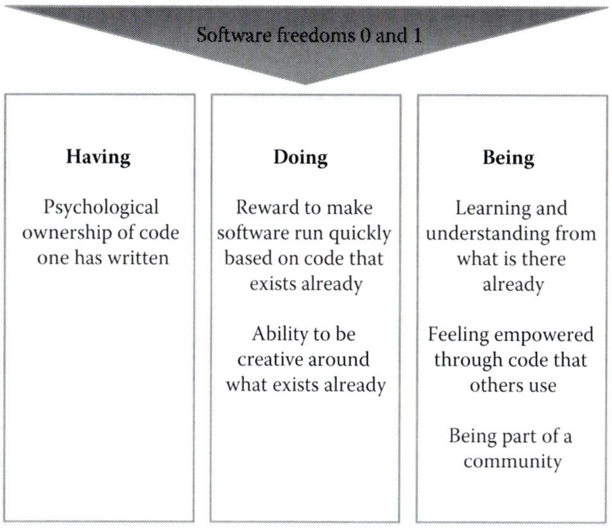

Software freedoms 0 and 1		
Having	**Doing**	**Being**
Psychological ownership of code one has written	Reward to make software run quickly based on code that exists already	Learning and understanding from what is there already
	Ability to be creative around what exists already	Feeling empowered through code that others use
		Being part of a community

Figure 11.4 Some selected effects of software freedoms on programmers.

General Public Licenses, it can do so for free, but it needs to distribute its derivative work under the same free conditions under which it got it. This practice is called "copyleft." For companies to benefit from the free codebase, they must also share their inventions back with the community. As I outlined earlier (and in Chapter 13), this sharing is important for building new systems. It is also good for the immediate gratification of programmers who can use an existing code base for making something work in a very short time frame.

A short final note though: The resharing of modified software worked well when modified software was still "distributed" or "redistributed" (i.e., sent on a CD to a recipient). This is what freedoms 2 and 3 say. But today's software provisioning is often not based on distribution any more. Instead, software runs as a service or service component on the servers of the modifiers. Take a search engine as a potential example. If the search engine provider used free software components for its search functionality and improved upon it, it would not be obliged under the GPLv2 or GPLv3 licenses to release the modified source code. It has the freedom to do so, but it does not have to, because the modified code runs on demand and is not "distributed." As a result, IT companies provide "software as a service" (SaaS) reaping the benefits of the free community work and can then decide to not give the code back to the community (Wolf, Miller, and Grodzinsky 2009). They can even "black-box" improvements of software that was initially open and free. The response of the FSF has been the introduction of the AGPL license (GNU Affero General Public License). AGPL says that when running (conveying) a modified version of a program as a service, then access to the source code must be offered to users. Its use would force application service providers to make their code base extensions accessible to the community.

11.6 Influence of Patents and Copyrights on Human and Economic Growth

Beyond free software, more exclusive and proprietary forms of property rights exist to commercialize the use of digital devices and digital content. First, there is the patenting system. When somebody invents a machine or service, he or she can apply for a patent with the national patent office if the proposed solution is new, not obvious and does not yet exist. *Wikipedia* defines a patent as "a set of exclusive rights granted by a respective sovereign state to an inventor or assignee for a limited period of time in exchange for detailed public disclosure of that invention."* Patents are openly available with

their full content.† Yet, the functionality they describe is not free to use. The *exclusive right* element of a patent means that inventors have the right to determine what is done with their invention. Inventors may prevent others from implementing the respective solution, sell the right to use it, or build the invention themselves. The latter of these three core rights is the reason why patents came into existence in the first place. They were a way to protect innovators from competition and give them time (typically 20 years) to harvest exclusive financial benefits from the market innovation. The legal fathers of the patenting practice thus wanted to incentivize innovation.

In the meanwhile, a large part of technology patents are used (unfortunately) to exercise the first two rights: Blocking patents are used as a competitive strategy to prohibit competitors from getting a foothold in one's market. As of 2015, major companies often get into costly patent wars to make each other pay for solutions they claim to have invented first or to block a competitor altogether. Many major corporations have started to pool their patents in order to avoid such patent wars or to form oligopolistic market structures, limiting a market to a controlled and small number of competitors. "Patent trolls" are a particularly negative abuse of the patenting system, to the extent that regulators consider limiting them (*The Economist* 2013). Patent trolls are commercial entities (often law firms) that file patents without ever intending to put the innovation into practice. Instead, they sell the rights to the patent to whoever wants to build something based on the technology. Typically, patent trolls plaster a potential digital service or machinery with patents from all imaginable technical angles; as a result, innovators are very unlikely to realize the technical solution in a sensible way without negotiating rights from the patent troll. Start-ups and small companies are often unable to innovate because they cannot afford royalty payments to patent trolls. They are also unlikely to get venture capital funding for a solution patented by others. Because of these patent practices, most technical innovation and funding efforts now start with an extensive patent search.

Against the background of this economically problematic situation, some companies question patenting practices. For example, the company Tesla recently released all its patents on electric vehicles, enabling free use by everyone. In his public blog, Tesla CEO Elon Musk (2014) writes: "Tesla Motors was created to accelerate the advent of sustainable transport. If we clear a path to the creation of compelling electric vehicles, but then lay intellectual property landmines behind us to inhibit others, we are acting in a manner contrary to that goal. ... Technology leadership is not defined by patents, which history has repeatedly shown to be small

---

* "Patent," *Wikipedia*, accessed September 19, 2014, http://en.wikipedia.org/wiki/Patent.

† Public patent libraries are made available, for example, through the United States Public Trademark Office (http://www.uspto.gov) or the European Patent Office (http://www.epo.org/index.html).

protection indeed against a determined competitor, but rather by the ability of a company to attract and motivate the world's most talented engineers. We believe that applying the open source philosophy to our patents will strengthen rather than diminish Tesla's position in this regard."

Some technology-driven companies like Tesla question patents (even though they own important ones themselves and could effectively block some competition) because IT markets are particularly prone to the phenomenon of network effects. This phenomenon means that the value of a market increases exponentially based on the number of market participants. Tesla sells electric vehicles and depends on the indirect network effect that people will buy electric vehicles only if there are enough fuel stations for them to refill their car. More electric fuel stations lead to more electric vehicles being sold. But fuel station owners only have an incentive to invest in servicing new vehicles if there are enough of the vehicles around. If Tesla uses its patents to block the entry of other electric vehicle players, then the overall market size for these devices may remain so small that fuel stations do not ramp up, damaging Tesla's own customer base. Economist Hal Varian explains this "information rule" as follows: "Unless you are in a truly dominant position at the outset, trying to control the technology yourself can leave you a large share of a tiny pie. Opening up the technology freely can fuel positive feedback and maximize the total value added of the technology. But what share of the benefits will you be able to preserve for yourself? Sometimes even leading firms conclude that they would rather grow the market quickly through openness, than maintain control" (Varian and Shapiro 1999, p. 199). This is exactly what Tesla seems to do.

I have shown so far that patents can be problematic because they block innovation, add cost to end products, and prohibit markets to grow. In addition, they also make it difficult to engage professionally with patented IT. In fact, patents often prohibit people from using the IT tools they own for their own creative entrepreneurial endeavors. Take Apple's QuickTime license as an example: As of 2015, owners of an Apple iPhone or iPad are not allowed to sell videos they create by using the embedded QuickTime software unless they pay for a license with the Motion Picture Experts Group Licensing Authority (MPEG LA).* Transferring this example to the offline world means to imagine that a carpenter who uses a hammer to put nails into the furniture of his clients would need to pay a license fee for every nail just because his hammer is patented. Does this influence the creativity of the carpenter or even a person's incentive to ever

become a carpenter? Probably yes. Patents are clearly questionable from the perspective of human growth, because they limit entrepreneurship and creativity. Because being entrepreneurial and creative is a form of being and doing (and later potentially having), I would argue that patents indirectly reduce many people's ability to build up self-respect and dignity.

The criticism that has just been described for patents has also been voiced for copyright protection schemes. A copyright is a legal right that grants the creator of an original work exclusive rights to its use and distribution. A copyright is intended to enable a creator (such as a photographer or author of a book) to receive compensation for his or her intellectual effort. Similar to a patent, a copyright is an intellectual property right. However, a copyright is not a technical mechanism but an idea or information that is substantive and discrete.† Copyrights are important for authors of creative works who need financial compensation for their publications. For example, a book author who invests many months or years in a book would like to receive appropriate financial compensation for the work, at least for some limited time, as the original copyright laws foresaw it. The United States therefore embraced a copyright protection scheme as early as the eighteenth century. This scheme gave authors the right to protect and receive royalties from their work for 14 years. After that time, the work entered the public domain and could be used by anyone for free as long as they cited the original author.‡ Now, copyrights span the entire life of an author plus 50 years after death. During this time, the content is protected and can be used only if royalty fees are paid to the publishing house that holds the respective rights. Only short snippets can be used for free.

From an individual growth perspective, copyrighted material can be problematic when the acquisition and ownership of copyrighted material does not allow buyers of the content to be creative with the material they own. For example, Digital Rights Management (DRM) software may not allow customers to listen to a piece of music bought from one publisher on a hardware device bought from a vendor. Similarly, customers are restricted in the way they are allowed to and able to take music bought from one publisher and mix it with music bought from another publisher. Such mixing and remixing of creative content has been recognized though as a major motor of innovation; applying copyright law too strictly can hamper "the future of ideas" as Lawrence Lessig (2001) has analyzed.

Because patents and traditional copyright schemes can significantly restrict innovation and hence humans' capability to build up property, be creative, become entrepreneurs,

---

* Apple Software License Agreement, http://store.apple.com/Catalog /US/Images/quicktime.html. The Motion Picture Experts Group Licensing Authority (MPEG LA) has a website with license terms (http://www.mpegla.com/main/default.aspx).

---

† Based on information provided by *Wikipedia*, http://en.wikipedia .org/wiki/Copyright (accessed September 19, 2014).
‡ Ibid.

and so on, some parts of the software industry slowly start to embrace free schemes for licensing software, hardware, and content, such as the GNU GPL.

For digital content such as photos, books, or lecture slides, the Creative Commons scheme is a relatively new copyright scheme under which authors can share their creations with others and be recognized for them. Instead of an "all-rights-reserved" scheme, Creative Commons licenses promote a "some-rights-reserved" kind of thinking. Authors can give the public the right to share and use their creative works on individual conditions. These conditions are specified by an author, and a license is created on the Creative Commons website (https://creativecommons.org/). The following options are available to authors of original work: They might say that a user of their original content (1) must attribute their original in a specific manner (attribution: "BY"); (2) must not alter, transform, or build upon their original work (no derivative works: "ND"); (3) must not use the original for commercial purposes (noncommercial: "NC"); and (4) must—if they alter, transform, or build upon the original—only distribute the resulting new work under the same or a similar license than the original (share alike: "SA").

11.7 Outlook on Human "Doing": Impact of Machines on Work

One of the biggest challenges for human identity and self-respect in the machine age will be the employment changes caused by the ubiquitous automation of work processes. In his science fiction novel *Manna: Two Visions of Humanity's Future* (2012), Marshall Brain, a U.S. entrepreneur and writer, describes two scenarios for societal development. In the first scenario, which takes place in the United States, machines slowly but steadily take over blue-collar and then white-collar work. The central figure in his novel, Jacob Lewis, resumes how first his student job as a waiter in a burger place is successively automated. Starting with a better structuring and modularizing of the work tasks, advancing through ordering people what to do through headset commands, the computer system "Manna" finally deploys robots that replace waiters altogether.

But the "autonomous economy" does not stop there. As Carl Frey and Michael Osborne from the University of Oxford outline in their 2013 study on the future of employment, machines are increasingly capable of cognitive computing that enables them to do "thinking" jobs for people as well. According to the two British scientists, 47% of total U.S. employment is at high risk for being automated within a decade or two. Only tasks that involve high levels of creative and social intelligence, manual dexterity, or highly unstructured workspaces are difficult to automate (for instance,

researchers, artists, product developers; Frey and Osborne 2013).

Coming back to Brain's (2012) novel, humans are equally replaced by machines. Jacob Lewis finally ends up in "Terrafoam housing," a kind of slum where former middle-class people end up jobless, and detained, surveyed, serviced, and supervised by robots.

The decisive point in this dark scenario is a lack of redistribution of wealth. In Brain's (2012) story, the U.S. version of developments does not foresee a generous sharing of productivity gains. As a result, jobless people are driven into poverty and then cannot develop their capabilities and interests. The author literally extrapolates the current economic situation that is described by Massachusetts Institute of Technology (MIT) professors Erik Brynjolfsson and Andrew McAfee (2012) in their influential book *The Race Against the Machine*. Brynjolfsson and McAfee argue that the historic alignment of technical progress and societal wealth (through employment and income) may not hold in the future because technology might create a "great decoupling" between productivity and employment. Will we simply not need people for productivity any more? The MIT scholars also outline how real corporate profits in the United States have soared for the past 15 years, while real median family income is stalling. The IT industry creates some superstars at the top of the income pyramid, but the distribution of wealth resembles conditions last seen in the late 1920s. Based on data from Piketty and Saez, the authors illustrate how more than 60% of U.S. income gains are going to the top 1% of the people (Feller and Stone 2009; Figure 11.5).

In Brain's (2012) futuristic novel, the second scenario for automation, robot deployment, and wealth distribution is

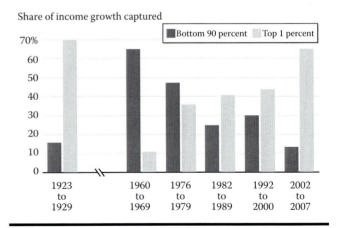

Figure 11.5 Over 60% of U.S. income gains are going to the top 1% of the people. (From Feller, Avi, and Chad Stone, 2009, "Top 1% of Americans Reaped Two-Thirds of Income Gains in Last Economic Expansion," Washington, DC: Center on Budget and Policy Priorities [CBPP]. With permission.)

called "The Australia Project." Here, machines are a public infrastructure, owned by everyone and servicing everyone, with income gains redistributed in society. The political and social setup in the Australia Project is obviously far from what we are heading toward today, at least in Western societies. But it is an interesting vision not only because of its economic setup but also because of the role robotics and automation play in the lives of people.

In the Australia Project, the role of machines is close to what Havelock Ellis, a British psychologist and author, once expressed: "The greatest task before civilization at present is to make machines what they ought to be, the slaves, instead of the masters of men" (1922, p. 129). In fact, machines could serve and relieve people. And the time and energy gains realized by delegating work to machines could free people to concentrate on tasks that they enjoy more than today's jobs. For example, people could spend more time with family and friends, innovate, learn, or become active in the community if they had the time to do so and received a good unconditional income (an income that could be paid out as a result of machines' productivity and not humans' own work). In such a positive scenario, people have the chance to engage in meaningful activities. By doing so, people could maintain and build their self-respect in new ways. What they are "doing" then may be more meaningful to them than what they are doing now as they work for wages. At least, this is the vision of Brain's Australia Project.

At this point, I want to return to value theory and the value pyramid I described in Chapter 4. If we deprive people of meaningful work and responsibilities, because we replace them with machines, then we deprive them not only of their financial basis to live but also of their basis for self-respect, self-esteem, and flourishing. Great thinkers have shared in this thinking before me. John Rawls argued that a lack of opportunity for meaningful work and occupation is destructive of citizens' self-respect (John Rawls in his *Political Liberalism*) (1921/2005). And Jeffrey Moriarty added: "To have self-respect, people must contribute to society. People contribute through work. So, if people lack access to meaningful work, then they may fail to contribute, and their self-respect may be damaged" (2009). In the face of automation, we must reshape our notion of meaningful work and must ensure that access to meaningful "doing" is ensured for everyone as well as the financial basis for it.

EXERCISES

1. Think of a system you interact with on a daily basis that currently embeds a primitive form of feedback. Using Figure 11.2, outline how the system could be redesigned to be more polite.
2. Think of a social network that you use regularly and all the personal data that you have on it. Using Figure 11.3, reflect on how much psychological ownership you feel for your information on that platform. What could social networks do to increase the perception of psychological ownership among their users?
3. Randomly chose 20 pictures and/or videos you like from various photo- or video-sharing platforms to build a collage. Once you made your choice and have a brief idea of what the collage could look like, analyze the license terms under which you are or are not allowed to share, change, mix, or resell the material you would like to use. What is your conclusion from the analysis? How much freedom do you have to be creative based on this media? Can you effectively create new media content with what you liked? And if yes, who are you allowed to share it with and under what terms?
4. Choose one of the new IT systems described in the scenarios in Chapter 3. For this IT system, identify a few major technological building blocks (components) you would need to make this system work. Use a public patent database to analyze to what extent each component is already patented. Who would be able to build the new IT system you chose?

References

Angwin, J. 2012. "Online Tracking Ramps Up: Popularity of User-Tailored Advertising Fuels Data Gathering on Browsing Habits." *Wall Street Journal*, June 18.

Ashcroft, R. E. 2005. "Making Sense of Dignity." *Journal of Medical Issues* 31(11):679–682.

Belk, R. W. 1988. "Possessions and the Extended Self." *Journal of Consumer Research* 15(2):139–168.

Black, M. J. 2002. "The Art of Code." Dissertation, University of Pennsylvania.

Botsman, R. and R. Rogers. 2014. *What's Mine Is Yours: The Rise of Collaborative Consumption.* New York: Harper Collins.

Boyer, A. 1702. *The English Theophrastus, or the Manners of the Age.* London: W. Turner and J. Chantry.

Brain, M. 2012. *Manna: Two Visions of Humanity's Future.* BYG Publishing Inc. Published online.

Brynjolfsson, E. and A. McAfee. 2012. *Race Against the Machine: How the Digital Revolution is Accelerating Innovation, Driving Productivity, and Irreversibly Transforming Employment and the Economy.* Lexington, MA: Digital Frontier Press.

Coleman, E. G. 2013. *Coding Freedom: The Ethics and Aesthetics of Hacking.* Princeton, NJ: Princeton University Press.

Dillon, R. S. 2010. "Respect for Persons, Identity, and Information Technology." *Ethics and Information Technology* 12:17–28.

The Economist. 2013. "Trolls on the Hill: Congress Takes Aim at Patent Abusers." December 7.

Ellis, H. 1922. *Little Essays of Love and Virtue.* London: A & C Black.

Etzioni, A. 1991. "The Socio-Economics of Property." *Journal of Social Behavior and Personality* 6(6):465–468.

Feller, A. and C. Stone. 2009. "Top 1% of Americans Reaped Two-Thirds of Income Gains in Last Economic Expansion." Washington, DC: Center on Budget and Policy Priorities (CBPP).

Franck, G. 1998. *Ökonomie der Aufmerksamkeit*. München Wien: Carl Hanser Verlag.

Frey, C. B. and M. A. Osborne. 2013. *The Future of Employment: How Susceptible are Jobs to Computerization*. Oxford: University of Oxford.

Fromm, E. 1976. *To Have or to Be*. New York: Harper & Row.

Furby, L. 1978. "Possessions: Toward a Theory of Their Meaning and Function throughout the Life Cycle." In *Life Span Development and Behavior*, edited by P. B. Baltes, 297–336. New York: Academic Press.

Hoff, J. 2015. "Contemplation, Silence and the Return to Reality," *The Way* 54(3):1–11.

Hofstede, G. 1980. *Culture's Consequences: International Differences in Work-Related Values*. Cross-Cultural Research Methodology Series. Newbury Park, CA: Sage Publications.

James, W. 1890. The Principles of Psychology–Volume 1, New York: Cosimo Classics.

Kant, I. 1784. "Beantwortung der Frage: Was ist Aufklärung." *Berlinische Monatsschrift*, December, 481–494.

Lessig, L. 2001. *The Future of Ideas: The Fate of the Commons in a Connected World*. New York: Random House.

Locke, J. 1689/1988. *Two Treatises of Government*. Cambridge Texts in the History of Philosophy. Edited by Peter Laslett. Cambridge, New York: Cambridge University Press.

Marx, K. 1867/1978. *Capital: A Critique of Political Economy*. Harmondsworth, England: Penguin.

Maslow, A. 1970. *Motivation and Personality*. 2nd ed. New York: Harper & Row Publishers.

McClelland, D. 2009. *Human Motivation*. Cambridge, UK: Cambridge University Press.

Mick, D. G. and S. Fournier. 1998. "Paradoxes of Technology: Consumer Cognizance, Emotions, and Coping Strategies." *Journal of Consumer Research* 25(9):123–143.

Moriarty, J. 2009. "Rawls, Self-Respect, and the Opportunity for Meaningful Work." *Social Theory and Practice* 35(3):441–459.

Musk, E. 2014. "All Our Patent Are Belong to You." In *Tesla Blog*, a blog by Tesla, June 12.

Nussbaum, M. 2004. *Hiding from Humanity: Disgust, Shame and the Law*. Princeton, NJ: Princeton University Press.

Pierce, J. L., T. Kostova, and K. T. Dirks. 2003. "The State of Psychological Ownership: Integrating and Extending a Century of Research." *Review of General Psychology* 7(1): 84–107.

Porteous, D. J. 1976. "Home: The Territorial Core." *Geographic Review* 66(4):383–390.

Rawls, J. 1921/2005. "Political Liberalism." In *Columbia Classics in Philosophy*. New York: Columbia University Press.

Rawls, J. 2001. *The Law of Peoples: With, "The Idea of Public Reason Revisited."* Cambridge, MA: Harvard University Press.

Reeves, B. and C. Nass. 1996. *The Media Equation: How People Treat Computers, Television, and New Media Like Real People and Places*. New York: Cambridge University Press.

Rogers, E. 2003. *Diffusion of Innovations*. 4th ed. New York: The Free Press.

Rudmin, F. W. and J. W. Berry. 1987. "Semantics of Ownership: A Free Recall Study of Property." *The Psychological Record* 37(2):257–268.

Sartre, J.-P. 1992. *Being and Nothingness: A Phenomenological Essay on Ontology*. New York: Washington Square Press.

Spiekermann, S., Korunovska, J. and B. Christine. 2012. "Psychology of Ownership and Asset Defence: Why people value their personal information beyond privacy." *International Conference on Information Systems (ICIS 2012)*. Orlando, Florida.

Stallman, R. 2015. "Free Software Is Even More Important Now." Free Software Foundation, https://gnu.org/philosophy/free-software-even-more-important.en.html.

Turkle, S. 2011. "Authenticity in the Age of Digital Companions." In *Machine Ethics*, edited by Michael Anderson and Susan Leigh Anderson, 62–76. New York: Cambridge University Press.

United Nations General Assembly. 1948. Universal Declaration of Human Rights.

Varian, H. R. and C. Shapiro. 1999. *Information Rules: A Strategic Guide to the Network Economy*. Boston: Harvard Business Books Press.

Venkatesh, V., M. G. Morris, G. B. Davis, and F. Davis. 2003. "User Acceptance of Information Technology: Toward a Unified View." *MIS Quarterly* 27(3):425–478.

Weil, S. 1952. *The Need for Roots: Prelude to a Declaration of Duties towards Mankind*. London: Routledge and Kegan Paul Ltd.

Whitworth, B. and T. Liu. 2008. "Politeness as a Social Computing Requirement." In *Handbook of Conversation Design for Instructional Applications*, edited by Rocci Luppicini, 419–436. Hershey, PA: IGI Global.

Wolf, M., K. Miller, and F. S. Grodzinsky. 2009. "On the Meaning of Free Software." *Ethics and Information Technology* 11(4): 279–286.

Chapter 12

Privacy and a Summary of the Value Fabric

"I don't see myself as a hero because what I'm doing is self-interested: I don't want to live in a world where there's no privacy and therefore no room for intellectual exploration and creativity."

Edward Snowden (2013)

In 2006 Daniel Solove, an American legal scholar, published an extensive taxonomy of privacy. Over 84 pages, he explained the concept of privacy by reviewing more than a hundred years of legal case studies concerning privacy harms in the United States. Based on this analysis, he summarized 16 privacy issues, as shown in Figure 12.1. He concluded that privacy is a "chameleon-like word." No one can define it precisely while covering all its facets. The term is relevant in so many contexts that when it comes to machine age computing, "privacy seems to be about everything" and therefore "to some it appears to be nothing" (Solove 2006, p. 479).

While writing this book, I have come to agree with this viewpoint in a very specific way: I fear that a long chapter on privacy in this book would have been a chapter about everything, because privacy issues are instrumental to almost all the intrinsic values that I covered here. Privacy is everywhere. If I wrote a detailed chapter on all of the privacy issues, none of the other values would have received the degree of attention they deserve. Finally, it is these other intrinsic values though—knowledge, freedom, security, trust, friendship, and dignity—that people care about perhaps even more than their privacy. I have therefore explained how various of Solove's concrete privacy dimensions (such as informed consent or surveillance) affect these ultimate intrinsic values. I now want to shortly recapitulate these findings, summarizing how privacy issues come into play at various levels of the value pyramid.

12.1 Privacy and Ethical Knowledge

When we align Solove's privacy harms (Figure 12.1) with the value pyramid (Figure 12.2), we find that most of the privacy issues he identified from U.S. legal history arise from knowledge being created about people. Allen Westin (1967) called this kind of privacy "information privacy."

Information privacy harm can be caused by *increased accessibility*. Increased accessibility means that public personal data is easier to access through the web today than it was in the past. If this accessibility is not handled in a careful way, a person's reputation can be damaged. Take the case of Mario Costeja González, who filed a lawsuit against Google in 2010. González accused the company of using its search service to publicize the fact that he had failed to pay social security debts in 1998. González asked Google to not display his behavior from the 1990s, because the incident occurred over a decade ago. He wanted the incident to be forgotten, arguing that it damaged his reputation. The European Court of Justice supported González.

Note that in the ethical knowledge chapter (Chapter 5, Figure 5.14), I described how a technical system (like Google's) can create transparency. Increased accessibility is the result of transparency. However, providing transparency in an *ethical* way means that only meaningful and appropriate information is published, and not any information that someone or something can acquire. If Google's search engine was technically optimized to provide *appropriate* information about González that creates a *truthful* image of his person, then at the very least the information provided by the machine would have needed to be timely. If that had been the case, the company would not have been sued. Looking at this example we see that transparency is deeply interwoven with privacy in terms of accessibility.

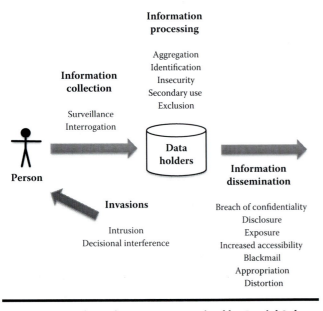

Figure 12.1 Privacy issues as summarized by Daniel Solove. (From Solove, D. J. 2006, "A Taxonomy of Privacy," *University of Pennsylvania Law Review* **154(3):490. With permission.)**

Interrogation is another form of privacy harm. The term originally referred to pressuring individuals to divulge information. Interrogation is different from surveillance in that it occurs with the conscious awareness of the subject and is not clandestine. Requesting to use customers' detailed personal information in the context of a service contract, combined with a denial of service if that information is not provided, can be considered a modern form of interrogation. Some national data protection laws therefore foresee a prohibition of coupling service use with personal data provisioning.* This prohibition of service coupling combined with informed consent procedures (Chapter 5) can ensure ethical information collection practices. Through informed consent that is voluntary, people can maintain control over data collection both technically and psychologically.

Solove outlines that privacy threats can be created through *aggregation* of personal information and how *distortion* of a person's image can result from analyzing aggregated data. Data can be legitimately aggregated if the data aggregator gets a person's explicit informed consent and keeps the data under that person's control (such as agent Arthur accumulating data about Sophia in Chapter 3). But distortion can still result from mistakes during the aggregation process. Therefore, a real challenge for companies is that their data quality needs to be very good for aggregation purposes (Section 5.4.1). Yet, as of 2015, data quality was not always good enough, leading to distortion of people's images during aggregation processes.

---

* For example, in German data protection law there is a so-called "Koppelungsverbot."

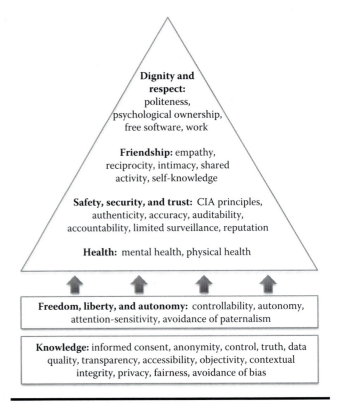

Figure 12.2 A pyramid of values important in future IT environments.

Distortion can be seen as a transparency issue. Transparency aims to reveal truth and avoid confusion, distortion, and pain as a result of unobserved errors. When data is aggregated in a transparent way, there is less risk of distortion of truth because a community of people can potentially look into the quality of data and aggregation practices. Another way to mitigate the negative consequences of distortion is to anonymize the original data and prevent it from being linked to specific individuals altogether (see Chapter 5, Box 5.1).

Finally, privacy is about the ethical use of knowledge about people. I have described how unauthorized secondary uses of data, a breach of confidentiality of information, exposure, public disclosure, and appropriation of data cause privacy harms (Chapter 5). Helen Nissenbaum's concept of "contextual integrity" can be used to think about the ethics and legitimacy of information flows that extend beyond an agreed context, allowing for such harms to happen. Beyond contextual integrity, I also described in Chapter 5 how and why people consider some data uses to be unfair and how data may be abused to create bias (see Figure 5.22).

12.2 Privacy and Freedom

Privacy dimensions are not only an issue for ethical knowledge creation and use. They are also relevant for our freedom.

In particular, surveillance as a special kind of privacy harm can undermine our freedom. The "right to be let alone" is harmed in a negative libertarian sense (Warren and Brandeis 1890).

A positive libertarian reason why surveillance has been said to reduce freedom is that it makes people behave in a restrained way. Scholars argue that we feel or are aware of being watched, and we therefore adapt our behavior to the expectations of those who watch us. The origin or source of our actions is therefore not our free will any more but the presumed expectations of our guards. I outlined this line of panopticon argument in Chapter 6 on freedom. However, there is little empirical evidence at the moment that citizens do in fact feel consciously constrained in their actions due to machine surveillance (such as video cameras). It seems that as long as surveillance data is not notably used against a large part of a country's population, people seem to accept and slowly adjust to the surveillance practice. Most people do not perceive the surveillance infrastructure as a threat. In contrast, some people even seem to desire being seen. A recent advertising video by a major cosmetics company flirted with the idea that a pretty woman is "admired" by a shop's surveillance camera.

One reason why the Panopticon effect is not obvious for many of us at the moment is because humans tend to systematically underestimate risks (Kahneman and Tversky 2000). "It will not happen to me" is a typical statement. We all think that we personally will not be negatively impacted by the surveillance infrastructure. If people are informed about surveillance they argue that they have nothing to hide and therefore do not care about being watched. They do not recognize the scope of today's surveillance infrastructure and massively underestimate it when being asked about it (Bizer, Günther, and Spiekermann 2006). This also explains the big surprise about Edward Snowden's revelations of massive National Security Agency (NSA) surveillance of citizens around the world. Against the background of these arguments, I hypothesize that the positive liberty of most ordinary citizens is not yet infringed by surveillance to the extent often argued. If we do not consciously feel the grasp of our invisible manacles, then at least our positive liberty is not affected. After all, positive liberty requires consciousness. This argument is, of course, no justification for building manacles. If it is not for freedom of thought that we should avoid surveillance infrastructure, it is certainly for the reason of avoiding power asymmetries between governments and citizens. In Chapter 8, Box 8.2 I described how we should strive for a more balanced planning of surveillance infrastructure in places where we really want it (like in dark parking lots at night). Such wise planning is a matter of leadership and foresight on the side of infrastructure investors.

However, our freedom is strongly impacted in another way: Nowadays, machines control a large part of our attention

and consequently free thought. Solove (2006) calls this privacy harm "decisional interference" and "intrusion." In Chapter 6, I outlined how information technology (IT) push architectures for messaging services lead to constant interruptions of our activities. As of 2015, people can hardly finish a train of thought without being interrupted by some kind of pop-up window, advertising display, or other form of incoming communication. We can hardly control this constant inflow of attention-grabbing machine messages. Attention-sensitive design of machines is therefore highly relevant form of ethical computing.

Attention-sensitive systems are built on the idea of information pull instead of push. Pull architectures for messaging and information retrieval are not only means for a renewed right to be let alone (Warren and Brandeis 1890). In terms of an information search on products and services, pull architectures would also allow us to better compare information and freely make up our minds around our own interests (Searls 2012). In an ethical machine design, this free thinking would take the place of today's setup, where we are kept in filter bubbles and bombarded with predictive advertising messages or search results (which infringe our positive liberty).

12.3 Privacy Trade-Offs at All Levels of the Value Pyramid

Privacy comes into play in various forms and guises when we relate it to the values at different levels of the pyramid. Unfortunately, however, the desire for privacy often seems to be accompanied by some value trade-offs. Take the case of using health data to better monitor medical histories. Patient monitoring is not only done for a person's proper benefit but also for higher social reasons. Large pools of health data offer better insights into the paths illnesses can take. Moreover, such Big Data allows us to watch the geographic spread of diseases, share experiences about the performance of doctors and hospitals, understand human genetics, and more. In Chapter 7, I described some of the benefits the health industry expects to get by collecting and sharing health data. At the same time, health data is highly sensitive personal data. It is sensitive not only because of its bodily intimacy but also because of its extraordinary potential for misuse. If health data got into the wrong hands, unfair treatment and bias would become a norm for people in all kinds of life situations, from looking for health insurance to searching for a new job. So is it good to collect, aggregate, and use people's health information?

At the next higher level of the pyramid, we see the widely discussed trade-off between privacy and security. Most governments and many fearful individuals argue that public security demands surveillance. There is a strong belief that surveillance infrastructure impedes crime and facilitates crime conviction. At the same time, the surveillance infrastructure undermines our right to be let alone and creates power asymmetries. As I outlined earlier, many argue that surveillance undermines our positive liberty to speak and act as freely as we would without being observed. And so people ask: Do we need to give up some privacy to promote public safety?

With regard to friendship, in Chapter 10 I showed how anonymity and invisibility increase online deliberation. People are less inhibited when they can shelter their identities. Many open up more and tell more secrets. Perhaps they can get closer to their true selves if they can be more anonymous than when they are identified. But at the same time, real friendship requires identification. True reciprocity, feedback, and learning from others can occur only when people know each other for real. So how can virtual worlds strike the right balance between identified selves and anonymous encounters? Should they take measures in favor of one form of self-representation? To what extent should virtual world operators themselves know about the true identities of their players? On the one hand, virtual world operators should know the "true names" of their players so that they can maintain order in the virtual world when players abuse their anonymity. On the other hand, the very fact of being completely anonymous—even to the service operators—allows for true "online deliberation." So what is more sensible from an operator's perspective: to maintain access to players or to allow them to be completely unobserved and open up?

Finally, dignity and respect (Chapter 11) cause potential privacy trade-offs. If people are fully respected, others should not be systematically surveying them. However, to build evaluative respect into machines—to make them polite—machines must be able to monitor people's preferences and try to understand them.

There is no easy answer for how we can resolve these trade-offs. In some cases we do not even know whether the trade-offs truly exist beyond theory. Take the example of public surveillance: So far, we do not have large-scale data to prove that surveillance infrastructure actually reduces and predicts serious crime. If we had such data, we could analyze how to scale surveillance to minimize crime while maximizing people's privacy. As I will show, this scaling is a process that is highly context specific. It is a process in which the real threats in a given context, the probabilities of these threats, and the amount of potential damage are combined to understand risks, such as the risk of crime. These risks are then addressed through controls and mitigation strategies like surveillance. The mitigation strategies correspond to the concrete threats identified and are evaluated by experts. By comprehensively weighing threats to values and enablers of values, we can resolve trade-offs, identify compromises, and thereby take ethical responsibility (see Chapters 14 through 18).

12.4 Summing Up: Values in Future IT Environments

I have identified and analyzed a number of core intrinsic values that are shared by all people around the globe. Knowledge, freedom, autonomy, security, health, friendship, and dignity are undoubtedly important for everyone to grow and feel good as individuals (Figure 12.2). Privacy is an important extrinsic value enabler that is essential to all these values. It is therefore of utmost importance that we protect these values in future IT environments. Moreover, we must build machines that actively embrace, embed, and foster these values.

I began with the *knowledge* value, outlining how building information for the machine age requires ethical conduct at all stages of the knowledge creation process. For people to trust machines and machine operators, data and information must be collected in a legitimate way. We can do this by implementing technically facilitated procedures for informed consent and by fostering a psychology of control in people around the data collection process. Data quality and transparency are also important for data aggregation. In the machine age, we want machines to build knowledge that we as humans can understand and hence trust. To accomplish that goal, we need to be sure that the machines release truthful information. By definition, there is no knowledge without truth. However, as of 2015, the data quality and transparency of data-processing activities are a challenge. Today's first-generation machines have not been built to ensure high enough levels of data quality and transparency. Consequently, the knowledge that is aggregated is often not reliable.

One of the reasons that current machine-generated knowledge is not reliable is that the quest for transparency is at an early stage. Transparency means that any knowledge we create is meaningful and appropriate. Yet, determining what is meaningful and appropriate requires judgment. And making good judgments is a trait that only humans have and that machines have yet to *learn* (if they are ever capable of it). For machines to make good judgments and learn from humans, the machines must not try to dominate human decisions as much as they do today. Humans are already in filter bubbles that obfuscate and distort the complex reality in which we live. Of course, easy use, time savings, or so-called efficiency are achieved when machines simplistically sort the world's information for us and nudge us into decisions. However, as I have outlined, training our judgment was at the core of what we call our own "enlightenment," and we need to be careful that the machine age does not take this capability away from us. We need to build hybrid machines that help us to sort out things but leave ample room for self-experimentation and discovery.

Finally, I wrote about the ethical use of information. I presented Helen Nissenbaum's concept of contextual integrity for data, information, and knowledge use. In doing so, I covered many privacy issues that arise today in the machine world as a result of ethically dubious data flows: secondary uses of data, unwanted appropriation of personal data, breaches of confidentiality, and even exposure have become an unfortunate norm. I therefore expanded on the ethical use of information, noting that we create bias when we categorize people and treat them according to such categorizations. Many users welcome personalized information, but this personalization needs to be perceived as fair in order to be trusted in the long run. Current economic rationale in our service designs often tends to prioritize short-term cost minimization and profit maximization over fairness. I encourage reflection on the ethics of such corporate practices.

Freedom, *liberty*, and *autonomy* are the philosophical building blocks of current Western political systems. To ensure that people remain free in the machine age, machines must be built with dynamic levels of automation, allowing people to manipulate and control the machine as needed. The industrial model of total automation that we observe in manufacturing today may not be the ideal solution for consumer-facing computing devices. In consumer markets where user demand determines the success or failure of technology, automation could backfire if it is too paternalistic or too simplistic. Machines must be accessible in such a way that they allow for manipulation on several layers. I distinguished between easy-to-use higher-level access to application layer dynamics and deep access to lower layers of a machine's functioning. Today accessibility or "openness" is noted as a fundamental software freedom. However, with a move to business models that provide software as a service, the fundamental freedom to access is threatened. Additionally, we must consider how such openness is used. Although we are still in the onset of the machine age, openness may sometimes be a way to pander to the curiosity and pride of software engineers, hackers, and a youth culture that wants to understand technology. At the same time, however, machine accessibility becomes crucial for balancing power between service providers, their machines, and the people.

A challenge in this power play is the protection of human mental skills and cognitive capabilities. If we want more than a tiny elite to be able to access, manipulate, and control machines, then we need to develop and strengthen our cognitive skills. We must train our cognitive abilities to understand the functioning and limitations of machines. Most important, we also need a healthy level of independence in decision making from our machines. Perhaps we do not always take their advice; we might override them and still feel good about our decision. Yet we will be able to develop these human capabilities only if we have sufficient time and free attention resources. A great digital divide has appeared

between the few people who can protect their scarce attention resource, stay focused, and make decisions autonomously, and those who do not have the mental strength to do so anymore. The digital attention divide can be overcome if we switch to "information pull architectures," which create something that Doc Searls calls "intention economy." In the section on attention-sensitive system design in Chapter 5, I described how information pull architectures support natural human attention allocation and how careful interruption design can help us to refocus and be less disturbed.

Being able to control our attention is also relevant for our *health* in the machine age. Many people today suffer from problematic internet use. Becoming addicted to machines or too absorbed in virtual worlds can lead to not only bodily pain but also to stress in everyday life followed by mental health problems (Chapter 7). Machines could certainly foster our health in many direct and indirect ways. New devices like the Talos suit or life-logging apps may bring people back into nature and motivate them to care for their bodies. I did not provide general guidelines in the health chapter on how to build "healthy machines." The subject domain of health is much too broad for that, and every bodily function may have its own supportive machine service at some point. But I did discuss the short-term and long-term effects of machines on our mental and physical health, and I outlined ways in which machines relate to today's phenomenon of burnout.

When we talk about burnout, it becomes clear why Maslow regarded knowledge and freedom as prerequisites for other basic needs in the pyramid. For example, Figure 7.3 (Chapter 7) shows how mental health in the form of burnout can be indirectly triggered by a lack of computer self-efficacy and job control. Ethical knowledge creation provides employees with a widely usable and legitimate database that they can understand (transparency), access, and use for fair purposes. This kind of ethical knowledge, as well as the autonomy to manipulate the machines they use at multiple levels, can foster employees' perception of efficacy and control. Employees can creatively meet the demands of number-driven jobs. People who feel empowered and in control will probably perceive a healthier balance between the demands of their jobs and their control. In contrast, employees feel out of control when they cannot access the machines they use in their jobs, cannot understand the numbers the machines produce, cannot alter these numbers nor the machines, and do not have documentation to understand how these machines work. This negative feeling is exacerbated when employees are forced to use the numbers and the machines they do not understand to meet job demands. The steep increase of burnout in companies today might be caused in part by machines that deprive people of ethical knowledge and autonomy vis-à-vis machines.

Besides health, another strong motivator to work on an ethical knowledge base for machines is the *safety* and *security* of these machines. In security projects, companies work toward more confidentiality, availability, integrity, authenticity, and accuracy of their data. They improve the auditability of their systems and take measures for better accountability. By doing so, they actually feed into a process for ethical knowledge creation and knowledge use.

But when ordinary people speak about "security" today, they often mean more than the securing of corporate data assets. In their mind, security is equated with safety. And security is also often equated with surveillance infrastructure. These simplistic equalizations are unfortunate because they lead security investments to be channeled into a surveillance infrastructure. It makes it easy to argue that one has done a lot to improve people's safety and security by increasing the budget for surveillance. In truth, however, security goals, auditability, and accountability are hardly improved by more surveillance. And the safety of an infrastructure, such as the quality and reliability of products, services, and assets is not enhanced by surveillance either. I therefore plead for a more stringent and precise use of terms when it comes to security and safety, and I propose a more reasonable, data-driven "golden mean process" to decide on surveillance investments.

The last sections of this chapter deal with the social need for *friendship* and the individual need for dignity and respect. The computer science world has barely addressed these last two human needs even though machines dramatically influence them. Machines alter the way we live and build relationships; three examples are our first-generation media (i.e., social networks, mobile phones, e-mail), virtual worlds, and interaction with artificial beings. In reviewing these influences, I found that Batya Friedman's value-sensitive design methodology is limited: We cannot simply build characteristics of friendship into robots or agents. It is ethically problematic to conceptualize and decompose the friendship value and then identify requirements for friendly machines. In contrast, if we build machines that become our friends, we face the ethical issue of replacing human touch with cold, lifeless, and uncaring marionettes. These objects may be very attentive and courteous with us. They might be easier to handle than unpredictable human characters, but they also make us accustomed to superficial, conflict-free relationships that are far from the demanding human encounters of the real world. Our ability to develop virtue and learn from the hard feedback of real human friends may be diminished as a result. So building the friendship value into a digital system has the counterintuitive effect of potentially destroying that same value in the offline system.

Many machine ethicists would probably argue that machine friendship is not as lifeless and dangerous as I presented it. First, not all cultures regard machines as lifeless. Some Buddhist cultures embrace the idea that every thing has some spiritual essence, including robots or other lifeless objects. Second, machine ethicists argue that machines can outperform humans in some respects. For example, machines could teach us ethics. Machine ethicists aim to build machines that embed ethical reasoning and that can inform humans about higher forms of philosophical knowledge and give support. Robots and agents could provide us with a knowledge base that has never before been accessible to humans. Personally, I am not sure whether this vision will deliver on its promise. Can a potential loss of humans' mutual socialization and self-development be countered by machines that embed ethical protocols? Today, little is known about whether humans' intrinsic knowledge and learning does not depend on human interaction and empathetic resonance. I reported on the importance of our bodies and their mirror neuron system for creating empathy and truly understanding what is happening in one's environment. Unless machines embed similarly powerful biological mechanisms, can they ever teach us much?

Finally, in the last section of this chapter, I reflected on how machines can influence our *self-respect* and the respect we receive from others. We can take a construct like *politeness*, decompose it (as I did in Chapter 11, Figure 11.2), and build machines that treat us politely. In fact, politeness would be a great new requirement for engineers to think about, since a lot of machines today tend to treat us as a cattle rather than humans. But a deeper reflection on higher-level needs in the value pyramid requires moving beyond interaction requirements. We need to think more holistically about the role machines play in human lives. And as we do so, we see that machine design, as well as the business models and legal frameworks created around machines, alter the power balance between people and machine owners. Requirements engineering for machines will become social engineering.

Let us take the psychological ownership value as an illustrative example. As shown in Figure 11.3 (Chapter 11), machine owners can design machine services in a way that systematically fosters perceptions of psychological ownership in customers. This perception benefits self-respect because people like to "have" and "possess" things. But fostering such a value through service design is a double-edged sword for companies. If customers use machines that strengthen their feelings of ownership, it is hard to take legal ownership and technical control away from them. For example, restrictive copyright laws and remotely controlled machine architectures challenge customers' psychological ownership perceptions in the long run. Recall the difference between the robot manufacturer Future Lab and Robo Systems in the scenarios. The robots built by Future Lab are completely owned and controlled by their users, whereas the robots built by Robo Systems are remotely controlled. Which one of the two companies seems more attractive to us as customers if all other performance variables are kept constant? Personally, I would choose the Future Lab robots that I could fully own and control. But is this the solution companies will prefer? If they truly cater the design of machines and business models to higher-level individual needs such as ownership, user control, and power, they will certainly gain a competitive edge and market share. But as they do so, they will also need to give up some of their control over the machines and their users. They will need to forgo personal data assets and knowledge about customers. Will they do so? What values will drive IT companies' investment in requirements engineering? I suppose that the struggle for power and control will become an important driver for IT companies' requirements engineering, perhaps even more than financial benefits. If this prediction comes true, people's higher needs will clash with the established corporate machine world. Exceptionally wise leadership is needed at that point.

EXERCISES

1. Using Solove's taxonomy of privacy depicted in Figure 12.1, identify an example you have heard of or experienced yourself for each type of privacy breach.
2. Take a social network platform like Facebook and reflect how the platform's functionality could harm the privacy of its users. Reflect on what could be done or is done already to avoid these privacy harms.
3. Draw the value pyramid with all of the values that were summarized in this chapter (see Figure 12.2). Then, align Solove's privacy harms shown in Figure 12.1 with the value pyramid, and discuss whether and how privacy harms may be created at various levels of the pyramid.
4. Take a new IT service from any of the scenarios described in Chapter 3 and apply it to the value pyramid depicted in Figure 12.2. What values are at stake with the new IT and why?

References

Bizer, J., O. Günther, and S. Spiekermann. 2006. TAUCIS: Technikfolgenabschätzungsstudie Ubiquitäres Computing und Informationelle Selbstbestimmung. Edited by Bundesministerium für Bildung und Forschung. Berlin, Germany: Humboldt University Berlin, Unabhängiges Landeszentrum für Datenschutz Schleswig-Holstein (ULD).

Kahneman, D. and A. Tversky. 2000. *Choices, Values, and Frames*. New York: Cambridge University Press.

Searls, D. 2012. *The Intention Economy: When Customers Take Charge*. Boston: Harvard Business Review Press.

Solove, D. J. 2006. "A Taxonomy of Privacy." *University of Pennsylvania Law Review* 154(3):477–560.

Warren, S. D. and L. D. Brandeis. 1890. "The Right to Privacy." *Harvard Law Review* 4(5):193–220.

Westin, A. 1967. *Privacy and Freedom*. New York: Atheneum.

Chapter 13

Ethical Value-Based IT System Design: An Overview

"Design is not just what it looks like and feels like. Design is how it works."

Steve Jobs (1955–2011)

Chapter 3 contained many examples of future information technology (IT) environments. Taking these scenarios as a baseline, we will see IT design projects focusing on the following kinds of systems:

1. *Embedded systems*—ID technologies (such as radio-frequency identification, RFID), sensors, cameras, actuators, and so on are embedded into the objects around us that so far are analog. Traditional products and structures may then look familiar from the outside, but they are internally enhanced with digital functionality. This digital functionality enables objects to collect information from humans, to communicate with each other, and act through their embedded "intelligence." Objects may also have a virtual representation of some sort on the Internet, a kind of website or dossier for each product. The owner of a PC, for example, could look up his PC online, see the warranties attached to it, where it was build, sold, and so forth. The term *Internet of Things* has been framed to describe this phenomenon.
2. *Material-enhanced systems*—Beyond digital functionality, nanotechnology will change the objects around us, giving materials new properties. As a result, new value propositions are enabled by the novel materials' characteristics. Roger's Talos suit and Sophia's glove are examples for people gaining bodily strength through the material. But smart textiles could also allow additional qualities, such as energy harvesting.
3. *New, yet unseen interactive devices and services*—The 1980s had the PC, and the 1990s had the mobile

phone. The next devices could be robots, drones, holographic figures, and so on.

4. *Highly integrated digital control systems*—These might be operated by governments or enterprises that pull information from and apply algorithms to data collected from myriad distributed objects and digital workflows. These systems then control institutional and corporate processes; even eventually manufacturing of products, sales and after-sales services. Agent Hal from the robot scenario in Chapter 3 is an example for this kind of system.
5. *Virtual reality (VR)*—VR could take on a different level of importance in future IT environments in the sense that it may be used for many more purposes than today, such as professional meetings. One variant of VR evolution is that it melts into the real world through augmented reality (AR) services.
6. *Existing (legacy) systems*—Such systems will need to constantly evolve to integrate with the new digital systems landscape. They will need to expand their current capabilities to embrace new service opportunities.

As companies prepare for these new IT opportunities, they run through new product development projects or product evolution processes.

In this chapter, I describe how companies can embrace and handle these technological changes while maintaining a focus on values. I first outline how innovation management departments and senior management traditionally approach new product development (often abbreviated as "NPD" in the management literature). I then align this innovation-management thinking with the computer science world and how IT departments have traditionally pursue IT system development life cycles (often abbreviated as "SDLC" in the information systems and computer science literature). Finally, I look at the computer ethics and social computing community and how scholars in these fields have thought about integrating values into IT design. The chapter closes with an integrated ethical system design process based on values that I call the *ethical* system development life cycle, or "E-SDLC" for short. This E-SDLC serves as a baseline for the subsequent chapters in the book.

13.1 Management View of New Product Development

All companies confront the modern challenge of constant change in different ways. In recent years, the acceleration of IT capability and the ensuing competitive pressure has forced companies to professionalize their innovation practices. One well-known and exemplary way to do so has been Robert Cooper's Stage-Gate® process. Stage-Gate is a conceptual and operational map for moving new product projects from idea to launch (Cooper 2008). Many major companies (e.g., 3M, Procter & Gamble, Hewlett-Packard and Rolls-Royce), including some from the IT industry, have used the process to manage innovation efforts. In the following I will use the Stage-Gate process model to explain how new product development (NPD) is typically viewed from an innovation management perspective.

Stage-Gate splits the innovation process into six phases and five stages that are separated by gates. Gates are a kind of checkpoint where the decision is made as to whether to continue a project (see Figure 13.1). The Stage-Gate process begins with a phase of discovery where ideas for innovation are collected from inside and outside the company. New technological capabilities may be screened, competitors and start-ups observed, and partners and customers questioned for ideas. All of these ideas are then evaluated at gate 1 and

benchmarked with a view to success criteria toward the company's operations and identity. At gate 1, companies look into the strategic fit of an idea with the existing product portfolio, its high-level feasibility in terms of available resources, and its main technical and financial risks. They then weigh these aspects against the idea's market potential and financial attractiveness. In theory, the ideas that pass gate 1 should match all of the company's success criteria. So far ethical criteria or social values have not been a regular part of these success criteria.

In the next stage, called "scoping," a cross-functional team substantiates the ideas that passed gate 1. In a relatively short time frame (for example, a month), promising ideas are evaluated in more detail. The market potential, technical and legal risks, actual resource commitments, and additional company-specific criteria, such as sales-channel availability, are analyzed. The true scope of a project is revealed. At gate 2, potential product innovations are filtered again for their feasibility. At this gate, the screening must be tough because the resource-intensive business planning stage that follows can be done only for a few select and truly promising leads. A company should therefore use relevant internal success criteria to judge a new idea's feasibility. Cooper notes that it is very important to rigorously apply success criteria. A major problem associated with product innovation processes at this stage is that too many product ideas are pursued for too long and too few are killed early. Cooper writes that companies easily forget that "the idea-to-launch process is a funnel, not a tunnel" (2008, p. 218).

Once new product ideas have passed gate 2, managers enter the stage where the business case is planned. Building a true business case for a product is a highly complex endeavor. It requires an in-depth market analysis that is used to gain an understanding of customer needs, values, sensibilities, and limitations. In the business case planning stage, a product or service is precisely defined. Technical requirements should be

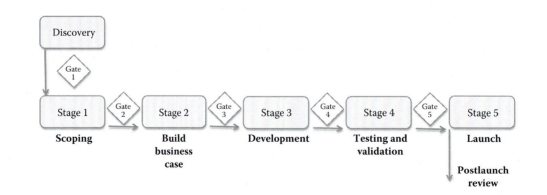

Figure 13.1 The Stage-Gate Idea-to-Launch Process is a trademark of the Product Development Institute. (Based on Cooper, Robert G., 2008, "Perspective: The Stage-Gate Idea-to-Launch Process—Update, What's New and NexGen Systems," *Journal of Product Innovation* 25:213–232.)

identified, because to a large extent, technical requirements determine how much the product will cost. (As I will show in Chapters 17 and 18, non-tech-savvy managers in IT companies often do not realize that a successful scoping phase is not enough to put them into the position of detailed IT business planning that can be completed at this early stage.)

A major challenge in the business planning phase is the overoptimistic and emotional involvement of project teams, which often want their idea to fly. Sometimes managers develop pet projects. There is therefore a tendency to overestimate market potential and underestimate barriers to market success, such as the limitations of a technical solution. Customer demand is often anticipated on the basis of analyst forecasts. At least in the IT industry, these forecasts are, however, normally overoptimistic (see the discussion of hype cycles in Chapter 2). Many forecasts suggest an unrealistic market take-up that is either too much stuck in our linear way of thinking or that promises rather questionable exponential growth. Some costs for products are easily ignored; for example, nonfunctional ethical requirements that pop up in this phase of product design might be categorized as nice-to-have traits and excluded from cost estimates even though they are important in later product development. Finally, context factors are often simply ignored in the business plan. An example is the necessity of an installed technical base that may be required for consumers to start using a product or a legal restriction that can impede a market altogether. In short, business plans risk envisioning an isolated reality that ignores everything the project team does not want to see. Experienced gatekeepers must see the bigger picture and determine the true feasibility and context of a new product or service.

After the business planning stage (gate 3), product development starts. This phase requires a constant dialogue between innovation managers and engineers. Today, many IT solutions are developed iteratively and with the help of *agile* development methods (see Section 13.2.4). In agile development, products evolve incrementally, often on a weekly basis, through constant feedback from customers, innovation managers, and other stakeholders. Early prototypes, mock-ups, and modeling diagrams can facilitate this communication. IT systems are not only implemented in this stage but also designed. Managers must therefore ensure that engineers are not left to make unsustainable and unchallenged product decisions by themselves.

In parallel to supervising the product's design and implementation, responsible innovation managers refine the financial analysis of the product and plan market introduction, distribution, and logistics. They may also need to support external intellectual property (IP) sourcing, for instance, when a software solution embeds components that need to be licensed from another IP owner. During product development valuable IP might also be created; this IP can be patented and sold beyond the initial product idea. Managers should track these opportunities and influence how any new IP is licensed.

When product development is ready, the project passes gate 4, where gatekeepers evaluate the final operational prototype before it enters the testing phase. Eventually, testing has two dimensions: First, the technical engineering team investigates how the solution integrates into a broader technical environment. Often, it happens that technical faults or interdependencies are discovered at this stage. Second, the innovation management team observes how customers in the field receive the prototype. If a product or service does not fly as expected, the team might need to revise the business plan. Depending on how cost-intensive large-scale manufacturing and rollout is, a new project may even be canceled after this stage at gate 5. If a product shows promising test-market results, and integrates and scales well from a technical perspective, gate 5 is passed and the product enters the market. Figure 13.1 summarizes the Stage-Gate Idea-to-Launch Process with its five stages and gates.

13.2 Engineering View of New IT Product Development

The management perspective on how to develop new products overlaps with the process view taken by IT system engineers (see Figure 13.4). *System engineers* are IT project leaders or senior IT executives who think holistically about machines as systems. They integrate a wide spectrum of requirements for hardware and software as well as the ergonomic, architectural, and human context factors in which IT artifacts operate. They should have a view of business processes, the organization at large, and the ways that society will have to deal with the machine. System engineers are not the same as *software engineers*, because the latter group focuses only on the software, the operating system on which the software runs, communications management, and data management. In his textbook on software engineering, Ian Sommerville (2011) illustrates the crucial difference of these roles: A system engineer is responsible for all technical dimensions, including software engineering, user interface design, ergonomics, architecture planning, and civil engineering, as well as electronic, electrical, and mechanical engineering. In contrast, a software engineer focuses only on his smaller share in the overall system.

The difference in roles and responsibilities for systems and software is particularly important because they are often mingled too much. In small-scale IT projects for instance (such as start-ups), the head of IT development or the CIO (chief information officer) after serves both roles. He or she is responsible for the software and the system. But because software engineering is so labor intensive and influential for

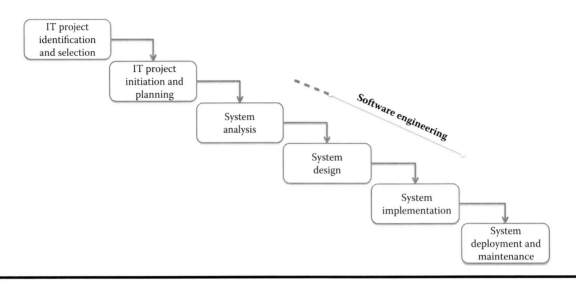

Figure 13.2 The waterfall model for IT system design.

product functionality, it (unfortunately) often dominates the IT project to the extent that relevant considerations of overall system design can suffer. The confusion of roles is reflected in the common mistake of equalizing *system development* life-cycle models such as the waterfall model (Figure 13.2) with software *engineering* models, such as the spiral model (Figure 13.4). The linguistic convention of calling an isolated piece of software a "system" makes the confusion even worse. In the following I will clearly distinguish between system and software engineering, because this is vital for value-based system design.

The most well-known and widely used model for system design is the waterfall model depicted in Figure 13.2 (Hoffer, George, and Valacich 2002). The waterfall model has a long history of practice in IT deployments. It is conceptually rooted in general engineering processes and is a plan-driven way to design and develop systems.

The waterfall model is a system development life cycle (SDLC). It prescribes that IT projects should run through six phases: (1) identification and selection, (2) project initiation and planning, (3) analysis, (4) design, (5) implementation, and finally (6) deployment and maintenance (Hoffer, George, and Valacich 2002). The waterfall model originally did not foresee any feedback loops and interaction between phases, which gained it the reputation to be very rigid. Companies can, however, opt to live this system design process in a more flexible way and many do so.

No matter how strongly a company sticks to the waterfall model or diverges from it embracing other system design approaches, the core of tasks required by this SDLC model are coming back in all IT projects. I will, therefore, use these waterfall model phases as the structural baseline to explain how to accommodate ethics in system design.

13.2.1 IT Project Identification and Selection

The first SDLC phase, *project identification and selection*, is similar to what Cooper calls the discovery phase. An organization determines whether to invest resources into a system at all. The development or enhancement of one out of what may be several IT investment alternatives is determined. At the end of this phase, an organization ranks system development projects and decides which IT project(s) to begin with, at least in terms of an initial study. Today, this phase is often influenced by hype cycles (Chapter 2). Technical departments and the media bring technical opportunities to the attention of management. Alternatively, companies may look to their competitors or supply chain partners to see what kind of IT they develop or need.

As I will argue in the next chapter, this first phase of IT development needs much wiser leadership in the future than is the case nowadays. More focus is needed on the values that a company wants to create or protect. It is in this phase that senior executives can identify whether a project is ethically compatible with their social or organizational goals. They must understand the benefits and harms that can result from a technology and how these map to values promoted or threatened by it. For example, what benefits and harms are created when RFID readers are deployed in retail shops and read out the belongings of customers? What values are impacted through such practices? Or, should an organization automate certain processes or stick to human work? As we will see, responsible and wise leaders must ask and answer these kinds of diverse questions. Management guru Ikujiro Nonaka argues that wise leaders are required to determine whether to pursue system development efforts from a moral perspective (Nonaka and Takeuchi 2011). In doing so they need to formulate value priorities for their company and must think about broad IT solution ideas that are able to respect and foster these values.

As senior executives seek to determine the value proposition from the new IT and to avoid negative externalities created through it, it is helpful for them to engage with stakeholders. Stakeholders are "people who will be affected in a significant way by, or have material interests in the nature and running of the new computerized system" (Willcocks and Mason 1987, p. 79). These can be customers, idea generators in the company, potential developers, end users, and vendors (Gotterbarn and Rogerson 2005). The stakeholders can also include outsiders who might be experts in the field.

13.2.2 *IT Project Initiation and Planning*

If management decides to consider a new technology, projects move to the second phase of the SDLC, which consists of *initiation and planning*. In this phase, a cross-functional IT project team is formed and sets up the project. IT project team members work with internal and external stakeholders to understand system requirements. The team engages in project planning, which includes a description of the project scope, goals, schedule, budget, and required resources. Most important, project planning includes a feasibility analysis so that different implementation alternatives can be assessed. Different implementation alternatives lead to varying projections for the total cost of ownership (TCO) an IT system will finally cause.

The tasks in this initiation and planning phase of the SDLC correspond to the scoping stage of the NPD process in that both models involve evaluating a promising idea in such detail that the team can make a prudent decision about whether to move forward. To gain this understanding, a scope, schedule, and resource plan is needed, and the project's feasibility needs to be clarified. A so-called baseline project plan is created. The two innovation approaches diverge though when it comes to mindset: Whereas the NPD process aims to systematically funnel (and therefore kill off) project proposals, the waterfall model traditionally does not foresee such rigorous gate-based thinking. Once an IT project is selected in the SDLC, the value and viability of a project is rarely challenged. At the very least, this practice is a common pitfall. IT system engineers tend to be more bullish than managers about what projects really require and how markets evolve. This tendency sometimes leads them to underestimate the scope of their projects in this phase. Most important, a study of the economic, technical, operational, legal, political, and ethical feasibility should be done to understand the nonfunctional requirements and true boundaries of the project. These limiting aspects go much further than engineers tend to consider today. In short, subsequent SDLCs benefit from a more rigorous gate here.

As I will show in Chapter 15, ethical issues in particular influence all feasibility dimensions of a project. For example, if a company wants to develop a drone that walks a child home from school, the drone has the potential to increase a child's safety and health. But the drone might also seriously undermine a child's freedom in terms of autonomy and intrude on its privacy through constant surveillance. In this second SDLC phase, such ethically risky issues or, in contrast, value-based design opportunities can be looked at in detail. Project team members and external stakeholders can use a conceptual value analysis to understand the dimensions of the values promoted or threatened by the technology. The team also needs to understand legal restrictions and investigate whether they can effectively handle ethical issues associated with the technology. If the organization or society is not ready for the respective project, the organization should not pursue it.

13.2.3 *System Analysis*

The third phase in the SDLC is the *analysis* phase. Here, the team determines concrete system requirements. What will the system physically look like? What technical architecture will it have? How will it interact with human beings and integrate with established human practices? The system requirements that are developed in this analysis phase must correspond to the physical, technical, organizational, and sociocultural context in which the system operates. For instance, when RFID technology in retail shops is met with consumer fears of being surveyed and scanned, then system analysis must come up with system development goals to address such fears.

Ideally, prospective users are empirically studied in this phase so that engineers can understand their concrete expectations for the new system. The requirements derived from contextual and user studies can result in process and data flow models as well as UML use-case and sequence diagrams (UML stands for Unified Modeling Language) that later guide development. Most important, a plan of the system's architecture is made in the analysis phase.

Unfortunately, the importance of rigor in this phase of system design is easily underestimated. IT project teams often do not like to see how important the details of various contextual factors are for the success of their solution and to what extent these factors can destroy the business case (see Chapter 17). Only if IT project teams engage in context and user analysis can they truly understand what a system should look like, how it should work internally, and how it should interact with its periphery.

Software engineers start to play a crucial role in the project in this phase of analysis. They normally have an intuitive idea of the final solution architecture, which is often based on existing software and service components. In fact, object-oriented programming has led to sophisticated libraries of readily retrievable code components. Similarly, commercial off-the-shelf software offers complete technical solutions that may have been difficult to build in the past. Service-oriented architectures can integrate existing and new components more readily. By considering the reuse of these existing components, software engineers can rapidly develop ideas about how to build a new IT system in a short time frame.

Managers tend to readily embrace engineers' "intuitive solutions" without much questioning. They do so because the project will consume less money and time when more software and service components are available and do not need to be coded from scratch. So this greatly aids the business case. Unfortunately, the tendency to rapidly embrace prefab architectural components from shared code libraries or existing web-services has its own pitfalls; especially when these are take-as-is components that operate like black boxes. Systems are often cobbled together from the bottom up, creating unforeseen dependencies. In addition, reference architectures or patterns are often embraced as blueprints for what a new system could look like without questioning limitations (i.e., too little security or privacy).

I do not want to dismiss the utility of existing system components or reference architectures. Using preexisting software components makes the creation of new services much easier and boosts the speed of innovation. Especially when opensource software components are used, the code quality of these can be superior to newly written software, because the code has undergone some group scrutiny. Interoperability problems can often be better avoided. And the probability is increased at the code- and user-interface level that engineers can build on existing knowledge and experience. But the downside of this approach is that such systems regularly bear technical compromises that can frustrate the final users of a system. Open-minded engineers and IT teams can uncover such shortcomings in the analysis phase through modeling. UML diagrams, for instance, can help to identify what a system really needs from a process organizational and social perspective. The models clarify users' roles in system interaction (Chapter 17). Based on this knowledge of the requirements, engineers should pick only those ready-made code structures that fit their case and are free and open to control (see Section 11.5 on software freedoms).

A visualization of a system's internal workings and interactions with its environment (i.e., based on UML) can also help the IT team to understand ethical system requirements. In the analysis phase, the team needs to judge the importance and likelihood of ethical risks and analyze how these risks can be broadly addressed. In security terms, the team is conducting a "threat analysis." Understanding the extent and probability of damage from potential ethical breaches helps the team prioritize the risks that need to be addressed in a system's design. The team refines the positive ethical value

proposition of a machine. If a machine can compete on values, the team now decides how to include these extra value propositions in the technical product design.

In this phase, it is useful to supplement intuitive judgments from the team and internal stakeholders with empirical user investigations. The team must identify and then test the relative importance of "value flows" (ethical enablers) and "value dams" (ethical barriers) to be included in the system (Miller et al. 2007; see Section 17.3.2).

13.2.4 System Design, Implementation, and Software Engineering

In the next phases, engineers convert the recommended system models into a logical and physical system specification which are then implemented. Elements such as input and output screens, dialogs, and interfaces are specified and programmed. In this phase, modern system developers often work iteratively with stakeholders and evolve systems until they achieve a final operational system status that can be tested in the market. Boehm's (1988) spiral model of the software development process, shown in Figure 13.3, is a well-accepted representation of how software engineering teams work in this phase. The model begins with a requirements plan that stems from the preceding analysis phase. The requirements are then built into the new IT system in cycles. At the beginning of each cycle, developers identify what objectives they can achieve in the next prototype and make multiple implementation plans for those objectives. The team challenges each implementation plan, systematically evaluating its risks.

Historically, spiral model engineering has focused on project-related risks only, such as staying within budget, meeting deadlines, and delivering specific functionality. Social risks that might later affect people or the company were not extensively considered. Limiting system risks to project, function, and usability issues is, however, much too narrow. In fact, Flanagan, Howe, and Nissenbaum (2008) have noted that values often "emerge in specifying (gritty) instrumental design features" (p. 334). To understand the true risks of an IT system, the team must therefore consider nonfunctional value impacts and the ethical sustainability of a system as well. For example, the data collection part of a system can be implemented according to several possible protocols. A protocol that records more data details could entail more privacy risk for later system users than a protocol that minimizes data collection. In the risk identification and resolution stage of the spiral model, the IT teams must therefore understand how all of the holistic system risks identified earlier can be controlled through appropriate measures or design alternatives. These design alternatives can be of technical nature as well as business process or governance related. The team's conclusions are implemented in the IT system and in the organization.

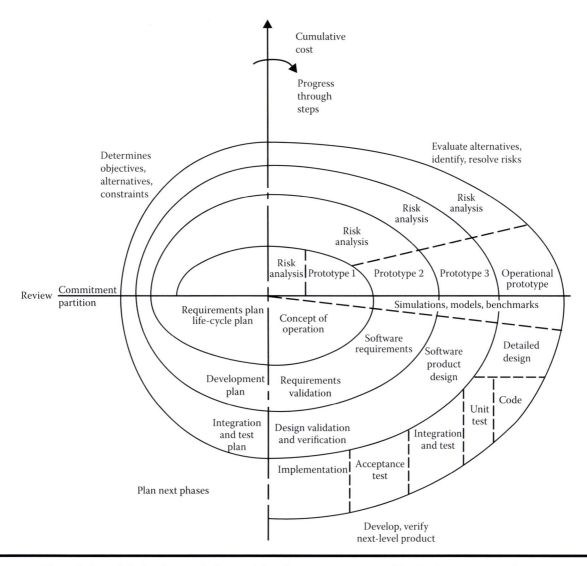

Figure 13.3   The spiral model of software design and development as proposed by Boehm. (From Boehm, Barry W., 1988, "A Spiral Model of Software Development and Enhancement," *Computer* **21(5):64.)**

The implementation and refinements of technical and organizational requirements should be accompanied by regular testing. Especially when systems are critical (potentially causing high social risks), companies have adopted a so-called V-model for software validation, which foresees a consecutive testing of the effects of systems' design decisions (for further information, see Friedrich et al. 2009, and Sommerville 2011, p. 43).

Whether the IT team takes risks or avoids them, risk mitigation is highly relevant in system design. The choices made in the design and implementation phase will constitute the code base of the later system. Product marketing managers should be closely involved in this system design phase, ideally full time and as part of the software engineering team. Simple twists to data flows, changes to algorithms, or user feedback mechanisms can dissolve ethical risks to a large extent, but they can also dramatically impact the business case.

The spiral model as depicted in Figure 13.3 is a *plan-driven* method of software engineering that promotes the use of successive prototypes. "A prototype is an initial version of a system that is used to demonstrate concepts, try out design options, and find out more about the problem and its possible solutions" (Sommerville 2011, p. 45). Prospective users can experiment with the new system before delivery and help to refine its requirements. Prototyping has proven very successful in practice, especially for achieving higher levels of system usability and better user experience. Prototypes can also help to stimulate stakeholder discussions about ethical system implications. But managers have to be careful in this phase to not mistake prototypes for final systems. Often, when product managers see prototypes, they believe that they are viewing the final system. Prototypes can look so sophisticated to managers that they underestimate the time and effort required to turn prototypes into final systems.

When software or system delivery is delayed, managers can pressure software engineers to release ramped-up prototypes rather than fully finished solutions. Time pressure also leads to a lack of documentation about what was done during the design and implementation phase. This practice causes problems for the later transparency of a system, its data quality, and the ethical use of the knowledge created by it.

Plan-driven methods for software engineering, such as the spiral approach, have been complemented by incremental forms of software development. Incremental delivery means that software solutions are delivered to customers in increments. Sometimes only one functionality is initially delivered to customers, and more functionality is added later. Customer feedback for increments or real-time experimentation with alternatives can be used to prioritize subsequent release versions of software. Agile development methods such as extreme programming (Beck 2000) or the scrum approach (Cohn 2009) are well-known incremental software engineering methods.

Unfortunately, agile software development is not always well suited for ethical system implementation or must at least be critically viewed for two reasons. First, agile development methods can become too narrowly focused on specific software applications and functionality. This narrow focus comes at the expense of integrating a holistic view of an overall system, its boundaries, and nonfunctional requirements, all of which are very important for ethical system design. Second, the agile development of software code stresses that working software is more important than comprehensive documentation. As a result, necessary documentation for later system maintenance is not created (Prause and Durdik 2012). Systems easily become opaque black boxes that can be understood only by delving into the details of program code, which is often not as well written or structured as the proponents of agile methods envision. This problem also affects the maintenance and evolution of IT systems.

That said, both plan-driven and agile software development can be used in ethical system design. The only thing that needs to be fulfilled is that earlier system design phases get the requirements and architecture right up front.

For the IT world, the last phase of the waterfall model, system operation and maintenance, is highly important because the majority of overall cost for IT systems is incurred in this phase. About two-thirds of software costs are evolution costs (Sommerville 2011). Despite their relative importance, operational maintenance and evolution of a system are not covered in later sections of this book because ethical IT innovation and design takes place mainly when a system is first built and in the early phase of the SDLC. I will show how the architecture of an IT system, its form factor, and its internal data flows determine whether it is ethical. Decisions on these factors are made during the project initiation and planning phase as well as in the analysis and design phase.

Companies are sometimes forced to "bolt on" ethical system traits after the system is already rolled out. For example, many companies have engaged in major IT security projects to protect their infrastructures from cybercrime and to not lose customer trust. But the necessary program and data restructuring as well as architectural changes that aftermath ethics require are the most cost-intensive forms of reengineering. *Aftermath ethics* goes beyond normal software maintenance activities such as correcting defects, adapting IT environments, or adding functionality. I therefore believe that, especially for cost reasons, the main avenue to create "good" machines should be an investment of time, thought, and money into ethical requirements engineering when systems are first conceived and rolled out. Unfortunately, most software engineering models do not advocate this practice. Instead, models suggest that feasibility analysis should be "cheap and quick" (Sommerville 2011, p. 37). It is commonly held that the requirements specification is only a "description of functionality" (Sommerville 2011, p. 39) and that only "critical systems" need precise and accurate descriptions (Sommerville 2011, p. 40). The problem with this view is that, in Internet economics, small-scale applications can unexpectedly become global solutions. Consider Facebook, Google, Skype, eBay, Microsoft, Apple, Amazon, WhatsApp, Doodle, Dropbox, and so on. Small services suddenly serve billions of people and do exercise a critical influence on people's private and work lives. The same phenomenon could happen to all the new services described in the scenarios. Software engineers should therefore embrace the notion that what they do could be critical on a global scale. As Jaron Lanier notes: "It takes only a tiny group of engineers to create technology that can shape the entire future of human experience with incredible speed" (2011, p. 6).

Once a system is in the field, it needs to be monitored, and its effects on people need to be understood. IT systems are often put to unintended and unexpected uses, some of which may even undermine their initial purposes. Also, new ethical issues and values may pop up. Unexpectedly, some values may turn out to be more important than initially thought, and other values may be questioned. Even if a system is developed with ethical values in mind and has been subject to rigorous ethical risk analysis, the system might turn out to be not that perfect. But perfection is not a reasonable expectation. If the products of our minds, the machines, turn out to have faults once they are in the field, we must recognize these faults and openly communicate them. We can try to correct the faults, incurring the necessary costs, or we can explicitly document machine limitations and treat the IT system's output with the appropriate awareness of these limitations. This latter point—an open and deep acknowledgement and acceptance of the

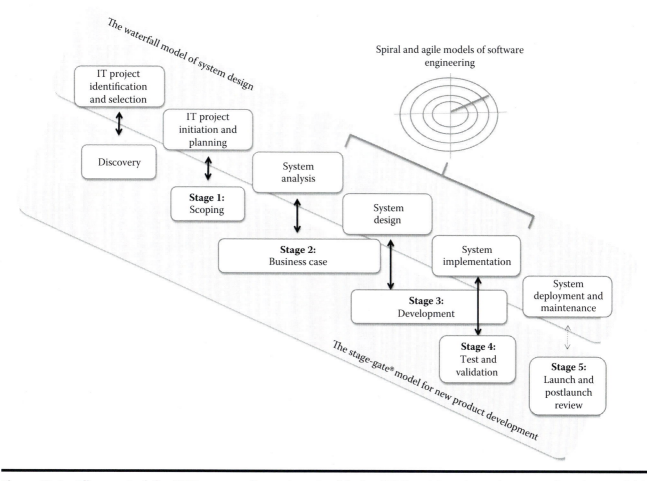

Figure 13.4 Alignment of the NPD process (lower boxes) with the SDLC and iterative software engineering model in spiral form.

limitations of IT systems—is an important psychological twist in the man–machine relationship.

Figure 13.4 summarizes how the waterfall SDLC can be aligned with the NPD process as well as approaches to software engineering. I adopt the spiral symbol in Figure 13.3 to illustrate the dynamic and iterative way in which most software is engineered today.* Software engineers' work starts in the business case stage when they bring in architectural views and influence system analysis. Together with product managers they then work iteratively on concrete system

design alternatives and the implementation. Ideally, software engineers go beyond feature-based thinking in these phases of design and implementation, and consider a more holistic system view. To facilitate this view, teams must invest the appropriate time and reasoning into earlier waterfall stages, proper requirements engineering, and documentation.

A final note of caution needs to be made: The three aligned models in Figure 13.4 are ideal and theoretical representations of IT product development projects. In reality, companies organize innovation processes in very different, unique ways. Because companies pursue varying strategies of how and where they source or build software, every project and every company is different. Also, all the phases overlap and interact with each other to some degree. That said, the tasks that need to be handled and mastered in NPD and IT system design recur and build on each other and can be seen as the kind of theoretical sequence chosen here. The ethical SDLC (E-SCLC) that I will describe in the subsequent chapters of this book integrates the perspective of NPD and the classical SDLC, and adds to it what we know from the value-sensitive design literature.

---

* Originally, *software* engineering was not modeled as an iterative spiral but as a waterfall model in itself. When machines resembled huge monoliths and took a very long time to develop, each phase of that waterfall model had to be completed before the next phase of system construction could begin. No interaction or feedback between phases was foreseen. Rigor and prudence reigned over requirements engineering to ensure that no mistakes happen when building systems. Some machines took years to build. And this model is still followed for some critical systems today (Sommerville 2011). But the waterfall model of the *software lifecycle* (see, e.g., Boehm 1988, p. 62) should not be mistaken with the higher-level *system design* waterfall model.

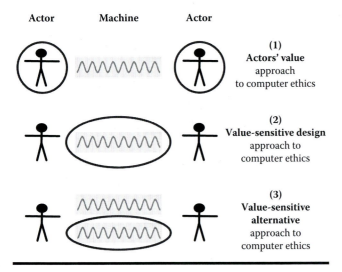

Actor   Machine   Actor

(1)
Actors' value
approach
to computer ethics

(2)
Value-sensitive design
approach to
computer ethics

(3)
**Value-sensitive
alternative**
approach to
computer ethics

Figure 13.5 Three classes of ethical discourses on computer ethics.

13.3 Computer Ethics and IT System Design

From the aforementioned descriptions of NPD and system design the question arises as to what contribution the field of computer ethics can make to ethical IT innovation. Computer ethicists, who are often philosophers by background, formed a scientific community in the mid-1990s.* Looking at the body of computer ethics literature, one can broadly discern two main avenues of discourse, which are illustrated in the top two parts of Figure 13.5. The first and most common approach to computer ethics focuses on the ethical consequences of a given technology for people. Examples of this kind of research include analyses on the loss of privacy through the Internet or copyright infringements. These questions focus on how a given technology changes a person, how persons may create new kinds of morally questionable behavior by misusing or abusing technology, and how social interactions change as a result of technology. In a metaethical debate, philosophers wonder whether computer systems create unique moral problems that have not existed before or whether old moral problems come in a new guise (Moor 1985). Philosophers analyze how people's ethical expectations and behavioral norms are formed and changed through IT. To a certain extent, philosophers are dealing with the question of how peoples' values change or are impacted. This form of computer ethics discourse has two key characteristics: It focuses on the human actor as the object of analysis, and it takes the machine as given.

Because this part of computer ethics literature focuses on people, I have circled the stick figures in Figure 13.5 and I call it the "actors' value approach to computer ethics." Others have called it the "standard approach" to computer ethics (Johnson 2009).

In a second class of computer ethics discourse called "values in design" (VID; Knobel and Bowker 2011) or "value-sensitive design" (VSD; Friedman, Kahn, and Borning 2006), scholars do not accept information technology as a given. These scholars whose disciplinary background is mostly human–computer interaction and computer science recognize that machines can be actively designed and shaped to respect and foster humans' values and moral norms. For example, systems can be built to respect people's privacy and give people control over system actions. In fact, I argue that systems can be designed to embrace all the values I described in Chapters 5 to 12. Value-sensitive design scholars have proposed an initial methodological framework for how we can build values into systems and hence build systems that are "good" or at least better than they are today. The scholars outline how values play different roles in technological design (Yetim 2011). Values underlie the system development process as a priority and requirement, they are considered during development for embedding in the artifact, they are (often visibly) present in an application, and they influence the application context. I will show how VSD can become part of system design and NPD (see Section 13.4). Strengthening and extending our knowledge about VSD is a major concern of this book.

A third path for ethical IT innovation becomes apparent when reading the scenarios in Chapter 3. I call this path the "value-sensitive alternative" approach to computer ethics. Pursuing value-sensitive alternatives means that we do not embrace an initial technological solution or service with all its ethical pitfalls trying to patch the drawbacks. Instead, we radically seek out a third way to create the same customer value. For example, in the gaming scenario, people want to play and to build friendships through virtual reality technology. The standard way to create this value today is to build virtual worlds for people to inhabit. But the gaming scenario describes an alternative approach in which a game brings virtual reality artifacts into the real world:

> Stern sits down and listens to the presentation of the invited lead user, who has been playing the new game for the past 3 months and is obviously thrilled. *Reality's* game content is cast as an overlay onto the real world through players' AR (augmented reality) glasses. As players move through real urban and rural space, they can meet fantasy characters and discover 3D mysteries that are sheltered in different geolocations.

---

* In 1985 James H. Moor published his seminal paper titled "What Is Computer Ethics?" The first Ethicomp conference was held in 1995, and in 1999 the journal *Ethics and Information Technology* published its first issue.

Thinking up more ethical and more valuable *alternatives* to standard technology is probably the most radical, disruptive, and competitive way of thinking about value creation in our digital future. Following this line of thought, some systems may not be built at all or may quickly be dismissed. For example, the robot scenario describes a robot manufacturer that has produced impressive humanoid Alpha1 robots for the police force (similar to Google's humanoid robot called "Petman" that is currently in development [Edwards 2013]). Such devices have the potential to threaten people and spread fear. They can lead to a loss of trust in governments if governments ever choose to deploy them. Political tension between citizens and their governments and silent or open rebellion could ensue:

> Stern's responsibility was to manage the Alpha1 series, a humanoid that Robo Systems sold to support the police force. Alpha1 systems look like tall steel men. They can act autonomously and in response to their immediate environment. If needed they can also be taken over remotely by an operator who then embodies the machine. Initially, the business went really well. Crime rates and social unrest, which had shot up so rapidly due to rising unemployment, dropped dramatically in areas where the robots were deployed. However, people hated them and were afraid of them. Many people did not want to go out at night in fear of meeting such a device. Retailers and restaurants complained. Gangs attacked the devices and broke them. The public indirectly supported these violent actions, even when human police officers accompanied the robots. They were seen as enablers of state control. As a result, Robo System's sales went down.

The question raised by this robot example is whether certain machines should ever be built. That's a question we already know from other areas such as nuclear power or genetic engineering. If such IT systems are built, we can question whether and under what conditions their usage should be allowed. The value-sensitive alternative approach calls for us to forgo the use of ethically problematic systems and to actively seek ethical alternatives to create value. In the robotics scenario, an alternative robot system called "R2D2" is presented that does not resemble or replace human beings at all and does not threaten anyone. This alternative embraces the important innovation potential of robotics but forgoes a humanoid design:

> The idea of the human–robot hierarchy (with humans always on top) is deeply embedded in Future Lab's design process. For example, robots manufactured by Future Lab never look like human beings.… However, Future Lab's robots are embedded with powerful artificial intelligence (AI) technologies including voice, face, and emotional recognition. But these AI functions run independently in dedicated sandboxes contained in the device that does not need to be networked to function.… With these user-centric control principles, the decentralized architecture as well as the recycling strategy, Future Lab gained tremendous ground in selling its robots. In particular, elderly individuals have started to buy the R2D2 model.

The practice of waiving potential IT innovations while seeking alternatives will be an important topic in Chapter 15, which is about wise leadership in IT project identification.

In the following sections, I discuss how computer ethicists have thought about integrating ethical value thinking into the concrete design of IT systems.

13.4 Value-Sensitive Design

In his book *Information Technology and Moral Philosophy*, Terrell Ward Bynum (2008) digs into the origins of computer ethics that were laid out by Norbert Wiener (1954). Bynum describes how in the 1950s Wiener had already identified several values as relevant for machines' design and use. In particular, Wiener thought about the impact of machines on freedom, equality, benevolence, security, happiness, and opportunity for the handicapped.

Many decades later, in the early 2000s, Batya Friedman and her coauthors proposed a methodological framework that helps to systematically embrace values in IT system design (Friedman and Kahn 2003): value-sensitive design (VSD). Friedman identified an initial set of values for machine design including human welfare, ownership and property, privacy, freedom from bias, universal usability, trust, autonomy, informed consent, accountability, courtesy, identity, calmness, and environmental sustainability. She acknowledged that this value classification is not comprehensive. In addition, Friedman contributed a methodology that calls for IT teams to consider values in system design and development. They can do so by first discovering and prioritizing value threats and benefits created through a system. In a second step, these values are conceptualized in the context of a particular system. Then, empirical and technical investigations serve to fine-tune value-specific adjustments to the technology. Friedman's methodological framework is broadly depicted in Figure 13.6.

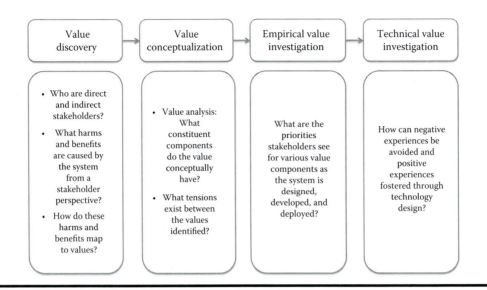

Figure 13.6 Value-sensitive design methodology.

VSD has been applied to a range of technologies including groupware, human–robot interaction, large display deployments, urban simulation, and browser security (Friedman, Felten, and Millett 2000; Friedman, Smith et al. 2006; Miller et al. 2007). Since its inception, the methodology has been discussed, expanded, and complemented (Miller et al. 2007; Le Dantec, Poole, and Wyche 2009; Yetim 2011; Friedman and Hendry 2012; Shilton, Koepfer, and Fleischmann 2014).

13.4.1 Value Discovery

When a new IT system is planned, the first step is to identify the harms and benefits it could create for direct and indirect stakeholders, and to map these harms and benefits to their underlying values.

Take, for instance, the privacy concerns of airline passengers about having their naked bodies exposed to security personnel through full-body scanners at airports. In the mid-2000s, many airports installed full-body scanners at security gates to increase passenger safety and mitigate potential terrorist attacks. In advance of the first deployments of these scanners, several technological alternatives were circulated in the press. As Figure 13.7 shows, such scanners could harm privacy if they exposed people's bodies to others. At the same time, the privacy concern was also an opportunity to differentiate one's solution in competition with other scanner providers. As Figure 13.8 shows, a person can be scanned for security without necessarily exposing intimate body details.

At the outset of scanner development, the potential privacy issue was probably obvious to all scanner manufacturers, whether they considered it as a threat to their business or

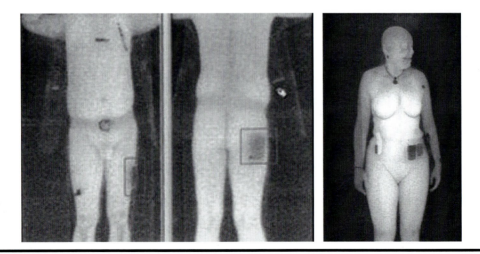

Figure 13.7 Human representation without value-sensitive design, not privacy preserving. (From United States Department of Homeland Security, 2007.)

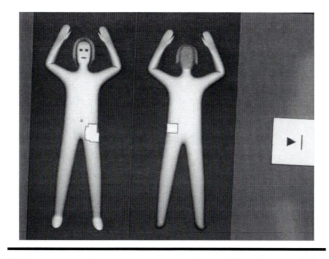

Figure 13.8 Human representation with value-sensitive design, privacy preserving. (© CC BY-SA 3.0 Berlin 2010.)

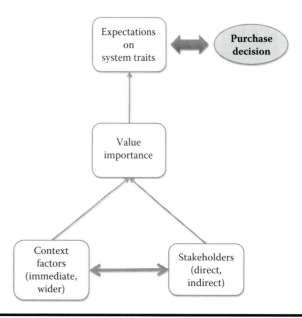

Figure 13.9 The importance of context factors and stakeholders for expectations on system traits.

as an opportunity. But scanner companies' IT design teams drew different conclusions as to the importance of the privacy value in the deployment context, and some misjudged or ignored stakeholder power. Here, one immediate context is the physical environment of scanners at airports. Many people stand in line and watch their neighbors in line with them. In this environment, seeing one's naked body parts exposed is extremely embarrassing to all *direct stakeholders* involved: oneself, the other passengers, and the security personnel. On the other hand, the broader context of the technology is that airline passengers are forced to pass the security gates, and they have no say in the technology that is used to screen them. So some scanner operators' IT teams probably thought that they did not need to consider human privacy; if passengers want to fly, they need to consent to scanning. The wider legal context allowed for such a conclusion. However, what these latter IT teams probably underestimated is the power of indirect stakeholders. *Indirect stakeholders* are those who might not be direct users of a system but who are nevertheless affected by the technology. In the body-scanner context, indirect stakeholders included security personnel working with the scanners, airport operators who address customer complaints, airlines that could see a drop in airline passengers as a result of scanning practices, and local politicians whose constituents could hold them responsible for the infrastructure. Because these indirect stakeholders were afraid of the privacy backlash, they ensured that airport operators bought only privacy-sensitive systems.

As this short case study shows, the context of a technology interacts with the direct and indirect stakeholders of the IT system. Both stakeholder groups and the context influence how important a value becomes for technology deployment. And if a value is important, it becomes an implicit part of the expectations buyers have on systems (see Figure 13.9).

For the scanner case study, a wise executive should have had the foresight to understand the importance of privacy in a body scanner's design. The executive should have pressured his or her organization to understand and tackle the privacy question before building scanners. Yet, the pressure to develop sufficient privacy-sensitive scanners was not high enough. Most scanner companies in Europe came up with intrusive technological proposals, as shown in Figure 13.7. Consequently, their solutions did not sell, and business was lost to one U.S. company that offered the privacy-friendly solution.* What went wrong within the European manufacturers who thought they could get away with the privacy-intrusive version?

In fact, identifying relevant values and anticipating their true importance and priority is not an easy exercise. Batya Friedman and I both propose value lists (or classifications) and thereby give some guidance on ethical issues that could be at stake. But this does not mean that any of these values we propose are relevant in a particular context. In each IT deployment context, a distinct value discovery phase must be done by the IT project team to understand the concrete contextual value expectations of direct and indirect stakeholders in a system (Le Dantec, Poole, and Wyche 2009). In this value discovery phase, harms and benefits are identified first. Only when these are known, can relevant values be derived.

---

* For a background news article on body scanners, see Sarah Spiekermann, "Privacy-by-Design and Airport Screening Systems," *derStandard*, March 15, 2012, accessed March 10, 2015, http://derstandard.at/1331779737264/Privacy-by-Design-and-Airport-Screening-Systems.

Engaging heterogeneous stakeholder perspectives is key in value discovery because stakeholders help to identify and prioritize values in a way that is often different from the view of IT project teams (Shilton, Koepfer, and Fleischmann 2014). For example, in the body scanner context, the issue was not just about the sheltering of genitals, as some manufacturers' project teams initially thought; stakeholders cared about the exposure of their entire naked figure. Stakeholders may also be in different roles vis-à-vis a system, using a system for different purposes, in different manners, and with different frequency. As a result, they have varying expectations and sensitivities. For example, an airline passenger has a different expectation for body scanners than a security guard. In fact, whereas the passenger wants maximum privacy, the security guard wants to uncover as much as possible of the person scanned. In such cases, a value conflict can arise. A frequent traveler with a time constraint may also have different expectations and value priorities vis-à-vis body scanners than a tourist who has plenty of time. Against this background, harms and benefits, and their respective values should be listed in a detailed manner for each type of stakeholder.

I will expand more on how stakeholder processes can be organized in Section 13.5. Habermas's "Discourse Ethics" has provided us with a valuable set of rules that can help to increase the effectiveness of stakeholder discussions. I also discuss Werner Ulrich's work on critical pragmatism, which can help us to identify the right stakeholders that should be around the table.

13.4.2 Value Conceptualization

Once an initial set of relevant values are discovered and prioritized, IT project teams then break each value into its constituent parts. The teams conduct a "conceptual analysis" to understand what the respective value is all about from a human user perspective, or from a legal or philosophical perspective. Consider the transparency value. If a company wants to create transparency, then this value implies that the information that a system provides is meaningful, truthful, comprehensive, accessible, and appropriate (see Section 5.4.3, Figure 5.14). A project team must identify and understand these constitutive dimensions of a value and then think about how to translate them systematically into the technical design.

The values software engineers typically learn to care for are efficiency, dependability, safety, and security. They refer to these values as "nonfunctional requirements" in the computer science literature (Sommerville 2011). This is not enough though. Bynum (2008) has recommended that traditional philosophical or also legal interpretations of a value could be used to understand values at stake. Unfortunately, IT project teams rarely have the time and educational background to find and interpret existing material on values. Project teams can try to understand value foundations, but a consistent challenge for VSD is that few value conceptualizations are available for IT project teams to apply.

Friedman and co-authors have started to change this. They illustrated conceptual analyses for the values of informed consent (Friedman, Felten, and Millett 2000), autonomy (Friedman and Nissenbaum 1997), and computer bias (Friedman and Nissenbaum 1996). Expanding on this work, I have provided in Chapter 5 to 12 conceptualizations for the values of user control, transparency, accessibility, objectivity, fairness, machine paternalism, attention, safety, security, trust, reputation, friendship, respect, politeness, psychological ownership, and privacy. Other highly relevant sources for these value conceptualizations have been historic summaries of case law (such as Daniel Solove's (2006) conceptualization of privacy); legal texts, such as the European data protection directive 95/46/EC (European Parliament and the Council of Europe 1995); and the computer ethics literature.

13.4.3 Empirical Value Investigation

As soon as values are well understood and conceptualized, *empirical investigations* can be used alongside system analysis, design, and rollout to further understand how stakeholders perceive the values' subsequent unfolding and importance. The earlier conceptual analysis can be substantiated by an empirical approach. After stakeholders have commented on their value preferences (i.e., in qualitative focus groups), a project team can quantitatively investigate the importance of the respective value perceptions and dimensions to learn about people's true priorities. For example, Friedman's VSD team tested a "dams and flows" methodology with more than a hundred prospective users to understand the relative weight of various values and their components in a groupware system (Miller et al. 2007). Value dams are technical features or organizational practices that hinder a value from unfolding. Value flows, in contrast, are technical features or organizational policies that support the unfolding of a value. The researchers decided that if more than 10% of system users were concerned about a potential harm, the VSD team would build a dam and not include the corresponding system feature in the new groupware system or find a solution to avoid the harm. If more than 50% of system users welcomed a certain system trait, the VSD team would include it, creating flow. The chosen percentages were arbitrary and must not be taken as a general rule.

To conduct a broader empirical investigation of value constituents it can be helpful to translate them first into corresponding technical features (or development goals). Consider the body scanner again. Here, the privacy threat of being exposed was translated into the development goal of prohibiting the exposure of genitals, prohibiting the exposure of skin surface, and prohibiting the exposure of one's figure. Only when development goals (or value dams and flows) are thus concrete, a wider audience can provide an informed and empirically collected judgment on what is most important to them. Chapter 17 provides a deeper discussion of available methods.

Empirical "reality checks" make system design (and new product development) a highly dynamic and open process throughout. Empirical research normally embraces both qualitative and quantitative approaches. In fact, value discovery is already an empirical investigation step of a qualitative nature. Qualitative empirical investigation is followed by quantitative empirical research in the system analysis phase. Qualitative empirical investigations come back during interactive IT system development work, when prototypes are continuously tested and reworked based on user feedback. Also, when systems are deployed and maintained, teams can continuously monitor users to understand how the system is really used and how it can be improved. Figure 13.10 shows how empirical investigations accompany the IT system design process.

13.4.4 Technical Value Investigation

Value discovery, conceptualization, and empirical investigation lead to a refined understanding of how stakeholders perceive future IT systems. After the team has completed these steps, they understand which values should be respected, which ones should be fostered in detail, and how value constituents translate into specific system development goals. For instance, it may be clear at this point that people do not want their figure, skin surface, or genitals exposed by scanners. Now, the team must translate these concrete user requirements into a technical specification. Engineers who employ security risk analysis would say that, once they understand the risks inherent in an IT system, they can specify controls to mitigate those risks. This specification of controls and weighing of various design alternatives, which VSD scholars call "technical investigation," is done in the system design phase.

In this phase, the team might find that a system property can support one value while undermining another. For example, the privacy threat of exposure may call for not capturing genitals with body scanners. Yet it is the genital area that is often used by criminals to hide and transport illegal

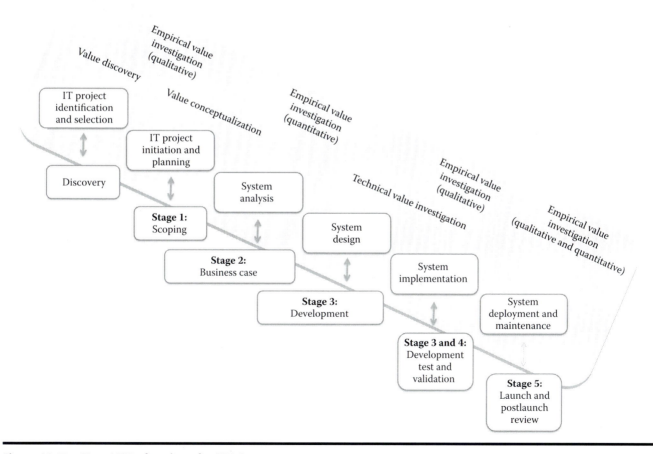

Figure 13.10   How VSD plugs into the SDLC.

objects. So a blurring of genitals in scanner representations or a complete forgoing of scanning these body areas would undermine the very function for which body scanners are built. In the technical investigation phase of VSD, IT project teams must therefore balance values or find a third way to resolve conflicting interests. What appears to be a trade-off at first sight may later be technically resolved by smart IT design. Just think of body scanners again as they are used today with stick figure representations. It is no problem to scan genitals and indicate that someone is hiding dangerous objects in that region. Privacy is preserved through the stick figure, and the scanners still fulfill their purpose. Finding such solutions is the purpose and result of technical investigations. Clever IT design often solves what seemed to be insurmountable value conflicts at the beginning of a project.

13.5 Stakeholder Involvement in Ethical IT System Design

Embedding values into systems is not an easy endeavor, but it is a huge responsibility. As a result, no one can make these kinds of decisions in isolation. Active stakeholder involvement in all phases of system design can be tremendously beneficial for project teams and company management. Stakeholder involvement should not be seen as a time-consuming nuisance. Instead it is fruitful, informative, and relieving to apply outside views to future technologies and to

one's own work. In today's consumer goods industries it is already well-known practice to engage in extensive qualitative and quantitative market testing and research before delving into any product manufacturing and launch.

Figure 13.11 summarizes the questions for stakeholders, which can be tackled throughout the SDLC. In the project initiation and selection phase, senior executives can work with stakeholders to understand the potential benefits and harms of new IT. What is the key value proposition of a new product or service? And where may a value be undermined? Once this is known and the project moves on, IT project teams can work with stakeholders to better understand the relevant values conceptually and how value constituents play out in the anticipated interactive contexts. The project teams try to understand whether the proposed IT solution is feasible given the values it potentially undermines. Once this process is done, project teams can analyze the market of users (prospective customers) in a quantitative empirical way to understand the absolute priorities of certain values over others. The teams can explore value dams or flows, which result from context factors that need to be discussed with stakeholders in this analysis phase. When these requirements engineering steps are finished, the project is ready to enter the product/service/software engineering phase. Here, stakeholders can participate in the iterative and progressive development of the system through various prototypical stages. In this phase, value trade-offs sometimes need to occur. For example, the security and usability of systems is often seen as a challenge

IT system design and selected stakeholder questions

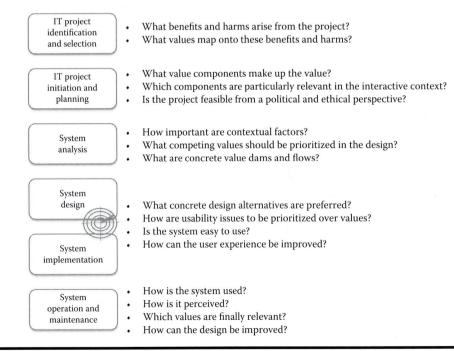

| IT project identification and selection | • What benefits and harms arise from the project?
 • What values map onto these benefits and harms? |

| IT project initiation and planning | • What value components make up the value?
 • Which components are particularly relevant in the interactive context?
 • Is the project feasible from a political and ethical perspective? |

| System analysis | • How important are contextual factors?
 • What competing values should be prioritized in the design?
 • What are concrete value dams and flows? |

| System design
 System implementation | • What concrete design alternatives are preferred?
 • How are usability issues to be prioritized over values?
 • Is the system easy to use?
 • How can the user experience be improved? |

| System operation and maintenance | • How is the system used?
 • How is it perceived?
 • Which values are finally relevant?
 • How can the design be improved? |

Figure 13.11 Selected stakeholder questions during the SDLC.

for which the right balance must be found. Finally, once a system is deployed and operated, the team can analyze usage statistics and user questionnaires to see how the system is perceived and used. These analyses may also identify how the system can be improved, especially when some values turn out to be more relevant than initially thought.

13.5.1 Challenges and Rules in Stakeholder Processes

In the first phase of system design, when senior executives meet stakeholders to first understand the potential harms and benefits of a system, the goal is to develop a design ideal. "A *design ideal* is simply the description of the ultimate values and standards that define the ultimate "good" at which [the] design project aims" (Klein and Hirschheim 2001, p. 76). One way to find the design ideal for a given technology is to identify the harms and benefits of the technology, map the harms and benefits onto values that a project could pursue, and define the resulting *ideal product* by carving out the concrete value dimensions engineers should focus on during development.

In this early phase of discussion about a system, conflicts can already arise because senior managers are often pressured to focus strictly on revenue and profits (financial values) instead of ideal designs. They may press for cost reduction through IT, low-cost solutions, automation (job replacement), or ethically ambiguous systems. Other stakeholders, meanwhile, may find such proposals difficult or unacceptable from a moral perspective. Besides such varying economic premises, stakeholders often bring different worldviews to the table. Attributes such as their personal values, political attitudes, virtue capacity (i.e., their ability to act in accordance with Table 4.2 [see Chapter 4]), intellectual strength, and farsightedness normally vary tremendously. Finally, on top of this individual diversity, organizational politicking is a common practice. "Politicking" refers not only to personal power plays but also to the fact that stakeholders may represent different interest groups (such as company shareholders or workers) when they discuss the deployment of a new IT system. Box 13.1 outlines Klein and Hirschheim's (2001) summary of four barriers to the rationality of organizational decision making that influence stakeholder discourse.

Because of these common communication barriers, stakeholders should define rules for discussions before they start. One rule can be a commitment to democratic stakeholder involvement. Klein and Hirschheim (2001) speak of "prior commitment to organizational democracy." This commitment does not necessarily mean that every decision or conclusion should be determined by democratic vote. Senior executives normally have a final say in projects because executives are held responsible

BOX 13.1 FOUR BARRIERS TO THE RATIONALITY OF ORGANIZATIONAL DECISION MAKING

a. Social barriers to rationality exist because of inequalities in power, education, resources, culture, etc. Social inequalities lead to bias of perception and presentation, blockage of information, and conscious or subconscious distortions. An example of the latter is that group decisions often are willing to accept more risk than individual ones. Cultural differences lead to misunderstandings and misperceptions.

b. Economic and motivational barriers exist because time constraints make it impossible to deal with all possible participants, arguments, and counterarguments as suggested above. In practice, the psychological and economic costs of debate often preclude the debate from even occurring. People are simply not motivated to argue for a long time to deal with all possible objections and implications. Furthermore there are social norms that discourage disagreement and thereby breed conformity. Hence, the decision making process is foreclosed prematurely and speed rather than quality becomes the primary determinant of the preferred solution.

c. Personality barriers exist because individuals differ in their decision-making styles and competencies. Biases may result from the mistaken belief in the law of small numbers and failure to consider the regression towards the mean. Objectives motivate estimates, e.g., sales managers underestimate costs and overestimates sales, while cost analysts do the reverse. Subconsciously many choices are value driven and once the choice is made, it needs to be justified with good reasons.

d. Linguistic barriers exist because the rationality of human communication tends to suffer from conflicting and ambiguous meanings, difficulty in expressing complex matters, limits of the human brain to comprehend lengthy reports and other factors which impede mutual understanding.[1]

NOTE

1. Heinz K. Klein and Rudy Hirschheim, "Choosing between Competing Design Ideals in Information Systems Development," *Information Systems Frontiers* 3(2001):79.

for a company's actions. Instead, the term "democratic" refers to a genuine willingness to listen to and include the views of everyone involved in the stakeholder process. Senior executives must also be willing to change their initial position. Enid Mumford (1983) referred to "consultative" or "participative" forms of discourse. As I will show in Chapter 15, senior executives are not always open to embracing such participation. There is a huge difference between paying lip service to stakeholder involvement versus really considering the arguments of others. I have often witnessed how higher-level managers pretend to listen without really wanting to understand what a stakeholder has to say. It is a matter of personality and continuous character improvement for senior executives to overcome such initial biases and practice the patience of true listening. Stakeholder processes can enforce true exchange when they are designed around "consensus," meaning that IT projects are supported only when they are accepted by all stakeholders.

That said, the organizational context of the IT system's design and deployment can make it more or less easy for senior executives to embrace the views of stakeholders. For example, a well-funded start-up's decision about how to build a new IT system is completely different in comparison to the decision of an established company to change established work processes by adding IT. Listening and embracing stakeholder processes is much more difficult when a company has established practices. Once an organization or system is established, it develops an inertia that is very resistant to change. Fear of failure or job loss has an emotional effect that can prohibit constructive stakeholder exchange. Listening to others and accommodating ideas is much easier when the external environment involves fewer conflicts.

Against this background, it is helpful to consider a rational way to conduct stakeholder processes. "Rational" means that the value discussion is driven by knowledge and cognitive rationale rather than emotion or political agenda. The literature offers an enormous amount of insight into the effective management of stakeholder processes and discussion. I will focus on three helpful concepts: choosing the right stakeholders, creating ideal speech situations, and constructing arguments logically.

13.5.2 Choosing the Stakeholders

When new IT systems are built or embedded in organizations, a wide range of stakeholders could potentially be involved. Today, a common practice is to involve end users of an IT system in the interactive and incremental design and implementation phases. The goal here is to obtain user feedback about system models and prototypes in order to increase the usability and user experience. However, as Gotterbarn and Rogerson noted: "Many

project failures are caused by limiting the range of possible system stakeholders to just the software developer and the customer" (2005, p. 742). For this reason, it has become more common in recent years to involve stakeholders in earlier phases of IT system design as well. The goal is to understand their expectations of a system and explore new market opportunities with the help of outsiders' views. In fact, outside of the IT industry, company stakeholder involvement has become a widespread innovation strategy: Companies often use crowdsourcing mechanisms for early phases of new product development. For example, toy manufacturer Lego invites customers from around the world to submit their ideas for new models. The company then selects the best ideas for production. Threadless, a t-shirt manufacturer, not only allows people to submit ideas for new designs but also allows the community to choose which designs will be printed.

But crowds may not be the best solution for ethical IT system design. In Figure 13.11, I outlined how stakeholders can contribute to an ethical, value-sensitive design of IT systems. A large battery of demanding questions needs to be answered throughout the entire SDLC. To answer these questions in a meaningful and complete way, it is not important to involve everyone but rather to involve the right people. So how can the right people be identified?

Werner Ulrich, the originator of "critical systems thinking" (CST), pointed to a number of helpful "boundary" questions that can help us think about who ought to be involved in a stakeholder process and what ought to be the relevant scope for the people involved. *Boundaries* in critical system thinking are established at the edge of what is and what ought to be. The concepts of what "is" and what "ought to be" serve as a point of reflection on a likely result versus an ideal. Figure 13.12 illustrates this thinking. It contains a checklist of boundary questions that support a CST practice in civil society (Ulrich 2000).

The first type of boundary question concerns the motivation for a new IT system: Who is or ought to be its client? For example, are airport operators the clients of body scanners because they pay for them? Or are the passengers the clients because their safety is at stake? The second type of boundary question concerns the sources of power. There is sometimes a difference between the ideal decision maker and the actual decision maker. For instance, when new IT systems are rolled out in various regional subsidiaries of a global company, the decision maker is often the CIO at the headquarters of that company. But the units affected by the system are based in each region. Should each unit be part of the decision-making process?

The third type of boundary question concerns the sources of knowledge. IT projects often involve outside experts who do not take a critical perspective on a new system. They are either paid to support the new solution or they

Sources of motivation
- Who is (ought to be) the client?
- What is (ought to be) the purpose?
- What is (ought to be) the measure of improvement?

Sources of power
- Who is (ought to be) the decision-maker?
- What resources are (ought to be) controlled by the decision-maker?
- What conditions are (ought to be) part of the decision-environment?

Sources of knowledge
- Who is (ought to be) considered a professional?
- What expertise is (ought to be) consulted?
- What or who is (ought to be) assumed to be the guarantor of success?

Sources of legitimation
- Who is (ought to be) witness to the interests of those affected but not involved?
- What secures (ought to secure) the emancipation of those affected from the premises and promises of those involved?
- What worldview is (ought to be) determining?

Figure 13.12 Boundary questions for critical system thinking. (Adopted from Ulrich, Werner, 2000, "Reflective Practice in the Civil Society: The Contribution of Critically Systemic Thinking," *Reflective Practice* 1(2):247–268.)

are not willing to point to its limitations. The final boundary type concerns the source of legitimacy: Are the stakeholders who are affected by a new system at the table? Do they at least have representatives? Mumford distinguishes between direct and indirect participation (Mumford 1983). She witnessed company projects where stakeholder representatives were elected by employees to participate for them in the IT decision-making process. Today, representatives from workers' unions are frequently included in discussions about large IT rollouts.

The boundary questions lead an IT project team and senior executives to include people in stakeholder groups who will not necessarily and unanimously support all aspects of a new product idea. They hence consciously introduce challenge to the group, which can make some people feel uncomfortable. Senior executives and IT project teams must be able to develop strong arguments for why and how they want to pursue certain ideas. The next section will expand on how to structure such discourses.

How well stakeholder groups work together can also be influenced by their selection and appointment process. Once we understand who should be involved, a second question is how the appropriate individuals are brought into the group. External end users of future systems can be easily recruited for focus groups, workshops, or JAD (joint application development) sessions by market agencies. Choosing stakeholders from inside a company is trickier. Here, individuals can be selected by upper management or an independent body, or they can be elected as representatives of interest groups.

Finally, even outside experts can be hired to represent internal groups.

13.5.3 Rational Arguments and Ideal Speech Situations

Once stakeholders are wisely chosen and the scope of stakeholder involvement is clear, the next question is how to realize effective communication. As Box 13.1 illustrates, barriers between people can hinder effective communication. At the same time, philosopher Jürgen Habermas noted that the most fundamental characteristic of human beings as a species is that we are able to coordinate our actions through language and communication. In his *Theory of Communicative Action*, Habermas argues that the primary function of communication is to achieve a common understanding. This understanding is then the basis on which shared activity can be built (Habermas 1984, 1987). Habermas sees communication as being oriented toward reaching agreement, and he proposes that the principal means of reaching agreement is through rational discussion and debate—the "force of the better argument"—as opposed to the application of power or the dogmas of tradition or religion (Mingers and Walsham 2010). With this kind of thinking, Habermas is very compatible with the belief in meritocracy that is shared by many members of today's technological elite and the programming community (Coleman 2013).

In his *Discourse Ethics*, which is highly intertwined with his *Theory of Communicative Action*, Habermas thinks about preconditions for rational discourse. Two main strongholds of Discourse Ethics have resonated widely: the nature of a rational argument and the nature of an ideal speech situation. A rational argument is truthful, right, and sincere. Truth means that a fact used in the discourse corresponds to the actual state of affairs. Rightness means that the argument includes only claims about norms of behavior and principles that are established in our social world, for example, through an open and democratic debate. Sincerity means that a speaker's arguments should arise from an honest perception of the world and be without guile. Some authors have pointed to intelligibility as a further trait of a rational argument (Zinkin 1998). Intelligibility refers to the assumption that the meaning of what is said is clear to all concerned (Klein and Hirschheim 2001). Intelligibility is at the core of understanding because simple differences of meaning associated with the same words often make communication difficult.

For a discourse to achieve a valid outcome, discussions should focus on rational arguments and ensure that rational arguments hold more sway than any distorting aspects of the people involved or the sociopolitical situation. Mingers and Walsham (2010) summarize the traits of such an "ideal

Discourse Ethics

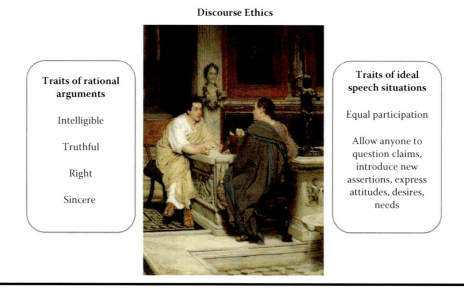

Traits of rational arguments

Intelligible

Truthful

Right

Sincere

Traits of ideal speech situations

Equal participation

Allow anyone to question claims, introduce new assertions, express attitudes, desires, needs

Figure 13.13   Core traits of a discourse ethics usable in stakeholder processes. Painting: *The Discourse* by Sir Lawrence Alma-Tadema.

speech situation" as proposed by Habermas. Habermas believed that these traits needed to be more than conventions; they needed to be inescapable presuppositions of rational arguments themselves (p. 840):

- All potential speakers are allowed equal participation in a discourse
- Everyone is allowed to:
  - Question any claims or assertions made by anyone
  - Introduce any assertion or claim into the discourse
  - Express their own attitudes, desires, or needs
- No one should be prevented by internal or external, overt or covert coercion from exercising the above rights

Figure 13.13 summarizes the main building blocks of Discourse Ethics that can be used to facilitate stakeholder processes for value-sensitive IT design. Of course, following these guidelines cannot guarantee that all value conflicts will be fully resolved. What we can achieve is a better consciousness of the core values we want to create though IT and an awareness of the values that may be negatively affected. Moreover, stakeholders can participate in the careful and detailed crafting of the values that should be created, and they can help to mitigate potential value risks. Not all risks may be sufficiently controllable, and stakeholders may not fully agree on the

design ideals chosen. There may be value trade-offs. But at the very least, a communication process that includes all relevant parties offers the best opportunity to reduce value conflicts; it creates transparency and a rationale for the chosen design ideals. Following quality stakeholder processes results in better acceptance of new IT systems and greater commitment to the final system from the people.

EXERCISES

1. Think about the company Lego and research how it involves stakeholders in its product design strategy. Apply Ulrich's boundary heuristic to the Lego case.
2. Explain the difference between a system engineer and a software engineer.
3. Discuss how the NPD process and the SDLC overlap and where they differ in their approaches toward innovation.
4. Reflect on the benefits and drawbacks of using ready-made software components for IT system design. Under what conditions do project teams benefit from preexisting software components? What harms and risks may arise?
5. In every scenario from Chapter 3, you will find a value-sensitive alternative approach, as outlined in Figure 13.5. For each scenario, identify these alternatives and discuss why they represent a third way to solve ethical challenges.
6. There are four recognized barriers to rationality in organizational decision making. For an organization

you worked for or a team you were involved in, recap what barriers you witnessed and how these barriers influenced the outcome of the projects.

References

Beck, K. 2000. *Extreme Programming Explained.* Reading, MA: Addison-Wesley.

Boehm, B. W. 1988. "A Spiral Model of Software Development and Enhancement." *Computer* 21(5):61–72.

Bynum, T. W. 2008. "Norbert Wiener and the Rise of Information Ethics." In *Information Technology and Moral Philosophy*, edited by Jeroen Van Den Hoven and John Weckert, 8–25. New York: Cambridge University Press.

Cohn, M. 2009. *Succeeding with Agile: Software Development Using Scrum.* Boston: Addison-Wesley.

Coleman, E. G. 2013. *Coding Freedom: The Ethics and Aesthetics of Hacking.* Princeton, NJ: Princeton University Press.

Cooper, R. G. 2008. "Perspective: The Stage-Gate Idea-to-Launch Process—Update, What's New and NexGen Systems." *Journal of Product Innovation* 25:213–232.

Edwards, J. 2013. "This Video of a Robot, Hooded and Chained by the Neck as Humans Test It, Will Tug at Your Soul." *Business Insider*, April 6.

European Parliament and the Council of Europe. 1995. Directive 95/46/EC of the European Parliament and of the Council of 24 October 1995 on the Protection of Individuals with Regard to the Processing of Personal Data and on the Free Movement of Such Data. L 281/31. Official Journal of the European Communities.

Flanagan, M., D. C. Howe, and H. Nissenbaum. 2008. "Embodying Values in Technology: Theory and Practice." In *Information Technology and Moral Philosophy*, edited by Jereon van den Hoven and Weckert John. New York: Cambridge University Press.

Friedman, B. and H. Nissenbaum. 1996. "Bias in Computer Systems." *ACM Transactions on Information Systems* 14(3): 330–347.

Friedman, B. and H. Nissenbaum. 1997. "Software Agents and User Autonomy." Paper presented at Autonomous Agents 97, Marina del Rey, California, May 5–8.

Friedman, B. and P. Kahn. 2003. "Human Values, Ethics, and Design." In *The Human–Computer Interaction Handbook*, edited by J. Jacko and A. Sears. Mahwah, NJ: Lawrence Erlbaum Associates.

Friedman, B. and D. G. Hendry. 2012. "The Envisioning Cards: A Toolkit for Catalyzing Humanistic and Technical Imaginations." Paper presented at Computer Human Interaction (CHI), Austin, Texas, May 5–10.

Friedman, B., E. Felten, and L. I. Millett. 2000. "Informed Consent Online: A Conceptual Model and Design Principles." Seattle, Washington: University of Washington.

Friedman, B., P. Kahn, and A. Borning. 2006. "Value Sensitive Design and Information Systems." In *Human–Computer Interaction in Management Information Systems: Foundations*, edited by Ping Zang and Dennis F. Galletta. New York: M.E. Sharpe.

Friedman, B., I. E. Smith, P. H. Kahn, S. Consolvo, and J. Selawski. 2006. "Development of a Privacy Addendum for Open Source Licenses: Value Sensitive Design in Industry." Paper presented at 6th Conference on Ubiquitous Computing (Ubicomp), Irvine, California, September 17–21.

Friedrich, J., U. Hammerschall, M. Kuhrmann, and M. Sihling. 2009. "Das V-Modell XT." In *Das V-Modell® XT*, edited by Jan Friedrich, Ulrike Hammerschall, Marco Kuhrmann and Marc Sihling, 1–32. Berlin: Springer.

Gotterbarn, D. and S. Rogerson. 2005. "Responsible Risk Analysis for Software Development: Creating the Software Development Impact Statement." *Communications of the Association of Information Systems* 15:730–750.

Habermas, J. 1984. *The Theory of Communicative Action.* Vol. 1, *Reason and the Rationalization of Society.* London: Heinemann.

Habermas, J. 1987. *The Theory of Communicative Action.* Vol. 2, *Lifeworld and System: A Critique of Functionalist Reason.* Oxford: Polity Press.

Hoffer, J. A., J. F. George, and J. S. Valacich. 2002. *Modern Systems Analysis and Design.* Upper Saddle River, NJ: Prentice Hall.

Johnson, D. G. 2009. *Computer Ethics: Analyzing Information Technology.* Upper Saddle River, NJ: Pearson International Edition.

Klein, H. K. and R. Hirschheim. 2001. "Choosing between Competing Design Ideals in Information Systems Development." *Information Systems Frontiers* 3(1):75–90.

Knobel, C. and G. C. Bowker. 2011. "Computing Ethics: Values in Design." *Communications of the ACM* 54(7):26–28.

Lanier, J. 2011. *You Are Not a Gadget.* London: Penguin Books.

Le Dantec, C. A., E. S. Poole, and S. P. Wyche. 2009. "Values as Lived Experience: Evolving Value Sensitive Design in Support of Value Discovery." Paper presented at Computer Human Interaction (CHI) Conference, Boston, April 7.

Miller, J., B. Friedman, G. Jancke, and B. Gill. 2007. "Value Tensions in Design: The Value Sensitive Design, Development, and Appropriation of a Corporation's Groupware System." Paper presented at GROUP, Sanibel Island, Florida, November 4–7.

Mingers, J. and G. Walsham. 2010. "Toward Ethical Information Systems: The Contribution of Discourse Ethics." *MIS Quarterly* 34(4):833–854.

Moor, J. 1985. "What Is Computer Ethics?" *Metaphilosophy* 16(4): 266–275.

Mumford, E. 1983. *Designing Human Systems: The ETHICS Method.* Lulu.com.

Nonaka, I. and H. Takeuchi. 2011. "The Wise Leader." *Harvard Business Review*, May, 58–67.

Prause, C. R. and Z. Durdik. 2012. "Architectural Design and Documentation: Waste in Agile Development?" Paper presented at International Conference on Software and System Process (ICSSP '12), Zurich, June 2–3.

Shilton, K., J. A. Koepfer, and K. R. Fleischmann. 2014. "How to See Values in Social Computing: Methods for Studying Values Dimensions." Paper presented at Computer Supported Collaborative Work (CSCW '14), Baltimore, Maryland, February 15–19.

Solove, D. J. 2006. "A Taxonomy of Privacy." *University of Pennsylvania Law Review* 154(3):477–560.

Sommerville, I. 2011. *Software Engineering*. 9th ed. Boston: Pearson.

Ulrich, W. 2000. "Reflective Practice in the Civil Society: The Contribution of Critically Systemic Thinking." *Reflective Practice* 1(2):247–268.

Wiener, N. 1954. *The Human Use of Human Beings: Cybernetics and Society*. 2nd ed. Da Capo Series of Science. Boston: Da Capo Press.

Willcocks, L. and D. Mason. 1987. *Computerising Work: People, Systems Design and Workplace Relations*. London: Paradigm.

Yetim, F. 2011. "Bringing Discourse Ethics to Value Sensitive Design: Pathways toward a Deliberative Future." *AIS Transactions on Human–Computer Interaction* 3(2):133–155.

Zinkin, M. 1998. "Habermas on Intelligibility." *The Southern Journal of Philosophy* 36:453–472.

Chapter 14

Value Discovery for IT Project Selection

"In conventional economics, the ultimate goal of any company is to maximize profit. But in the knowledge society, a corporate vision has to transcend such an objective and be based on an absolute value that goes beyond financial matrices."

Ikujirō Nonaka (2011)

Value discovery is the first phase in the ethical system development life cycle (E-SDLC). Identifying the values inherent in a future information technology (IT) system is no easy task. To debate the impacts of a system on values in a meaningful way, the project team needs to envision the future.

The most intuitive way to come up with values is to brainstorm the harms and benefits created by a new IT. As the team thinks about a potential IT service, they anticipate the things they like or dislike about it, the things that could go wrong or be very positive once the system is in place. These harms and benefits are on a different logical level than values. For example, the fear of being tracked through radio-frequency identification (RFID) chips represents a harm that is related to the underlying value of privacy.

When tech-savvy team members think about harms and benefits, they naturally come up with ideas about how harms could be addressed technically or what features would be the most fun. These ideas should also be collected because they feed into the later vision for an ideal IT design.

Once harms and benefits are collected, the next step is to identify the human values that are underlying them. Figure 14.1 summarizes the value discovery phase. Note that the arrows do not indicate a fixed and rigid sequence of tasks. Sometimes a value comes to mind earlier than its source, the harm, or benefit it relates to. Often it is the other way round. Some people will think more about technical solutions than value terms. When stakeholder groups and IT project teams

are heterogeneous enough, they will come up with many aspects relevant for service design.

Envisioning harms, benefits, and values can be facilitated with a few key tools. For example, envisioning cards enable people to think through a future technology from various angles, such as time and pervasiveness. Alternatively, project teams can ask people to reflect on photos and videos that illustrate future scenarios. Innovation project teams can invite potential future users, customers, and direct and indirect stakeholders to participate in envisioning exercises. I hereafter call those people who take part in value analysis "participants."

During envisioning exercises, IT innovation teams should avoid using preconfigured value schemes and conceptual decompositions (as I provided them for many values in Chapters 5 to 12). Using such preconfigured value schemes too early leads participants to focus too much on the value concept rather than the concrete IT context and service under scrutiny. Creativity can hence be reduced, and participants miss the harms and benefits that often lie outside preconfigured value conceptualizations.

14.1 Envisioning with Cards

Envisioning cards are a toolkit developed by Batya Friedman, Lisa Nathan, Shaun Kane, and John Lin (Friedman and Hendry 2012). The cards help participants look at a new technology from four angles, which are represented by four types of cards. The first angle takes various direct and indirect *stakeholder* positions. Each stakeholder card shows a different stakeholder (for example, children, as in Figure 14.2). It then asks the participants to develop a scenario that portrays the respective stakeholder interacting with the new system. How might the system influence the stakeholder's

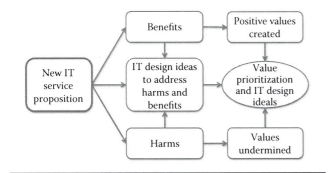

Figure 14.1 Overview of E-SDLC, phase 1: value discovery for IT selection.

personal growth and eudemonia? What harms, benefits, and values are touched upon and in what way? The second angle of the envisioning card toolset looks at the *time* criterion. Participants are asked to picture the system many years in the future. What would society look like if the new technology was fully adopted and integrated into society? How would values be impacted over time by the new system? How would intentional nonuse of the system affect a person's daily

life? The third angle of the cards is the *pervasiveness* criterion, which is somewhat linked to the time perspective because it asks participants to imagine pervasive use of the new system. Use could spread across geographic borders and across social subgroups. For example, suppose a social networking site that was exclusively used by teenagers is suddenly also used by adults. Or suppose it was used only in China and is now suddenly adopted throughout the United States. How would values change? What values would be spread? Are there harms and benefits we can imagine for another culture but not for our own? Finally, the envisioning cards help to identify and specify *new values*. One way to do so is to ask participants to create their own value card, where they write a brief definition of the value they deem important. Alternatively, participants can be asked to specify a value that is part of a higher value system. For instance, a participant could be asked to specify how a future system could impact environmental sustainability or privacy. The latter mechanism is important to ensure that major values are covered that an IT project team sees as having an impact. Figure 14.2 shows a sample of envisioning cards with image, title, theme, and activity.

Image	Title and Theme	Activity
	Crossing National Boundaries (Pervasiveness). Nations have different rules, customs, and infrastructure that affect use of technology. What challenges will be encountered by your system if it is used in other countries?	Choose three countries across the globe and envision challenges for your system if it was deployed in each of those countries. Label any common concerns across the identified challenges.
	Consider Children (Stakeholders). Children often appropriate systems originally designed for adults. How might this system influence a child's social and moral development?	Develop a scenario that portrays a 7-year-old interacting with the systems. How might the system influence the child's learning, or play with other children?
	Environmental Sustainability (Values). Many systems can be applied or extended to support a desirable environmental outcome (e.g., a system designed to support efficient printing from web browsers may lead to less use of paper and ink). At the same time, systems may have unintended negative effects on the environment (e.g., pollution and waste created in the production of electronics).	Specify the required resources needed to create and support your system, and the by-products of its production and use. Can your design be applied or extended to support a more positive environmental outcome?
	Choosing Not to Use (Time). Some people may decide to use your system, or may attempt to remove themselves from an indirect stakeholder role (e.g., choosing not to publish a telephone number). How might deliberate non-use of the system affect a person's daily life (e.g., employability, relationships, civic participation)?	Picture your system in use many years from now. Identify three ways in which an individual's intentional non-use of the system might affect that person's daily life or the system as a whole.

Figure 14.2 Figures from examples of the envisioning card set. (From Friedman, B., and D. G. Hendry, 2012, "The Envisioning Cards: A Toolkit for Catalyzing Humanistic and Technical Imaginations," in *Proceedings of the SIGCHI Conference on Human Factors in Computing Systems,* New York: ACM, p. 1147. With permission.)

14.2 Envisioning with Scenarios

To help project teams explore the innovation space, they can also construct and visualize scenarios for participants to reflect on. Scenarios have the following characteristic elements (Carroll 2000): (1) They include or presuppose a setting for the technology. Where is it used and when? (2) They specify the main and secondary actors (also called "agents") that use the system. (3) Each of these actors uses (activates) the system in one or several ways and (4) with one or several goals or objectives in mind. (5) Finally, in a scenario the goals of system usage could change, leading to actions, events, or uses other than those that were originally foreseen.

Scenarios can be pure text descriptions, but using one or more photographs often encourages participants' creativity. Photo elicitation is an ethnographic technique (Schwartz 1989) that leverages the "inherently ambiguous … meanings emergent in the viewing process … in order to elicit reactions and information … which might otherwise never become apparent" (Le Dantec, Poole, and Wyche 2009, p. 1144).

If a technology has to be embedded in a specific setting or context, one way to find values is to have participants (or project team members) photograph that setting and imagine the impact that the new technology will have on that setting. Such a "photo elicitation interview" is a particularly immersive experience for participants (Le Dantec, Poole, and Wyche 2009). They can reflect on why they took that particular photograph to discuss the technology and what was particularly important for them in the photo. These questions act as a starting point for participants to reflect on their own life and the technology envisioned.

Finally, teams can use animations or videos to demonstrate the look and feel of a new technology. When such material is used, it must not be biased toward or against a technology. The material should not include background music or emotional scenes. If there is a dialogue, it should be purely informative, without any adjectives characterizing the technology as good or bad. In short, the material should be as neutral as possible. Note that this goal is not easy to achieve. In practice, project teams tend to embellish the scenario, biasing participant discussions to focus on the benefits of the solution more than on potential harms.

14.3 Critical Information Systems and Reflective Design

A common pitfall in value discovery is to identify the obvious. Participants in a value discovery exercise often tend to only come up with obvious mainstream values deeply ingrained in our common culture and political system without questioning them. An example from one of my university classes illustrates this: Together with my students we had a discussion about the value of saving time through technology. The students came up with the idea for an intelligent software agent that would optimize travel planning. Initially, they argued—as expected—that the software agent could create benefits in terms of time savings during travel. The software agent could ensure that business people spend as little time traveling as possible on the overall journey. We then critically reflected on the meaning and value of time savings during travel. We speculated that business travelers might not want only to save time. In contrast: in some situations, they might want to spend more time! For example, when they fly to business destinations, they may appreciate optimal plane routing. But they might also wish to have some time to visit the towns they go to or see a friend who lives there. They would like to have the time to recover from the physical stress of traveling. Generally, people wish for more time in what they do and not less, but this desire is currently not supported by our business practices and our technology. So the students finally asked: Do we really want a software-based travel agent that always only minimizes the amount of time people spend on a business trip? From an ethical perspective (which looks at what is good and right for people) probably not. Instead, the students came up with a software agent that had "time-gain" logic embedded in its algorithms. This travel agent looked completely different than one that would have been built solely to minimize time. It would for instance check with the social network (i.e., on Facebook) to see whether any friend lives at the planned business-trip destination and recommend a dinner with that friend reserving the required time for such a break in the time-schedule.

A lack of *critical reflection* of common values is problematic, especially in the IT selection phase. The IT selection phase must give room to critical and challenging reflection about mainstream values as well as about a new technology. Why should a system only save time instead of giving time? It is ideal if someone in the team is playing the advocatus diaboli (devil's advocate) who questions everything. *Critical theory* and critical information systems (IS) are useful frames of mind in this endeavor. According to critical theory, societies advance because they do not unthinkingly accept the teachings of authorities but instead are open to critical reflection and change. The Western tradition of critical theory is embodied in intellectual strands such as feminism, racial and ethnic studies, and psychoanalysis.

Scholars in the field of critical IS (Sengers et al. 2005; Stahl and Brooke 2008; Myers and Klein 2011) call for a systematic uncovering and challenging of subconsciously held value assumptions when we build information systems. In their work on reflective critical IS design, Phoebe Sengers et al. (2005) write: "Reflection itself should be a core technology design outcome. … We define 'reflection' as referring to *critical* reflection, or bringing unconscious aspects of experience to conscious awareness, thereby making them available for conscious choice" (p. 50). The earlier example about the time value shows how vital critical IS thinking is for ethical system design. Critical thinking generates a more complete spectrum of relevant system requirements that lie out of today's corporate mainstream thinking.

A useful mechanism for an uncovering of critical values is the critical technical practice outlined by Phil Agre in his 1997 book *Computation and Human Experience*. Critical technical practice identifies the core "metaphors" of a field, which we may interpret here as the mainstream values immediately recognized. Critical technical practice then looks explicitly into what value aspects (metaphors) remain marginalized. These marginalized concepts are the ones that are then brought to the center of technology development. In the travel agent example, the originally marginalized idea is gaining time. With critical technical practice, this marginalized idea is brought to the forefront.

Once IT innovation teams have envisioned a new technology with a first mix of values to be created senior leaders in the company need to decide on whether to proceed. The next chapter gives insight into this decision-making process and what ethical pitfalls it can entail.

EXERCISES

1. Select a relatively new IT service that you or people you know have started using. Then, use the card system described in this chapter to engage in an envisioning exercise.

2. Take the smart coffee machine service that is described in the work scenario in Chapter 3. Discuss the harms, benefits, and values of such a service from a critical perspective. What value aspects might remain marginalized by such a service? How might the marginalized values affect the success or failure of such a service?

3. Discuss the benefits and harms caused by the attention-monitoring program that was described in the work scenario in Chapter 3. What values are impacted by such a program? Use the card system described in this chapter to engage in an envisioning exercise.

References

Agre, P. 1997. *Computation and Human Experience*. Cambridge: Cambridge University Press.

Carroll, J. M. 2000. "Five Reasons for Scenario-Based Design." *Interacting with Computers* 13(1):43–60.

Friedman, B. and D. G. Hendry. 2012. "The Envisioning Cards: A Toolkit for Catalyzing Humanistic and Technical Imaginations," in *Proceedings of the SIGCHI Conference on Human Factors in Computing Systems*, 1145–1148. New York: ACM.

Le Dantec, C. A., E. S. Poole, and S. P. Wyche. 2009. "Values as Lived Experience: Evolving Value Sensitive Design in Support of Value Discovery." Paper presented at Computer Human Interaction (CHI) Conference, Boston, April 7.

Myers, M. D. and H. K. Klein. 2011. "A Set of Principles for Conducting Critical Research in Information Systems." *MISQ* 35(1):17–36.

Schwartz, D. 1989. "Visual Ethnography: Using Photography in Qualitative Research." *Qualitative Sociology* 12(2):119–154.

Sengers, P., K. Boehner, D. Shay, and J. Kaye. 2005. "Reflective Design." Paper presented at AARHUS '05, Aarhus, Denmark.

Stahl, B. C. and C. Brooke. 2008. "The Contribution of Critical IS Research." *Communications of the ACM (CACM)* 51(3):51–55.

Chapter 15

Wise Leadership in the IT Project Selection Phase

"Courage, that is for sure, drives most essentially of all human characteristics the achievement of happiness."

Johann Heinrich Pestalozzi (1746–1827)

In a fast-moving environment, surrounded by information technology (IT) hype cycles, senior executives must regularly decide whether to invest in a new technology or not. Their investments can take multiple forms. A company may integrate a new technology into existing products; for example, embedding radio-frequency identification (RFID) chips or sensors into analog artifacts such as clothing or machine parts. A company might also decide to develop a whole new digital product or service from scratch. It might upgrade an older IT-based product or service. Or, it might change its operations with the help of IT functionality. In any case, senior managers should ask the following ethical questions:

1. What positive human values can be created by the new IT, and which of these are most important?
2. What harms can be caused by the new IT, undermining which human values? And which of these harms are most important to avoid?
3. Given the benefits and harms, is the resulting value proposition a strategic fit for the company's past and future narrative, stakeholders, and for society at large?

As of 2015, such questions are rarely asked. In the IT selection phase, the dominant rationale for a decision about new technology is the technical progress and financial benefit it promises. Progress is seen as a positive value per se that comes automatically with a new IT opportunity. However, the meaning of *progress* is hardly understood. A progressive concept or technology is often described as green or open

or smart, but the value reflections stop there. Instead, to have a chance for approval, idea promoters must present their IT projects in terms of financial benefits (Shollo and Constantiou 2013). So, besides a description of the technology, the goal of idea givers and innovation teams in this initial phase of the system development life cycle (SDLC) is to broadly anticipate how much income or cost savings could be generated from the new IT. Equally, the team might consider operational improvements from using the new IT, for example, greater process transparency, increased product quality, or more security. The team also anticipates the availability of resources, potential technical difficulties, and approximate project size and duration. All this material is combined to form a proposal that is presented to the company's leadership team. Proposals that are prepared this way allow leadership teams to judge a project with a so-called "limited economic view" (Waldman and Gavin 2008).

In contrast to this limited view, which emphasizes shareholder interests, recent management literature holds that companies should embrace an "extended stakeholder view" (Waldman and Gavin 2008); a view that includes the kind of ethical questions raised earlier. In Chapter 13, I explained the ethical and economic significance of extended stakeholder views. I used the example of body scanners at airports. I described how one US company clearly understood the ethical challenges and privacy harms inherent in body scanning and was therefore able to develop a stick-figure solution that now dominates the market. I also showed how some other equipment manufacturers ignored passengers' privacy concerns around nudity. Those manufacturers trusted that they could ignore the value *harm* caused as body scanning is mandatory. As a consequence they failed in the market.

The airport body scanner case is a very blunt example of how things can go wrong when IT idea givers and executives ignore value questions. But the example is quite representative of the typical situation today, where few laws regulate the new technical environments. Hence, managers must constantly use their own judgment on whether an IT investment is good from an ethical perspective or not. When they see a good bottom line for the company and feel sufficiently empowered, they often tend to ignore the human value perspective.

15.1 Why Should Leaders Care about the Ethical Design of IT?

Recent research in the field of responsible leadership provides theoretical and empirical evidence that executives should generally care more about ethical decision making in their companies (Pless, Maak, and Waldman 2012; Miska, Hilbe, and Mayer 2014; Stahl and De Luque 2014). The IT world in particular has witnessed many cases where a lack of concern had negative consequences.

The worst thing that can happen for a company is to pull a technology from the market after it has been introduced. A small-scale example is the retailer Metro, which recalled 10,000 RFID-tagged loyalty cards in 2004. It had used these cards to track its future store customers without their knowledge and consent. When a small nongovernmental organization (NGO) and the press found out about the practice and protests took place in front of their store, the company recalled the cards. Metro's IT management had completely ignored and underestimated the privacy backlash from secretly using chip technology.

Another consequence can be an unexpected technology overhaul or service interruption. A famous example is the hacking of Sony's PlayStation network, through which attackers obtained the personal data of 77 million players due to poor system security. Sony was forced to take the gaming platform offline for 24 days. The incident cost the company €128 million, not including a tremendous loss of brand value estimated by some to reach more than a billion euros (Rose, Röber, and Rehse 2012).

Taking a solution out of the market or improving a technology in operation is an extreme and costly step. A situation that is more common but less visible and measurable is when an IT product or service never reaches the level of acceptance or customer loyalty that it actually could reach. For example, some Internet users abstain from Facebook's social network service due to its poor privacy reputation. Many who do use the network do not trust it and are cautious about what they post and share; they create less information on the platform than others and may be less loyal. Strictly speaking, the company incurs an opportunity cost of lost income from ethically aware potential customers and makes itself vulnerable to competitors who are more concerned with privacy.

Beyond such return on investment (ROI) and competitive considerations, ethical issues can also influence the morale of a company's workforce. Management literature has shown that the success of products and processes depends on employee commitment to a company's strategy (Kaplan and Norton 1996). New IT can considerably influence internal processes and can alter the product strategy of a company. Employees' personal commitment to such changes and offerings influences the operational and financial success of an IT rollout.

Finally, an ethical breach can lead to governmental penalties. In the United States for instance, the government can sue companies whose practices harm consumers. Alternatively, the Federal Trade Commission can impose sanctions on companies that do not behave ethically. In 2012, Google paid $22.5 million in fines because it bypassed users' expressed privacy settings in Apple's Safari browser in order to display ads.* In the European Union, a data protection regulation is currently planned to force companies to pay up to 2% of their world turnover in fines if they breach customers' privacy (European Commission 2012).

Beyond such negative incentives to invest in ethical IT selection and design, leaders can and should take a positive perspective. Ethical buy-in from customers can create great business cases. The literature on mainstreamed corporate social responsibility (CSR) has accumulated many examples (Smith and Lenssen 2009). Mainstreamed CSR means that CSR is not outsourced to a "do-good" department with a philanthropy budget. Instead, companies create a competitive advantage by building social values directly into their products. These "good" product characteristics foster demand and drive long-term company value. Good examples include "organic food"-seals in the retail industry and hardware vendors' adoption of green (energy efficient) IT. Companies that differentiate their products for better quality and environmental sustainability gain competitive advantage due to higher margins and market growth, while those that do not are faced with tough price competition and decreasing ROI. "The business case (for CSR) at the level of the firm is becoming increasingly clear as more companies are coming to understand that, aside from any moral obligation, it is in their economic interest to address environmental, social and governance issues and in a manner that is integrated with their strategy and operations" (Smith and Lenssen 2009, p. 2).

Of course, financial and instrumental incentives to engage in ethical IT leadership should only be one aspect of a leader's decision. Another is the leader's own personal history, role models, values, and motivations (Maak and

---

* Gerry Smith, "FTC to Pay Record Fine over Safari Privacy Violation," *Huffington Post*, last modified October 9, 2012, accessed January 12, 2015, http://www.huffingtonpost.com/2012/08/09/ftc-google-fine-safari-privacy-violation_n_1760281.html.

Pless 2006). A leader might want to be remembered as a good leader instead of an inhumane or shortsighted executer. Unfortunately, wise and responsible leadership is not something every top executive can easily live up to. In Section 15.5, I show that wise leadership requires moral and decisional autonomy. Autonomy and courage is a gift of character as well as a zenith of personal maturity. Authenticity and integrity in decision making follows suit. But not everybody who has climbed the organizational hierarchy possesses these traits. Even if someone is wise enough to make good decisions, the next step is to have the courage to implement them. Responsible leaders need to have the courage to lead instead of running opportunistically after mainstream ideas, economic trends, and IT hype cycles. Unfortunately, many organizations today tend to discourage both wisdom and responsibility.

15.2 Why Many Executives Still Ignore Ethics

In today's business climate, being a wise and responsible leader who seeks to make ethical decisions can backfire. Ethical IT system design does not always pay off for companies, at least not in the short term. For example, IT products are often built to stop working after some years. This is not because they cannot naturally endure, but because some IT manufacturers intentionally build their systems so that they stop functioning after a while. This is surely not an ethical practice. It undermines environmental sustainability by creating unnecessary waste, and it betrays customers who believe that their product broke when it really comes with a limited lifetime that is build already into the product. From a business perspective, however, such a practice is excellent because it leads to replacement sales.

Another example of lucrative ethical ignorance is the unauthorized use of customers' personal data. Most IT services today collect personal data from their customers to deliver high-quality services. For instance, map services use location data to show their users where they are on a map. But in many cases, this personal data is not collected exclusively to optimize service offerings. The data is often also used as an asset in itself, sold to other parties for profit. This unethical practice is legitimized through lengthy, unreadable terms and conditions that customers are forced to sign if they want to use the service.

Executives can almost count on people's inability to properly judge risks (Kahneman and Tversky 2000) and on their desire for immediate gratification (Löwenstein and Prelect 2000). People normally underestimate and misjudge the degree of personal risk caused by using technologies. For instance, they underestimate the risk of personal data sharing online and how that data can be used against them. They also prefer immediate gratification, such as free service use, instead of thinking about eventual financial drawbacks in the future. Hence, ethical misconduct is facilitated by many people's "predictably irrational" and short-sighted behavior (Ariely 2009).

For all of these reasons—customer ignorance, bending and abuse of laws, customer inability to judge risks, customers' desire for immediate gratification, and so on— companies seem to get away with not caring about ethics in their product and service design. When unethical behavior goes unpunished or when ethical decisions are financially costly, value-based IT design and lengthy ethical decisions pose more challenges. In such situations, wise leadership becomes even more important.

15.3 Tough Decisions for Wise Leaders

In 1589, a man named William Lee traveled to London. Lee was in a good mood and full of confidence because he had succeeded in inventing a new machine: a stocking frame knitting machine that promised to relieve workers of their laborious hand-knitting tasks. Seeking patent protection for his invention, he rented a building in London for his knitting machine to be viewed by Queen Elizabeth I. When the queen entered, he proudly demonstrated his invention. To his disappointment, the queen was more concerned with the employment impact of his invention and refused to grant him a patent, claiming that: "Thou aimest high, Master Lee. Consider thou what the invention could do to my poor subjects. It would assuredly bring to them ruin by depriving them of employment, thus making them beggars" (as cited in Frey and Osborne 2013, p. 6).

Imagine yourself as a chief executive today sitting in your office with a consultant. The consultant tells you that you can change your shop-floor manufacturing operations by replacing workers with robots. Laying off hundreds of workers and using robots instead can save you millions. How would you react? Do you think the queen's reaction in 1589 was naïve? Or was it aristocratic in the sense that it expressed a deep care for her subjects? Would it be the appropriate way to react to such a situation today?

Michael Porter and Mark Kramer recently wrote that "the purpose of the corporation must be redefined as creating shared value, not just profit per se. This will drive the next wave of innovation and productivity growth in the global economy … learning how to create shared value is our best chance to legitimize business again" (2011, p. 64). The management literature now acknowledges how important it is to manage companies sustainably and to take responsibility for all stakeholders, including employees. In 2011, ISO 26000 defined CSR as "the responsibility of enterprises for their impacts on society." Would you, as an executive, have the courage to turn down the argument of the consultant and refuse to automate so that jobs can be maintained?

Despite wise theoretical insights in the recent management literature, it must be extremely difficult for senior managers to act accordingly. If the financial benefits of automation, data sales, throwaway products, and environmental abuse are high enough, top executives are under significant pressure to implement them and have a hard time justifying not optimizing the bottom line. Moreover, considerable competitive pressure is created through serious innovations as well as IT hype cycles. The problem is that managers act in a highly competitive environment in which "economic theory [has become] stripped of any ethical or moral dimension" (Mingers and Walsham 2010, p. 839). Often, when even well-meaning managers face situations such as automation, they think that if they do not automate, others in the market will; or if they do not collect personal data, others will soon have more knowledge about customers than they have. Managers today often do not risk challenging such general market assumptions or the goodwill of the stock exchange. The result is that "the market acts on its moral foundations, as the industry acts on its fossil fuels: They are being burned in the verge of expansion" (Dubiel 1986, p. 278).

Even if managers were willing to question technology investments and critically reflect on questionable business practices, they face the challenge that the ready embracing of any new IT hype has become something of a fetish. New IT functionality and the use of hyped terms such as cloud, Big Data, Internet of Things and green IT (see Chapter 2, Figure 2.2 on the hype cycle) is widely equated with progress. And absurdly enough, saying no to new IT systems or automation sounds like being critical of human advancement in general.

So how can managers act wisely and responsibly in this tense field? How can they justify tough and wise decisions against a hyped IT investment without being called a Luddite? How can executives investigate their own value base and decide what is good or bad for the company, the stakeholders, and for society at large? The way forward is to establish a process of more conscious ethical decision making around new IT investments and IT requirements (Jones 1991; Stahl and De Luque 2014). Corporate leaders' (and regulators') IT decisions are now so far reaching that they must be much more cautious and reflective when they decide whether to roll out a system or decide what values the system should embed. "Cautiousness is becoming the core of moral action," Hans Jonas already wrote in 1979 with regard to managing modern technologies (p. 82). This does not mean that leaders should be timid or slow, but they must be extremely mindful in deciding what is right or wrong and take the time to make the right decision. Management needs to go into "philosophical mode" (Flanagan, Howe, and Nissenbaum 2008).

15.3.1 What Is Philosophical Mode?

About 2500 years of philosophical heritage have provided us with a great compass for thinking about what is right or wrong. In the following sections, I will introduce the most prominent theories of ethical reasoning: the teleological theory of utilitarianism, deontological theory, and virtue ethics.

Teleological theories argue that what is good can be determined by the science (in Greek: *logos*) of telos, which means "the end." Thus, all is good that leads to a good end or good consequences. Scholars sometimes refer to teleological theory as "consequentialism." In contrast, deontological theories believe in the science of duty (in Greek: *deon*). Here, it is the characteristic of an action itself that determines its goodness. An example is the duty to not lie to anyone. A deontologist would not lie no matter the situation. A consequentialist, in contrast, would consider the effects of lying. In some cases, he or she could decide that lying is the right thing to do, for example, to save someone else from pain or fear.

I will start with the philosophical stream that underlies today's business theories: utilitarianism. I will then show how deontologists would think about ethical management dilemmas. Both of these theories are limited though and are recently giving way to a third, potentially more powerful avenue of ethical thinking: virtue ethics.

15.4 Utilitarian Reasoning

When Jeremy Bentham (1748–1832) and John Stuart Mill (1806–1873) proposed utilitarianism in the nineteenth century, it was a reaction to the strict and top-down ethical paternalism of the churches at that time. Philosophers of the enlightenment and those who followed them embraced the idea that the right actions and paths to take are not necessarily God-given; instead, people can decide for themselves what is good or bad. They believed in people being moral agents, autonomous individuals, and decision makers. And they wanted to provide guidance on how good judgments can be achieved and argued for.

The teleological philosophy of *utilitarianism* they founded holds that the standard of right and wrong can be found in the principle of utility, which says the end to be sought in our actions should lead to the greatest possible balance of good over evil. Mill wrote: "The creed which accepts as the foundation of morals, Utility, or the Greatest Happiness Principle, holds that actions are right in proportion as they tend to promote happiness, wrong as they tend to produce the reverse of happiness … and that standard is not the agent's own greatest happiness, but the greatest amount of happiness altogether" (1863/1987, pp. 278, 282).

Utilitarians of the time proposed considering the consequences of one's decisions by weighing—in almost mathematical terms—their positive and negative outcomes. That said, positive and negative consequences of decisions in Mill's and Bentham's thinking were not restricted to money, as today's economic utility models suggest. Instead, Mill and Bentham embraced myriad forms of "utils." Bentham suggested that we should think of utils or utility in terms of pleasures and pains, which can be ranked according to their intensity, duration, and certainty. A decision is good if it maximizes pleasures while minimizing pains. This thinking is not too far from today's "economics of happiness," where scholars go beyond financial utility and include qualitative utils such as employment, health, environment, relationships, control, and freedom (Frey and Stutzer 2001).

15.4.1 Act Utilitarian Reasoning

Different kinds of utilitarianism include act utilitarianism (AU), general utilitarianism (GU), and rule utilitarianism (RU) (Table 15.1). As the term suggests, act utilitarianism applies the principle of utility to a specific action. One must ask "What effect will *my* doing *this* act in *this* situation have on the general balance of good and evil?" (Frankena 1973, p. 35). Let us take as an example a difficult IT investment decision similar to the ancient knitting machine case from earlier: the decision to automate flight check-in counters at an airport. Imagine that an airline is confronted with the decision to replace its airport ground staff with machines. In AU this decision will be made for each airport individually.

Table 15.1 Forms of Utilitarianism as Described by Frankena

Forms of Utilitarianism	Explanation
Act utilitarianism (AU)	What effect will *my* doing *this* act in *this* situation have on the general balance of good and bad?
General utilitarianism (GU)	What would happen if *everyone* were to do so-and-so in such cases?
Rule utilitarianism (RU)	It is for the greatest general good if we *always* act in the way the rule dictates.
Primitive-rule utilitarianism	The rule is deduced from GU reasoning.
Actual-rule utilitarianism	The rule is based on accepted and prevailing moral rules.

Source: Frankena, William, 1973, *Ethics*, 2nd ed., Upper Saddle River, NJ: Prentice-Hall, p. 34 et seq.

Now let us assume that we are talking about 100% automation and the layoff of all check-in staff at one airport. On the positive side, we have the traditional economic argument that the monetary gains from layoffs minus the investment and service cost of the new IT will probably increase profit for the airline. This profit is considered a good consequence. Shareholders' happiness will be increased by a larger dividend. In management theory, we would talk about the creation of "shareholder value" (Hillman and Keim 2001). Utilitarianism would furthermore recognize the quality-of-service benefit; if ticket machines are well designed and usable, customers will be able to get their tickets more quickly. An often-cited benefit of automation is also that fewer errors are made.

At the same time, quality of service could be reduced. If the airline staff is gone, nobody will take care of the airline's check-in area, and customers who are not technically savvy may feel helpless or incapable of using ticket machines, finding out about schedules, and so on. Beyond these arguments, a utilitarian would also consider the reduction in happiness of the check-in staff who are laid off and both their loss of income and the potential loss of identity that some will suffer, as employment is tied to identity formation (Gini 1998). Finally, the airline will lose knowledge about the physical check-in experience of its passengers, because ticketing machines cannot give feedback about customer complaints or desires. In an AU calculation, all these negative consequences of automation are considered and contrasted with the positive effects. Table 15.2 shows a summary of this broad AU calculation. It suggests a score of 4:5 against automation.

Of course, this line of reasoning can be challenged in many respects. First, the number of points given to each consequence could be more granular; a 100-point scale could be used to weigh different consequences differently. For example, such a scale would allow for a finer-grained consideration of the effects of the decision on the happiness of different stakeholders, potentially including more stakeholders than Table 15.2 contains. The scale could consider different durations and intensities of happiness, or unhappiness among shareholders, customers, and employees. And, of course, these weighted happiness scores are also a judgment of the very specific context at the distinct airports under scrutiny, at a particular point of time. Let us assume, for example, that this airline has a small number of shareholders who are very rich. So the marginal utility they gain or marginal happiness they perceive from the extra profit they make from the layoffs at one airport will be relatively low. In contrast, the ground staff members who lose their jobs are deprived of their entire income. So the decrease in happiness will be immense for them, and suffering may be even greater for those who are personally identified with the job.

Table 15.2 Simplistic Exemplary Act Utilitarian Calculus on an Automation Decision

Positive Consequences of Automation	Negative Consequences of Automation
+1; Shareholder happiness increased due to extra profit	–1; Employee happiness decreased (for those who lose their job) due to loss of income
	–1; Employee happiness decreased (for those who lose their job) due to short-term or long-term loss of identity
+1; Quality of service increased for those who are tech savvy and can use more rapid ticketing machines for check-in	–1; Quality of service reduced for those who are not tech savvy and need more time to check in
+1; Fewer operational errors are made	
	–1; Knowledge lost for airline due to a future lack of customer feedback on check-in quality
+ 1; Staff that is laid off may have the chance to find a more fulfilling job elsewhere	–1; Staff that is laid off may not find a job that is as good from their perspective
+ 4	–5

In contrast, employees' loss of happiness in this particular situation may also be much less grave. Maybe the airline employees at the respective airport are old and wanted to retire anyway. It also might be that they never identified with the job and are happy to leave it. The company may compensate them with a very generous indemnity package or shares in the company so that their monetary losses are well covered until they start a new job. The social security system in the country could provide employees with a comfortable income from the state. And the job market in the region may be so good that the employees will find a comparable or even better job soon after, thereby avoiding an identity crisis or social decline.

All of these individual and contextual factors are unique for each airline, airport, shareholder, and employee. And the idea of AU-based decision making is to factor them all into the decision calculus. It is this detailed weighing of the local unique decision context that drives a good or bad decision. Wise leaders who pursue AU try to recognize the fate and dignity of all individuals who are impacted by a decision. To a certain extent, entrepreneurs who govern small and medium enterprises (SMEs) probably act with a similar kind of rationale.

But as this description makes plain, pursuing AU is quite complex and probably costly for large companies that operate across multiple regions and see myriad human context situations. Airlines with many airport ground operations and thousands of employees will argue that it is too much effort for them to consider each employee and stakeholder individually. Therefore, utilitarianism has embraced more generic ways to find solutions to moral problems. One of them is general utilitarianism.

15.4.2 *General Utilitarian Reasoning*

General utilitarianism (GU) does not ask, "What will happen if *I* do this action in this case?" but rather "What would happen if *everyone* were to do so and so in such cases?" (Frankena 1973, p. 37). GU is closer to today's reality, where companies make principled decisions for whole parts of the organization without looking into the detailed effects these decisions have for individual employees and the myriad of customer contact points they maintain.

Let us return to the airport check-in example. Leaders who adopt a GU approach would not look at one airport or the individuals employed there. Instead, they would ask, "What will happen if all of the airlines fully automate their check-in process?" Of course, the short-term profit and happiness of airline shareholders would strongly increase. Massive cost savings would drive dividends and stock prices. At the same time, however, airlines would have a harder time competitively differentiating based on quality of service, as automation streamlines operations and offers less of a human touch at check-ins. Tech-savvy customers might view airlines that use the latest technologies as innovative. The check-in process would also become generally quicker. In contrast, passengers would find airports to be largely vacant of competent staff. A whole branch of jobs at airports would be outmoded, with many employees and their families losing their income and identities. Generally, airlines would know less about what is happening on the ground, and the whole travel experience would change for passengers. But new service branches would probably rise, for example, services to supply and maintain the machines, and security companies and surveillance cameras to prevent vandalism and ensure passenger safety on the ground. These services would create new kinds of jobs.

The example shows that GU thinking requires leaders to reflect on the broader consequences of their actions: changes in the perception of airlines and airports, changes in the nature of competition in the industry, and changes in customer experience during travel as well as shifts in job market structure and unemployment. When GU is applied, executives are asked to think in such broader and longer-term societal and economic dimensions. To do so well, however, they need to have a very mature and wise view of the future and an ability to anticipate the wider implications of their decisions. This is not a talent every leader has. But if they have and do see the broader consequences of their decisions, GU leaders have great power to shape the future in line with their convictions.

A drawback of GU is that it enables leaders to justify exceptions. A GU decision maker may understand the general theoretic consequences of certain decisions for society at large. But the leader may still act against this theoretical insight if he or she doubts that the action has universal consequences. Imagine an airline executive who generally does not favor automation at airports. Her general utilitarian reasoning would lead her to conclude that check-ins should not be automated by every airline. Yet, in the case of her company, she would still automate, perhaps for financial reasons. She would do so under the assumption that automating her check-in areas would not lead everyone else to follow likewise. She likes to believe that her line of reasoning would not be followed by all of her competitors; that it is not universal.

15.4.3 Rule Utilitarian Reasoning

A way to avoid GU exceptions is to embrace RU. In RU, the question is not what *action* has the greatest utility but what *rule* has the greatest utility. A decision maker has to follow the rule that is accepted for her context. So the airline executive would look out for any rule for how to behave when confronted with the question of check-in automation. What is the accepted rule in the industry?

However, even if a utilitarian line of reasoning is used to define rules, rule making can follow different strategies. One form of rule setting in RU has been called *primitive-rule utilitarianism*. Primitive-rule utilitarianism simply accepts the conclusions from GU as rules for the behavior one should pursue. Another form of RU is rule setting according to *actual-rule utilitarianism*. Actual-rule utilitarianism views an action as right if it conforms to the accepted or prevailing moral rule. In doing so, it assumes that the accepted rule is conductive to the greatest general good.

Coming back to automation: As of 2015, it seems to be generally accepted that automation is a good thing to do. The good of automation is justified in part as technical progress, which is regarded as conducive to the general good of society. In addition, job markets are in some countries believed to be fluid enough for everyone to get a job. Hence, automation is accepted because it seems to drive profits and progress while creating limited harm for people who are laid off but quickly reemployed. The validity of these accepted assumptions has recently been questioned by leading economists (Wright 2004; Brynjolfsson and McAfee 2012). Rules must therefore be constantly reviewed, revised, and replaced based on their true utility (Frankena 1973).

15.4.4 Problems with Utilitarian Reasoning

All of the various forms of utilitarianism make some sense, but none of them is perfect or complete. An advantage of act utilitarianism is that it tries to look at the local context of a decision to optimize it for all direct and indirect stakeholders. But it is problematic to find the right weights and make correct judgments for all consequences.

To avoid arbitrary decisions, decision makers must have considerable information up front. For instance, information is needed on how the dynamics of a region's job market really is, and how financially and psychologically dependent people really are on their jobs. Decision makers must understand how much customers value automation. Good AU depends on a thorough collection of unbiased facts, which is difficult and costly to obtain.

The second challenge of AU is of a moral nature, that is, the readiness to weigh human destinies against management parameters, such as financial gain or quality of service. If one holds a deep belief in human dignity, then engaging in such weighing exercises can appear quite cruel and immoral. In fact, Mill recognized that we cannot just equalize pleasures and harms, and make them comparable on the basis of duration, intensity, and certainty. Instead, he argued that we should distinguish between higher and lower pleasures. He also believed that human happiness would ultimately be served by those higher pleasures that allow for an expansion of creative powers (MacIntyre 1984).

The third pitfall of AU is the general assumption of utilitarianism that decision makers are autonomous individuals. Remember that autonomy means the freedom to make self-regarding choices in which people express their authentic self (see Section 6.4). Yet, when local contexts are investigated and judged in practice, decision makers confront all kinds of influences and are not free to make independent decisions. For example, airline executives who try to understand the operations and employment situation at a local airport and who look at individual employees will hear about and understand the fate of the people affected. They will hear some stories but not others, which will bias their decisions. And even if an executive concludes that it would be better for the airport and the airline to automate operations, his or her judgment could still be influenced by the pity he or she has for individual employees who hope to keep their jobs. As a result of more or less biased information as well as emotional involvement, an executive may decide to not automate. The executive may even present the facts that influenced the decision in a light that turns the business case in favor of human employment. In short, truly autonomous judgment in a local context seems almost impossible. With AU, it is too easy for decision makers to accumulate and weigh facts and decision elements in a way that justifies personal decisions that have been made on other grounds. This does not mean, of course, that pity is not an excellent reason to tweak rational AU arguments in favor of human fates.

Given the challenges of AU, leadership forums, such as the World Economic Forum, support and engage in GU and RU reasoning. GU requires leaders to reflect and act upon the wider consequences of a decision for society. GU makes high-level executives think about what would happen if everyone (at least whole industries) decided in the same way. In the airline business for example, top executives certainly consider profitability, general service-level development, and employment in general. Unfortunately, GU equally encourages weighing profitability against human destinies. But the consideration of human destinies in GU happens at a higher level of abstraction. When decision makers apply GU, they do not look at the fate of individuals in a specific situation anymore. Instead, they think in an abstract form about the dynamics of job markets in general if everyone was to automate. They wonder whether job markets are robust enough to create new forms of employment at the same rate as jobs are lost (Brynjolfsson and McAfee 2012; Frey and Osborne 2013). They might consider John Maynard Keynes's theory (1933) that the "capitalization effect" of innovation (new jobs created) might fall behind the "destruction effect" (old jobs lost), meaning that our discovery of ways to economize the use of labor outruns the pace at which we find new uses for people. Philosophically, GU decision makers might argue along the lines of Schumpeter's idea of "creative destruction." Schumpeter described creative destruction as a process of industrial mutation where innovations incessantly revolutionize an existing economic structure from within, destroying the old economic structure but also constantly creating new ones (Nickles 2013). From this perspective, the loss of jobs and the creation of new ones is a natural and acceptable matter of change. And such a change may lead to a society where people simply have to work less and enjoy higher-quality tasks.

Besides such broad employment-related questions, the GU perspective would also require airline managers to consider various operational effects of automation. They would

need to decide whether the ubiquitous presence of automated machines, surveillance cameras, and maintenance companies is preferable to an atmosphere where the premises are run by responsible personnel.

As this example shows, GU confronts senior managers with grand political reflections as well as actions they need to take based on the conclusions they draw. Unfortunately, senior executives have not acted much upon GU reasoning (if they engaged in it). Instead, as Porter and Kramer pointed out, they have pursued "a narrow conception of capitalism [that] has prevented business from harnessing its full potential to meet society's broader challenges" (2011, p. 64). Many top executives may have seen and questioned the general and grand implications of their management actions for factors such as the environment and employment, but their concrete decisions have largely reflected a priority of company profitability.

In defense of this pattern, these leaders' failures are probably rooted to some extent also in the pitfalls of the GU philosophy itself. As I outlined, GU lends itself to justify exceptions. Even though one may see the negatives of some decisions on a general and universal level, the level of abstraction in GU enables leaders to separate their concrete decisions from the theoretical conclusions they draw. RU tries to step in here, advocating that GU conclusions should be acted upon as a rule (primitive rule utilitarianism). But even then another question is how GU conclusions or RU rules are obtained in the first place. Are the theories that are accepted at a certain time always the right ones?

A major challenge for GU and RU reasoning is that the accepted theories in economics and technology management are full of fashions in thinking. Economic theorists and business scholars tend to cultivate and propagate beliefs that become so dominant in an economy that senior executives have a hard time ignoring them, even if they have the gut feeling that some of these theories may be wrong for their particular companies. Prominent examples are textbook models of financial markets or the 1990s idea that companies should radically specialize in services rather than manufacturing. Besides such dangerous fashions, Western societies have started to cultivate a tech fetish, which seems to equate new technology with progress. The problem with GU and RU are their core assumption, which holds that such current rules and judgments are for the general good because they are *accepted* by or *conformed to* in a society. This acceptance can often be misplaced; though GU and RU are based on norms that might be common but not right.

For managers who are supposed to be autonomous, acting according to current trends or normative economic models can be a problem. Privately, executives may hold different opinions than what the rules of the day dictate. If they then make decisions in accordance with contemporary GU or RU thinking that are against their own personal judgment, they can become personally alienated from the realities they create.

Very often, however, executives do not hold different opinions from the mainstream. They simply embrace what is currently in vogue. The question here is whether they do so consciously and out of their own free will or whether they really are not mature enough to make independent decisions. As I will show, *wise* leaders are endowed with a rare strength: the moral autonomy and ability to judge independently from fashionable ways of thinking. Whether they can always act responsibly and in line with these judgments is another question unfortunately.

15.5 What Makes a Wise Leader?

The philosophies of GU and RU, which dominate managerial decision making today, do not make things easier for managers. The destiny of companies relies more than ever on the wisdom, foresight, and integrity of companies' top executives. As a result, leaders need to be independent and follow their own judgments instead of hypes and fashions. They need to distinguish hypes from reality. They ultimately determine whether a company acts as a responsible social entity vis-à-vis its employees, customers, stakeholders, and society at large. As Nonaka and Takeuchi wrote in their 2011 *Harvard Business Review* article on wise leadership: "Judgments must be guided by the individual's values and ethics. Without a foundation of values, executives can't decide what is good or bad" (p. 64). But how can such values be cultivated? How can executives be wise enough to make decisions on good and bad? And what is true wisdom? In the following, I want to outline some core aspects of personal wisdom based on what the Greek philosopher Aristotle and the contemporary philosopher Edward Spence said.

Edward Spence (2011) has defined wisdom "as a type of meta-knowledge that is used … to make right judgments … that are of value and good for us in our lives personally … and that are of value and good for others in their lives (ethically good) for the ultimate attainment of … eudemonia" (p. 266). So, wise decisions should've driven by what is good for our eudemonia and others' eudemonia in life (see Chapter 2).

But many modern executives in capitalist societies do not look directly at human life values or eudemonia anymore. Instead, they look at lifeless, abstract, and mediating economic constructs such as productivity, efficiency, unlimited growth, contracting debts, and new technology. They are justified to do so on the grounds of the currently accepted utilitarian theory in capitalist societies that a

healthy economy—measured in terms of these constructs— equals a good life for its citizens. If we take Spence's definition for granted, then leaders in capitalist societies will have a hard time being wise because the system in which they operate prevents them from looking directly at life or eudemonic values.*

Another dimension in Spence's definition of wisdom is "metaknowledge," which is needed for the right judgment. Aristotle made a similar claim, saying that a wise person needs "scientific knowledge, combined with intuitive reason, of the things that are highest by nature" (2000, VI, 1141b). Note though that unlike Spence, Aristotle did not reduce wisdom to knowledge, instead combining scientific knowledge with intuitive reason. So, in Aristotle's perspective, wisdom includes an informed intuition for the "highest in nature."

To distinguish what this highest by nature as opposed to what is lower in nature, we must be able to order things. Aristotle defined practical wisdom as sorting the specific from the general: "Now it is thought to be the mark of a man of practical wisdom to be able to deliberate well about what is good and expedient for himself, not in some particular respect, e.g., about what sorts of thing conduce to health or to strength, but about what sorts of thing conduce to the good life in general" (2000, VI, 1140a–1140b). So wisdom enables us to see order where others see only multiple bits of knowledge as if they were all of equal value. *Sapientis est ordinare* ("Wisdom means to create order") is a point that was later taken up by Thomas Aquinas (1261–1263/1976).

Both Spence and Aristotle see wise decisions rooted in care for the community. Spence talks about what is "of value and good for others in their lives." Wisdom in the Aristotelian sense equally manifests itself in the ability to consider the community of people. He referred to the term *nous* when speaking about wisdom. Note the linguistic overlap between Aristotle's *nous* and the French *nous*, which stands for "us." This recognition of community is an argument for why leaders may find stakeholder processes helpful in wise decision making (Chapter 13).

Terrell Bynum points to another Aristotelian dimension of wisdom. According to him, Aristotle believed that practical wisdom requires the physical appropriation of information within a person's body (Bynum 2006). Looking at Box 19.1 in Chapter 19, we find confirmation in neuroscience that the body plays a key role in understanding the emotional context in which we interact with others. Mirror neurons seem to play a role in our ability to have empathy that arises in response to relationships and the world around us.

---

* I emphasize "capitalist societies" here, because some readers of this textbook may not be from capitalist nations. I do not want to impose the particular challenges of leaders in capitalist societies on readers in nations with other dominant management philosophies.

Since decisions are typically linked to relationships, wise leaders are able to understand how a decision resonates with them. CSR scholars have talked about "gut feeling" (Smith and Lenssen 2009) or "emotional intelligence" (Goleman 2005) to make a similar point.

Finally, and most important, no wise decision can be made without autonomy. "Moral philosophy arises when, like Socrates, we pass beyond the stage in which we are directed by traditional rules … to the stage in which we think for ourselves in critical and general terms and achieve a kind of autonomy as moral agents" (Frankena 1973, p. 4). Unfortunately, autonomy in thinking is a gift not given to everyone. In fact, individuals have been shown to pass through several stages in their lives before they reach the necessary degree of maturity to become truly autonomous individuals. Many individuals never reach this stage. Sociologist David Riesman (2001) identified four maturity stages of an individual's autonomy:

1. The tradition-directed individual
2. The inner-directed individual
3. The other-directed individual
4. The autonomous individual

Riesman's model describes how people start out by being educated by society and groups around them. In stage 1, they absorb traditions and beliefs, and they repeat them mechanically without much reflection. Then comes a moment (stage 2) when these traditions are not mechanistic but internalized and understood. As people grow, some (but not all) will be able to not only accept what they learned, but they will also be able to consider why they hold particular beliefs and why they make judgments in line with them (stage 3). People at this stage become self-aware in a sense. But still they only repeat the traditional school of thought they have internalized. They don't diverge from these traditional rules. Due to this repetition of an outside ideology and a lack of independent thinking, stage 3 is still referred to as an irrational state of being. It is only transcended when people become truly autonomous in the sense that they make their own judgments, build their own theories, and act accordingly. At this stage, people may arrive at conclusions that differ from the mainstream tradition or current trends. For example, autonomous reasoning could lead an executive to not automate a corporate process even if this automation might maximize shareholder value, company profit, efficiency, or any of the traditional corporate management goals. But how can a leader arrive at conclusions or decisions that are outside the mainstream and outside of what utility theory would preach? The next sections will expand on the alternative philosophical concepts of deontology, justice theory, and virtue ethics. Figure 15.1 summarizes the aspects of wise leadership described here above.

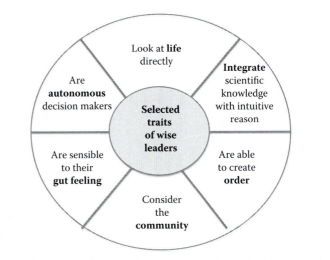

Figure 15.1   Selected traits of wise leaders.

15.6 Deontological Philosophical Reasoning

Deontological ethics is a theory of obligation, the idea that, regardless of any rational calculus, there are certain principles or duties that humans need to live up to. Deontological ethics looks at the internal character of an act itself rather than its consequences. William David Ross, a Scottish philosopher, is a famous proponent of this school of thought. Ross argued that all "plain men" have an intuitive sense of what is right or wrong to do in a situation, a common sense that they inherit and learn from the generations before them. Duties like telling the truth, keeping a promise, or being loyal are examples of what most of us have learned to be moral behavior. Ross believed that this common and learned sense of what is right or wrong is much more meaningful than the attempted systematization of decision making that the utilitarian philosophers proposed. Ross (1930) also thought that pleasure or happiness cannot be the only and ultimate good that determines the morality of an act.

Ross and other deontologists believed that good human relationships ultimately rely on a number of behavioral duties. Ross emphasized five such duties (Skelton 2012):

1. Fidelity (the duty to keep our promises)
2. Reparation (the duty to act to right a previous wrong)
3. Gratitude (the duty to return services to those from whom we have in the past accepted benefits)
4. A duty to promote a maximum of aggregate good
5. The duty of nonmaleficence (the duty to not harm others)

As part of the duty to promote a maximum good, he recognized a responsibility of justice (a responsibility to bring about a distribution of happiness between people in proportion to merit) and a responsibility of self-improvement in terms of virtue and knowledge. Ross called his duties "prima facie duties" instead of obligations. "Prima facie" because there may be situations in which conflicts among duties arise that sometimes prevent an actor from fulfilling all of them. But even if she is not able to fulfill it, her duty to do so still remains. And this duty remains as a fact and speaks against any other decision or behavior (Skelton 2012).

Returning to the airline automation example: According to Ross, an airline executive would have the duty to show gratitude toward those who have worked for her. She would need to weigh the services and benefits she (or the airline) received from those individuals who are on the potential lay-off list. As a result, she may decide to spare some employees. The idea of showing gratitude may sound antiquated and naïve to those who see a company's main goal as profit maximization. If one views a company as a social entity, though, in which people spend their lives and form their identities, then human relationships and the principles that enable them also count. Because gratitude, fidelity, and reparation rely on personal relations with others, Ross regarded these particular duties as even more important and weighty than the duty to promote an aggregate good.

Some critics of Ross argue that his duty theory is not systematic enough. How do we know that his list of duties is complete? And if it is not complete, what mechanism justifies additional duties? In fact, Ross himself did not claim that his list of duties was complete or final. Another deontologist provided us with a much more rigorous method to arrive at ethical judgments: Immanuel Kant.

Immanuel Kant (1724–1804) was a German philosopher who is regarded as one of the most influential thinkers of the Enlightenment. His way of thinking remains dominant in Europe. He wanted to create a universal justification for moral actions. In order for moral justifications to be rational, he anticipated that the consequences of an act, which are subject to volatile ideas of human happiness, cannot serve as a reliable moral guideline (see the earlier criticism of utilitarianism). A moral obligation, which he called a "categorical imperative," can be justified only by something that is a universal principle in itself. So Kant formulated one *Categorical Imperative* that all more specific actions should conform to: "Act only in accordance with that maxim through which you can at the same time will that it become a universal law" (1785/1999, p. 73, 4:421). Note the use of the word "maxim" here. For Kant, maxims are not just values that one can try to live up to. Instead, maxims are a kind of subjective law that can be universalized and upon which one has the duty to act. Take the example of lying to someone: Wanting to tell the truth would not be enough for Kant. In Kant's sense, I have the duty to never lie or to always tell the truth ("Act only according to that maxim").

Kant recognized that our actions are often means to an end, and that people have very different conceptions of what constitutes a valuable end. Therefore, he specified what could be a universal principle for his imperative. He argued that there could only be one universal end: that is human beings in themselves. And for this reason, he completed his categorical imperative with a second rule that stressed human dignity: "So act that you use humanity, whether in your own person or in the person of any other, always at the same time as an end, never merely as a means" (Kant 1785/1999, p. 80, 4:429).

Taken together, Kant's Categorical Imperative says that the duty to perform moral action arises out of the universal law of respect for other people. If I do not lie because I do not want to get caught, then my act is not morally worthy. Why is the motivation to not get caught not morally worthy in Kant's eyes? After all, do not many companies work in such a way today, ensuring only that their practices remain within the confines of the law? This kind of behavior is not enough because the motivation for action is not the right one. The motivation of action—according to Kant—must come from respect for human dignity.

Applying Kant's thinking to our airline case, we would first ask whether we want all airlines to automate all of their check-in systems. Potentially, this question can be confirmed. We may want all airlines to automate their check-in counters, because the machines are quicker and cheaper than human-operated check-ins. At first sight, speed and cost seem to work as universal maxims. Not in Kant's perspective though. If we replace the check-in staff only because automated machines are quicker and cheaper than employees, we would be treating the employees as means rather than ends. We would deprive them of their dignity by comparing the profit we can make with them to the profit we can make with machines. Taken together, Kant's categorical imperative would conclude that check-in automation is ethically difficult to justify. Based solely on an economic rationale, the layoff of staff would be morally reprehensible. Figure 15.2 summarizes Kant's main contribution to ethical thinking.

German philosopher of the Enlightenment (1724–1804)

His main contribution to moral philosophy is the **categorical imperative**:

- "Act only in accordance with that maxim through which you can at the same time will that it become a universal law."
- "So act that you use humanity, whether in your own person or in the person of any other, always at the same time as an end, never merely as a means."

Figure 15.2 Immanuel Kant and his main contribution to ethical thinking.

Interestingly, though, Kant's argumentation can be twisted: An executive could argue that the work at airline check-in counters is boring and demeaning, and thereby also justifies a layoff decision with the Categorical Imperative. The executive would show respect for the dignity of his people and their justified desire to do a more meaningful, responsible, and qualified work elsewhere. This second line of argumentation shows that Kant's Categorical Imperative lends itself to a subjective and paternalistic (if not abusive) argumentation in the name of human dignity. The potential abuse of the Categorical Imperative is one of the reasons why 20th century scholars such as Ross, Habermas, and McIntyre criticized Kant's theory. As previously noted, Ross, came up with very concrete duties that decision makers should follow. Habermas' Discourse Ethics forces individuals to at least defend their moral imperatives in a dialogue with others (see again Section 13.5.3).

An important critique of both Kant's thinking and utilitarianism comes from Alasdair MacIntyre, who published a seminal analysis of the history of ethics called *After Virtue* in 1984. MacIntyre noted that the central problem of moral philosophy is that it relies too much on the formulation of behavioral rules. Indeed, I have already presented several such rules: Bentham's and Mill's act, rule, and general utilitarianism' Ross's duty ethics; Kant's categorical imperative. But "how are we to know what rules to follow?" asks MacIntyre (1995, p. 314). If none of the rules are perfect and all of them lend themselves to abusive argumentation, then one particular virtue gains in relevance: justice.

15.7 Rawls's Theory of Justice and the Veil of Ignorance

Aristotle viewed justice as the most important virtue of political life (MacIntyre 1984). In Section 5.6.3 in Chapter 5, I introduced political theories of justice when I discussed just and fair use of personal information. I showed how the perspective of what is just varies in society. Some argue that the egalitarian treatment of all people is just. Others argue that it is just to treat people and distribute goods according to merit.

John Rawls (1971) developed a thought experiment to uncover just decisions in democratically organized institutions. Rawls asked us to imagine ourselves as a decision maker with the following traits: On the one side, we are rational and self-interested individuals who have some knowledge about human psychology and human nature. On the other side, we know nothing about ourselves: We don't know our sex, race, IQ, talents, social class, positions, and so on. Having these traits, we sit behind a "veil of ignorance" with regard to our own characteristics as we observe a society or a specific situation. We are asked to decide on the rules of the society we observe or decide on the specific actions to

take. Rawls argued that we could use this thought experiment to exclude normal self-interest from our decisions and be more just.

Furthermore, Rawls argued that observers in this position generally agree to two principles (Lamont 1994, p. 6): "Each person has an equal claim to a fully adequate scheme of equal basic rights and liberties, which scheme is compatible with the same scheme for all; and in this scheme the equal political liberties, and only those liberties, are to be guaranteed their fair value. Social and economic inequalities are to satisfy two conditions: (a) They are to be attached to positions and offices open to all under conditions of fair equality of opportunity; and (b), they are to be to the greatest benefit of the least advantaged members of society." These two general principles should constrain the formulation of more specific rules. As basic rights and liberties Rawls listed primary social goods, such as equal freedoms and opportunities, income, wealth, and self-esteem. For cases where rules conflict, principle 1 has priority over principle 2, and principle 2a has priority over 2b.

Rawls's theory of justice is much more extensive than what I can cover here. It was developed to critique and envision just institutions and not necessarily individual decisions. Normally it only applies to democracies; which does not necessarily correspond to today's organizational hierarchies. But even though this focus of analysis is on a higher political level and democratic context, I think we can still transfer some of his thinking to organizations, such as our airline company. Let us imagine ourselves sitting behind the veil of ignorance. We see an airline executive, airline shareholders, airline employees, and so on. We may see also external stakeholders that have been chosen according to Ulrich's methodology (2000) outlined in Section 13.5.2 and Figure 13.2. The airline executive needs to decide whether to automate. She has all the information at her disposal that I gave earlier about financial implications, quality implications, and so forth. Rawls's theory would first focus our thinking on the social and human implications of the layoff decision. Taking the first of Rawls's principles (which is prioritized here), we can argue that a fair decision should benefit all persons who are affected by the automation in an equal way ("the same for all"). Everyone who is affected should have an equal opportunity to benefit from the short-term (and long-term?) profits that are made from automation. Since managers are not affected, they should get no income incentive to automate. Furthermore, one way to view equal opportunity here would be to evenly distribute the full sum saved now and in the future from automation among shareholders and employees. Everyone involved would then receive the same amount of money made. Rawls himself never called for equal monetary rewards. In contrast, he was more concerned about equal opportunities. But one way to apply his thinking in this case could be to distribute financial benefits from automation

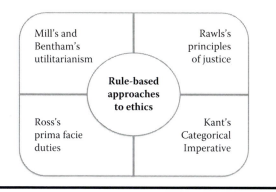

Figure 15.3 Overview of well-known rule-based approaches to ethical decision making.

more equally. The effect of such a financial sharing of wealth might then create better opportunity for those who are affected later.

Rawls's theory of justice is regarded as a contribution to social political liberalism. And MacIntyre (1984) has noted that even though the theory claims a neutral approach to justice (as suggested by the veil of ignorance), it really favors a definition of justice that is oriented toward the satisfaction of the needy. His egalitarian-based, distributive justice is only one approach to justice though. Other philosophers, such as Robert Nozick (1938–2002), did not support Rawls's view of justice. In a case like the airline example, Nozick (1974) would argue that if all people have equal starting positions in life, they should have a legal right to what they earned during their lifetime; people who achieved ownership or shareholder status should therefore have the right to benefit from all the profit that a change makes. Whose theory of justice is right? Again, we face the challenge that different rules can be applied to assess ethical behavior. Figure 15.3 summarizes the various rule-based approaches that I have described.

15.8 Classical Virtue Ethics

Where do we go from here? I will close this chapter on ethical decision making with the postliberal perspective of Alasdair MacIntyre, who has heavily influenced moral philosophy since the 1980s. Here, less emphasis is put on universal rules. Instead, MacIntyre offers an approach to ethical decision making that is based on "classical virtue ethics" as Aristotle conceived it.

MacIntyre (1984) based his analysis of ethical decision making on Aristotle's *Nicomachean Ethics* and philosophers of the classical tradition. I summarized the list of Aristotle's virtues in Table 4.2 (Chapter 4) and described that virtuous behavior aims for a golden mean in one's actions. So, a virtue like generosity finds the right balance between wastefulness and stinginess. And a decision like the one in the airline

example would be influenced by a desire to be generous in this sense.

But why should a decision maker want to be generous? According to classical virtue ethics, the goal of human beings is to be good. Virtue ethicists argue that it is not money or some utility that people want to maximize in life (a fictitious idea found in the *homo oeconomicus* assumption of many neoclassical economic models). Instead, they want to live a virtuous life; they strive for moral integrity. When they come home at night, they want to look in the mirror and feel good about themselves as people.

Now, feeling good about oneself can obviously mean different things to different people. The cardinal virtues that are formulated in classical antiquity and in Christian traditions are prudence, justice, temperance, and courage. Benjamin Franklin established a catalog of 13 virtues including cleanliness, tranquility, and frugality. In her novels, Jane Austen made her characters win through reliability and kindness. Recently, a fashionable virtue for rich people in the United States is philanthropy. So, just as we see a multitude of competing rules among utilitarian and deontological philosophers, we also see many manifestations and interpretations of virtues among virtue ethicists. How do we know what virtues to adapt for ourselves? Which ones are right? Are any right or wrong?

People's practice of virtues can also differ. Originally, in works like Homer's *Iliad*, virtues were characteristics that were attached not to a person but to a social role. Virtuousness required people to fulfill exact predefined social roles. We still find a lot of this thinking today, for example, when we expect members of a royal family to act royally and venerably, think that a professor should be punctual and knowledgeable, and want an accountant to be diligent. These role-attached virtues are complemented by the idea that virtues are attached directly to people and that practicing them is part of a good life. Franklin regarded virtues as useful personal characteristics that serve a means–end relationship to reach a successful and prosperous life (here and in the afterlife). However, as MacIntyre notes, Franklin's perspective on virtues is outward. People are virtuous because they expect some utility from it.

When Aristotle and MacIntyre write about virtuousness, they see it as an inward endeavor: Virtues are characteristics within a person that he or she builds and nurtures through practical experience. And the kind of virtues one should build and nurture are those in line with one's personal history. Aristotle believed that human lives or histories have a *telos*, a higher goal to which we naturally strive. This higher goal is not necessarily a material one but rather asks what kind of person you or I strive to be. Note that I do not talk about the kind of person you or I want to be. The pictures we have about ourselves and to which we want to live up to are often not in line with who we really are. Cultivating virtues in service of a narcissistic self-image is not what is meant here. "Know Thyself" was written on the forecourt of the oracle's temple at Delphi. And it means that a wise person needs to understand who he or she really is, what his or her true story is—not the story that parents, employers, media, and so forth have taught us to believe in and live out. From the perspective of classical virtue ethics, virtuousness means living a life and making decisions in line with who we really are. Contemporary philosophers have called this kind of living and acting in accordance with a personal *telos* "authenticity." In his book *A Secular Age*, Charles Taylor defines authenticity in line with MacIntyre's classical virtual ethics: "[By authenticity] I mean the understanding of life which emerges with the Romantic expressivism of the late-eighteenth century, that each one of us has his/her own way of realizing our humanity, and that it is important to find and live out one's own, as against surrendering to conformity with a model imposed on us from outside, by society, or the previous generation, or religious or political authority" (2007, p. 475).

Authenticity can have various motivations. George Orwell once remarked: "At 50, everyone has the face he deserves" (1968, p. 515). And Gerhard Richter created a portrait gallery where he tried to capture the essence of twentieth century leaders' character (see Figure 15.4). Do we want to have a face at some point in our lives that reflects our authentic self? Or are we going for a face that shows how we betrayed our own better judgments throughout our lives?

What do such wider reflections on virtue ethics mean for IT investment decisions and leadership? Leaders who think about an investment decision or values in their company should first think about who they are, if what they personally think is in line with their convictions and personality. Leaders leave an imprint on organizations. Their personal histories and virtues are embedded into the narrative of the organization they serve. Hewlett Packard and Apple are great examples. A leader like Steve Jobs fiercely shaped the values of Apple and drove its success by emphasizing criteria such as beauty, usability, and inner perfection.* Maak and Pless (2006) show how a sense of care and duty of assistance are two important virtues of responsible leaders today.

Organizations are also embedded in a wider social environment, where they fulfill a certain narrative. For example, Xerox is seen as a legendary Silicon Valley think-tank for true innovation. Companies like Xerox have such a strong history that consumers from all over the world can tell what virtues those companies stand for.

Coming back to my airline example, I had myself a great experience once when I checked in with Singapore Airlines. The competitively differentiating story of Singapore Airlines is that it offers passengers outstanding personal service. All staff on the ground and in the air are wearing exquisitely beautiful uniforms and offer a very welcoming attitude toward its passengers that can be noticed already at check-in. Checking-in to a flight is a truly fun experience, because most of the staff who serve and treat passengers are very beautiful. A top executive at Singapore Airlines who sees himself as part of such a company's tradition and feels right to further nurture and strengthen this airline history, would think twice about whether, how, and to what extent he would adopt check-in counters, no matter how much cost savings they offer. The example may sound simplistic in the face of the grand theories presented in this chapter, but remember that for the executives at Singapore Airlines such a decision costs millions.

Wise leaders understand their own narrative and that of their organizations. They try to make decisions that are in line with this narrative.

15.9 Ethical Decision Making in the IT Selection Phase

So how can these philosophical approaches to wise leadership be integrated into IT selection? The literature on responsible leadership often quotes James Rest's framework of ethical decision making, which is at the core of Figure 15.5 (Rest 1979; Stahl and De Luque 2014). Here, ethical decisions are divided into four phases: a first phase in which the moral

Figure 15.4 *48 Leaders* **portrayed by the German artist Gerhard Richter. (© Gerhard Richter 2015.)**

---

* There is a moving section in Steve Jobs's autobiography where he describes how his father taught him to repair cars and educated him to make a car and technical products as perfect inside as they look outside. Jobs then instructed his Apple engineers to do the same when designing the circuit boards of Apple computers.

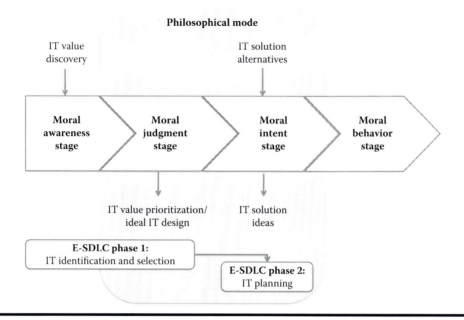

Figure 15.5   James Rest's framework of the ethical decision-making process and the E-SDLC.

significance of an issue is recognized (awareness stage); a second phase in which a moral judgment is made that determines the course of action (moral judgment stage); a third phase that prioritizes values and potentially favors moral values over values such as profit (moral intent stage); and a fourth phase in which moral values are implemented (moral behavior stage).

The moral awareness stage of ethical IT design is when values are discovered. IT innovation teams ask what benefits and harms arise from an IT project and try to understand what values these benefits and harms map to (see previous Chapter 14). Ideally, their ideas are challenged in moderated stakeholder discussions so that they can gain a heterogeneous understanding of the situation and begin to prioritize values. "Moderated" means that the IT innovation team does not conduct the stakeholder group meetings by themselves but instead engage external, professional, and neutral moderators to discuss the proposed IT idea. The innovation team can monitor the discussion by using a video recording or a professional focus group room, where the innovation team watches the group discussion from behind a mirror wall. The result of this phase should be a description of the core values to foster and protect with the new IT.

IT decision makers see the value spectrum and then judge whether and how to proceed. They prioritize the values for their IT solution and formulate expectations for an ideal design. The identified ideal IT alternatives are further scrutinized in the moral intent stage, which corresponds to the IT planning phase described in the next Chapter 16. Here, ethical feasibility analysis requires IT teams to enter philosophical mode again.

15.9.1 Mapping IT Effects and Values

The moral judgment of IT investments can be supported by various ethical approaches. New IT solutions can positively impact a company's bottom line while harming customers, society at large, or both. Socially responsible leaders try to do good and avoid harm through their decisions. But they must do so not only in the short term, for only their shareholders, or for only the operations within their corporate boundaries. Instead, they must reflect on the impact of their decisions for society at large. "A phronetic leader must make judgments and take actions amid constant flux. And he or she must do so while taking a higher point of view—what's good for society— even though that view stems from individual values and principles" (Nonaka and Takeuchi 2011, p. 61).*

When leaders judge investments, it can be helpful to think of IT's consequences in terms of the value landscape depicted in Figure 15.6. Various values might be harmed as a consequence of introducing a new IT system. Potentially, even the act of introducing a system could be wrong. At the same time, other values might be fostered and the act of introduction might be good for those values. Figure 15.6 isolates the way that a single system can have positive and negative effects. Note that the x-axis, which distinguishes positive and negative effects, is not a continuum; "avoiding harm" does not necessarily correspond to "do good." In contrast, the y-axis, which shows the affected parties, is a continuum

---

* Phronetic is based on the Greek word *phronesis*, which means "practical wisdom."

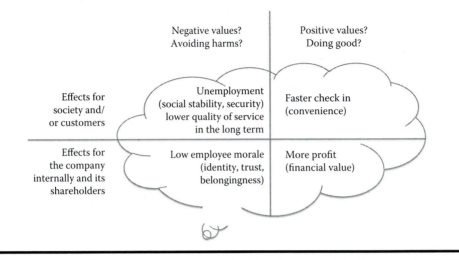

Figure 15.6 Mapping the value effects of an IT investment decision from an ethical perspective (with some exemplary value effects from the airline automation case used in this chapter).

in terms of the number of people affected. Even though we can initially distinguish the effects of a decision for a company's internal affairs from the effects for society at large, companies are—at the same time—part of society. In the long run, the value effects on society will pull companies in as well. For example, it may be a positive short-term strategy for a company to lay off its staff. But over time, unemployment causes instability and security problems in societies, which leads to higher costs for companies.

15.9.2 A Matrix for Ethical IT Decisions

Once decision makers understand the effects of their IT systems, as well as the values they create or destroy, deontological theory and utilitarian thinking might be combined to support IT investment decisions. Figure 15.7 illustrates how the two streams of normative ethical theory can be combined. On the x-axis the value consequences of an IT investment are plotted (as identified in Figure 15.6). On the y-axis the goodness of an IT investment itself is questioned (alongside the Categorical Imperative). If both of these ethical theories signal that a positive situation is created, values are created and the investment act itself is fine, decision-maker can go for the new IT. When both ethical theory approaches signal that there might be a problem with the new IT, then an IT leader should not invest.

On the deontological y-axis it can be hard to judge a new technology because deontological philosophies were originally developed to investigate only human relationships. When we introduce a new technology, this 1:1 human relationship dissolves. The act vis-à-vis another person becomes indirect; it is mediated by a machine. To maintain a deontological line of thinking, we need to ask: Do I want the kind of behavior and

the effects that I trigger through the new technology to become universal? If the technology was universally used in a certain way, would people be treated by the technology only as a means to an end? Or would the technology empower some people to treat others as means rather than ends? Given the importance of these questions, IT teams who make deontological judgments must consider long-term effects of up to at least 15 to 20 years.

An ethical investment dilemma arises when a new IT investment promises multiple positive value consequences, but the act of introducing it is morally difficult (Figure 15.7, lower right quadrant). For example, suppose that an airline could survive financially only if it automated its check-ins. Although the act of laying off employees for cost reasons is deontologically problematic (see earlier), the positive consequence of company survival speaks for the layoffs and

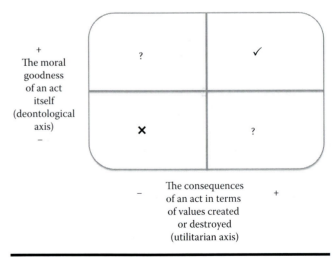

Figure 15.7 A matrix for ethical IT decisions.

automation from a utilitarian perspective. In such a decision dilemma, wise leadership guidance is badly needed. At this moment, the leader's wisdom and virtues come into play as described here above in Figure 15.1. Looking at the life and community of their operations, leaders must listen to their intuition and create order from the multitude of competing information they receive. In this situation, they must challenge the true value proposition of the new IT. Is it really good enough to justify a harmful act? Leaders must be ready to autonomously decide and potentially formulate new company values in line with their personality and the organization's narrative. When leaders commit to an act that is actually not good on moral grounds and can only be justified through positive ends, these positive ends need to be really good ones.

A similarly difficult situation arises when an act may be morally right but has negative consequences for the company (Figure 15.7, upper left quadrant). Here again, leaders need to challenge the nature of the values, which seem to be undermined by an act that is actually the right one to do. IT innovation teams will normally recognize that their initial idea can be refined from various angles. Typically, an IT idea touches upon multiple internal and external, positive and negative values. For instance, one new technology may lead to financial growth, efficiency gains, and more transparency for a company internally. But it may undermine customers' privacy and trust, lead to less secure operations, and put some employees' income as well as their identity at risk. To face such a diverse ethical dilemma, company leaders must decide what values rank first for them.

As a short note of caution: Deontological reasoning and utilitarian judgments overlap when utilitarianism requires a decision maker to judge the consequences of an act from a human dignity perspective. So it could be argued that the two axis in Figure 15.7 are not completely independent. I would still use the quadrant, because in an ethical IT decision context, the defense of human dignity cannot be overemphasized. And if this one value plays into both axes, then this ensures the moral emphasis of the decision-making process.

What is helpful in challenging decision quadrants is to not only formulate and concentrate on value priorities but also to envision broad IT solution alternatives. Because IT is something we create, we have control over how it is built and formed. Consequently, any IT solution can embed multiple values in various hierarchical arrangements. Through appropriate IT design, many harms can usually be avoided. Thinking in terms of alternatives, IT solutions can move a new idea to more favorable decision quadrants while retaining the benefits of the original idea. Think back to the gaming scenario in Chapter 3, where the company Playing the World Corp. placed a new virtual reality game in the real world and hence avoided many

traditional gaming problems such as negative health effects and physical isolation.

Once an innovation team has identified a preferred solution, wise leaders can challenge the solution one more time along the philosophical belief systems by asking the following questions (adapted from Barger 2008):

1. Would I be willing for everyone to be permitted to use the solution I choose? Why? Does this solution treat people as ends rather than as means? Why? (These questions look at the ethical impact of a decision from a deontological perspective.)
2. Would there be a majority agreement that this solution is the best means to an end? Why? Will it produce the greatest good for the greatest number of people? Why? (These questions look at the ethical impact of a decision from a utilitarian perspective.)
3. Is this solution the one I feel most committed to in my own conscience, regardless of whether it benefits me personally? Why? Did I choose this solution in an autonomous manner, as the final arbiter of the good, free from the influence of others? (These questions look at the decision from a virtue ethics perspective.)

15.10 Challenges for Leaders in the IT Selection Phase

The described insights, recommendations, and proposed methods for the IT selection phase are difficult to implement in organizations that swiftly follow hype cycles. The enthusiasm for new technical possibilities and the explosion of innovation opportunities is driving the spirit of IT innovation teams. Going into philosophical mode, in contrast, seems like stepping on the brakes. Yet, I argue that current organizational IT innovation practices often look like pinball machines. New ideas (pinballs) are shot up into the machine and then bump themselves down to failure. Even worse, they clog the pipeline for better ideas. Going into philosophical mode and questioning IT opportunities more thoroughly is therefore the wiser idea.

The main argument against philosophical mode and ethical decision making is time. Everybody in a company will tell you that they have too little time to do the kind of job described in this chapter. Running through various ethical-decision perspectives, organizing stakeholder groups, and mapping values to harms and benefits all takes effort and time. However, even more time and money are lost by pursuing a multitude of thoughtless projects.

When companies do decide to engage in more ethical and thoughtful IT innovation, they often meet various practical and organizational challenges.

15.10.1 Practical Challenges for Ethical Decision Making in the IT Selection Phase

Innovation teams face practical challenges before engaging in discussions with their leaders. The challenges directly relate to the procedures and methods that I described in this chapter. What can go wrong? Two main challenges must be expected: One is that the identification of true harms and benefits, and their mapping to values is not an easy task. Often, IT innovation teams tend to state the obvious and try to justify their IT ideas by aligning them with mainstream values. The teams do not understand the value they actually want to create. A second problem is that innovation teams fall in love with their ideas. They want to defend the ideas they own and tend to embellish them. This practice can lead to an unhealthy bias in project selection. It can even lead engineers to overstate what a new technology can do before they have tested it thoroughly.

15.10.1.1 Stating the Obvious: Catering IT Solutions to Mainstream Values

One of the most important practical challenges in the IT selection phase is to identify the true values created by a new solution. Working for companies and with my students, I have observed that mainstream values such as efficiency, time savings, or cost savings dominate people's thoughts. Most people are initially not critical (in the critical information systems sense described in Chapter 14). Sometimes they justify a new IT capability simply by citing a new buzzword. However, when challenged to explain why efficiency, time savings, cost savings, or the buzzword as such are better than existing solutions, they have a hard time making the argument. As MacIntyre (1984) argued on moral philosophy, people hold values and they argue for them, but they do not know why they hold them and why they are good. It is therefore important for IT project teams to really put themselves in the position of prospective users. Teams must try to emotionally reproduce and envision how the new IT will affect their personal lives and that of customers, stakeholders, and society at large. This exercise then allows them to identify the true potential of a new IT.

An example from my ethical IT class can explain this point in more detail. My students came up with the idea to build "DocBoxes": 2 m × 6 m white sterile cubicle rooms that could be spread out across cities to take blood and saliva tests from citizens. Small and highly automated lab equipment, automated chip-card-based access authentication, smart interface technology, and remote diagnostic services could be bundled to replace some typical nursing jobs. The students were enthusiastic about their idea and made a great effort to explain the technical functionality and appearance of the room. When asked for the value proposition of the DocBoxes, they argued that the benefit would be time savings for patients and better health. However, when I asked them to explain why blood and saliva tests would improve health per se and how time could be saved in a city with many doctors and low waiting times, they had a hard time justifying their solution.

The discussion eventually uncovered what had really moved them to propose the DocBoxes. In fact, one student said, he would be curious to learn about his physical health from time to time, and he would enjoy checking it at regular intervals even when he was healthy. He would feel embarrassed, he said, to take up a doctor's time for such a wellness purpose. He would also appreciate the flexibility of going to a DocBox, potentially using it spontaneously and without any appointment during idle times. It turned out that students wanted to use DocBoxes to raise health awareness and improve proactive health protection with the help of a solution that made people feel good about themselves. So the values they really wanted to create were not health or time savings, as they initially said, but a kind of bodily consciousness among a wider population, which would be supported by a more anonymous and accessible care infrastructure. When we jointly reflected on these true values, we realized that the box should be called "YourCareBox," should include many more tests than they had originally thought of, and should provide feedback with simple health recommendations (food, exercise, etc.).

When teams are not critical and only cite high-level mainstream values that are not connected to personal life experience, the IT that the team produces often fails to address the positive values the IT is really capable of creating. The example shows that innovation teams need to wisely identify the true life-values that are achievable through the IT. A new IT solution can then be conceived to correspond to these values rather than to superficial mainstream justifications.

15.10.1.2 Functional Enthusiasm Leads to Embellishing the New IT Idea

IT project teams often fall in love with their new idea. They are like artists who are extremely passionate about their creations and do not want to hear any criticism. This passion can be dangerous because IT project teams sometimes embellish their solutions and hence bias the feedback they get on it. For example, they may not present the full truth of technical drawbacks to stakeholders or even executives. Sometimes the testing phase for IT is not extensive enough or does not occur for long enough to understand the technology's effects in various contexts and circumstances. A famous case is the presumed benefit of rolling out RFID technology for all kinds of products. For some time, the retail industry believed that RFID tags should be included in all items. Over time, they realized that the benefits from the RFID investment did not pay off at an item level for most product groups.

Sometimes materials (such as photos, videos, etc.) about the new idea create a positive mood for the discussion among investment decision makers. One trick to create a favorable discussion mood is to show videos that dramatize the problem space that is tackled with the solution. Alternatively, stimulating background music or emotionally laden descriptions are used to present the new idea. These tactics are very common, but they are not expedient. They bias and inhibit wise decision making on IT investments.

Summing up, IT leadership in organizations suffers if the solutions presented to them (photos, videos, etc.) are embellished in presentation or based on shallow values. Leaders need to thoroughly challenge the values that IT project teams present to them and ask for presentation material that is neutral and factual. In the IT selection phase, leaders' main task is to identify the true positive and negative values that take priority in a project. They need to engage in what scholars have called a "frontloading of ethics" or "proactive application of ethics" (Manders-Huits and Zimmer 2009, p. 41) so that IT project teams and software engineers later know their priorities in system design. For this purpose, senior executives are advised to witness some of the stakeholder discussions themselves to get a feeling for the proposed solutions. Unfortunately, this kind of focused, substance-driven engagement of top executives is often inhibited by the organizational structures of today's companies.

15.10.2 Organizational Challenges for Wise Leadership

Nowadays, senior executives, or whom I have called leaders in this chapter, have the challenge that they often work in extremely complex, even anonymous organizational structures that are under external (i.e., financial-market driven) pressure to generate profits. In this environment, executives are constantly asked to justify their decisions on financial grounds rather than ethical reasoning. They are asked to execute rather than lead. In this environment, ethical reasoning can seem too scientific, and it takes a significant amount of time. In many companies, ethics has been outsourced or at least delegated to units that have little influence on the actual business, such as the CSR department.

15.10.2.1 Inflated Trust in a Scientific Method for IT Investments

Today's IT project prioritization largely follows an evidence-based paradigm. It condenses products into a presentation format that allows for financial comparison between project ideas. Typically, IT ideas are prioritized based on factors such as their net present value (NPV) and the number of full-time employees needed.

A problem with this approach is that intangible and nonfinancial benefits and costs, including value aspects, are inappropriately excluded from the IT idea's presentation. Furthermore, numbers for portfolio analysis can be tweaked, and they often reflect the enthusiasm of project teams or personal preferences more than the true potentials and pitfalls of a solution. As the conductors of an intensive long-term study of IT prioritization Shollo and Constantiou (2013) state: "The main problems with using financial data and facts as prioritization criteria, according to managers, are the inaccuracy of costs and the unreliability of benefits' calculations. The managers respond to costs and especially benefits' estimations with disbelief and a sense of irony, which become evident during the interviews.… IT project prioritization at the PG (Prioritization Group) meetings is viewed as a 'Turkish Bazaar.'"

15.10.2.2 Leaders' Lack of Time and Alienation from the Decision Base

A further problem is that of company size and the alienation of decision makers from baseline decision making. In a concrete company case that was investigated by Shollo and Constantiou (2013), an IT project team would show its work and ideas to an IT project prioritization group that would again present the work to the actual decision makers, an IT committee with budget responsibility. This three-layered approach is a common way of hierarchically organizing IT selection processes. However, the final decision makers normally do not investigate the project thoroughly enough themselves then. They rely heavily on what lower-level employees decide for them. As a result, leaders' proven experience and intellectual capacity is not fully leveraged. One reason for this is that leaders' schedules are often too full for them to get deep enough into individual IT project proposals. Large organizations demand their attention in too many places. From an ethical perspective, this alienation from the base and limited attention is a problem. Individuals are more willing to contribute to the welfare of others or to a community that is close to them personally and care less about those that are far away (Morris and McDonald 1995). Today's organizations are so large and spread among so many units and countries that leaders have less physical proximity resulting in limited care for employees and consumers.

15.10.2.3 The Idea That IT Is Merely a Support Function

A further problem of IT investments is that some executives still believe that IT is just a support function and not part of the core business. They underestimate how much their business processes and products embed IT, and delegate IT decisions to a technical department. Some of these beliefs are probably driven by executives' fear of learning about technology (see Chapter 2).

15.10.2.4 "Being Good" Is Delegated to the Corporate Social Responsibility Department

Finally, the activities around being ethical are easily put into the CSR box. Although a few companies have realized that CSR must be part of a company's day-to-day mainstream processes (Smith and Lenssen 2009), many companies do not see it this way yet. CSR and ethical thinking is still often outsourced to a CSR department. Potentially, the CSR department could send a representative to stakeholder meetings when new IT is planned. Sometimes CSR representatives work with product managers to create awareness for consumer concerns. But CSR representatives may not always have sufficient influence and knowledge to impact products' value propositions. And a separate CSR department may also raise the impression among IT project teams that ethics is somehow dealt with elsewhere and is not their responsibility.

EXERCISES

1. Based on the robot scenario in Chapter 3, write an essay on whether you would invest in the artificial-intelligence-based, automated sales force of Future Lab. Consider the ethical perspectives described in this chapter and weigh them against an economic and management perspective.
2. Discuss the different uses and definitions of the term *knowledge* that were explained in Chapter 5. Taking a utilitarian perspective, argue whether it is beneficial for a society to adopt an information processing perspective on knowledge.
3. From an ethical perspective, can a company justify using illegitimately collected personal data? Suppose that the company operates in the health industry and could use the data to gain research insight into the development of and propensity for diseases. Does your ethical judgment change?
4. Think of a leader you know whose judgment you value. Then, consider the description of wisdom I provided and review Figure 15.1. Describe the ways that Figure 15.1 relates to the person you thought about.
5. Write an essay in which you compare rule-based approaches to ethical decision making with virtue ethics.
6. Recap the Goethe schools described in the education scenario in Chapter 3. Imagine that Goethe schools were offered technology that would enable them to select their students based on Big Data, in the same way as Stanford Online. First, argue from a virtue ethical perspective whether Goethe schools should embrace the data-driven approach to selecting students. Then, apply utilitarian and deontological reasoning to the same question. Compare your results.

References

Aquinas, S. T. 1261–1263/1976. *Summa Contra Gentiles.* Vols. 1–4. Translated by A. C. Pegis. Notre Dame, IN: University of Notre Dame Press.

Ariely, D. 2009. *Predictably Irrational: The Hidden Forces That Shape Our Decisions.* New York: Harper Collins.

Aristotle. 2000. *Nicomachean Ethics.* Translated by Robert Crisp. Cambridge Texts in the History of Philosophy. Cambridge: Cambridge University Press.

Barger, R. N. 2008. *Computer Ethics: A Case-Based Approach.* New York: Cambridge University Press.

Brynjolfsson, E. and A. McAfee. 2012. *Race Against the Machine: How the Digital Revolution Is Accelerating Innovation, Driving Productivity, and Irreversibly Transforming Employment and the Economy.* Lexington, MA: Digital Frontier Press.

Bynum, T. W. 2006. "Flourishing Ethics." *Ethics and Information Technology* 8(4):157–173.

Dubiel, H. 1986. *Populismus und Aufklärung.* Berlin: Suhrkamp Verlag.

European Commission. 2012. Proposal for a Regulation of the European Parliament and of the Council on the Protection of Individuals with Regard to the Processing of Personal Data and on the Free Movement of Such Data (General Data Protection Regulation). 2012/0011 (COD).

Flanagan, M., D. C. Howe, and H. Nissenbaum. 2008. "Embodying Values in Technology: Theory and Practice." In *Information Technology and Moral Philosophy,* edited by Jeroen van den Hoven and Weckert John. New York: Cambridge University Press.

Frankena, W. 1973. *Ethics.* 2nd ed. Upper Saddle River, NJ: Prentice-Hall.

Frey, B. and A. Stutzer. 2001. "What Can Economists Learn From Happiness Research?" Munich: Center for Economic Studies & Ifo Institute for Economic Research.

Frey, C. B. and M. A. Osborne. 2013. "The Future of Employment: How Susceptible Are Jobs to Computerization." University of Oxford.

Gini, A. 1998. "Work, Identity and Self: How We Are Formed by the Work We Do." *Journal of Business Ethics* 17(7):707–714.

Goleman, D. 2005. *Emotional Intelligence: Why It Can Matter More Than IQ*. New York: Bantam Books.

Hillman, A. J. and G. D. Keim. 2001. "Shareholder Value, Stakeholder Management, and Social Issues: What's the Bottom Line?" *Strategic Management Journal* 22:125–139.

Jonas, H. 1979/2003. *Das Prinzip Verantwortung: Versuch einer Ethik für die technologische Zivilisation*. Vol. 3492. Frankfurt am Main: Suhrkamp Taschenbuch Verlag.

Jones, T. M. 1991. "Ethical Decision Making by Individuals in Organizations: An Issue-Contingent Model." *Academy of Management Review* 16(2):366–395.

Kahneman, D. and A. Tversky. 2000. *Choices, Values, and Frames*. New York: Cambridge University Press.

Kant, I. 1785/1999. "Groundwork for the Metaphysics of Morals." In *Practical Philosophy*, edited by Mary J. Gregor and Allen W. Wood. New York: Cambridge University Press.

Kaplan, R. S. and D. P. Norton. 1996. "Using the Balanced Scorecard as a Strategic Management System." *Harvard Business Review*.

Keynes, J. M. 1933. "Economic Possibilities for Our Grandchildren." In *Essays in Persuasion*, 358–73. London: McMillan.

Lamont, J. 2013. Distributive Justice, in *The Stanford Encyclopedia of Philosophy*. Stanford: The Metaphysics Research Lab. Online Resource. URL: http://plato.stanford.edu/entries/justice-distributive/ (last visited August 12th 2015).

Löwenstein, G. and D. Prelect. 2000. "Anomalies in Intertemporal Choice: Evidence and an Interpretation." In *Choices, Values, and Frames*, edited by Daniel Kahneman and Amos Tversky, 578–596. New York: Cambridge University Press.

Maak, T. and N. M. Pless. 2006. "Responsible Leadership in a Stakeholder Society: A Relational Perspective." *Journal of Business Ethics* 66(1):99–115.

MacIntyre, A. 1984. *After Virtue: A Study in Moral Theory*. 2nd ed. Notre Dame, IN: University of Notre Dame Press.

MacIntyre, A. 1995. *Der Verlust der Tugend: Zur moralischen Krise der Gegenwart, Wissenschaft 1193*. Frankfurt: Suhrkamp.

Manders-Huits, N. and M. Zimmer. 2009. "Value and Pragmatic Action: The Challenges of Introducing Ethical Intelligence in Technical Design." *International Review of Information Ethics* 10(2):37–44.

Mill, J. S. 1863/1987. "Utilitarianism." In *Utilitarianism and Other Essays*, edited by Alan Ryan. London: Penguin Books.

Mingers, J. and G. Walsham. 2010. "Toward Ethical Information Systems: The Contribution of Discourse Ethics." *MIS Quarterly* 34(4):833–854.

Miska, C., C. Hilbe, and S. Mayer. 2014. "Reconciling Different Views on Responsible Leadership: A Rationality-Based Approach." *Journal of Business Ethics* 125(2):349–360.

Morris, S. A. and R. A. McDonald. 1995. "The Role of Moral Intensity in Moral Judgments: An Empirical Investigation." *Journal of Business Ethics* 14(9):715–726.

Nickles, T. 2013. "Scientific Revolution." In *The Stanford Encyclopedia of Philosophy*, edited by Edward N. Zalta. Stanford, CA: The Metaphysics Research Lab.

Nonaka, I. and H. Takeuchi. 2011. "The Wise Leader." *Harvard Business Review*, May, 58–67.

Nozick, R. 1974. *Anarchy, State and Utopia*. New York: Basic Books.

Orwell, G. 1968. *The Collected Essays, Journalism and Letters of George Orwell: In Front of Your Nose*. London: Secker & Warburg.

Pless, N. M., T. Maak, and D. A. Waldman. 2012. "Different Approaches toward Doing the Right Thing: Mapping the Responsibility Orientations of Leaders." *The Academy of Management Perspectives* 26(4):51–65.

Porter, M. and M. R. Kramer. 2011. "Creating Shared Value." *Harvard Business Review* 89(1).

Rawls, J. 1971. *A Theory of Justice*. Oxford: Oxford University Press.

Rest, J. R. 1979. *Development in Judging Moral Issues*. Minneapolis, MN: University of Minnesota Press.

Riesman, D. 2001. *The Lonely Crowd: A Study of the Changing American Character*. 2nd ed. New Haven, CT: Yale University Press.

Rose, J., B. Röber, and O. Rehse. 2012. "The Value of Our Digital Identity." In Liberty Global Policy Series, edited by Boston Consulting Group.

Ross, W. D. 1930. *The Right and the Good*. Oxford: Oxford University Press.

Shollo, A. and I. D. Constantiou. 2013. "IT Project Prioritization Process: The Interplay of Evidence and Judgment Devices." Paper presented at European Conference on Information Systems (ECIS), Utrecht, Netherlands, June 6–8.

Skelton, A. 2012. "William David Ross." In *The Stanford Encyclopedia of Philosophy*. Stanford, CA: The Metaphysics Research Lab.

Smith, N. C. and G. Lenssen. 2009. *Mainstreaming Corporate Responsibility*. Chichester, England: John Wiley & Sons.

Spence, E. H. 2011. "Information, Knowledge and Wisdom: Groundwork for the Normative Evaluation of Digital Information and Its Relation to the Good Life." *Journal of Ethics in Information Technology* 13:261–275.

Stahl, G. and M. S. De Luque. 2014. "Antecedents of Responsible Leader Behavior: A Research Synthesis, Conceptual Framework, and Agenda for Future Research." *Academy of Management Perspectives* 28(3):235–254.

Taylor, C. 2007. *A Secular Age*. Cambridge, MA: Harvard University Press.

Ulrich, W. 2000. "Reflective Practice in the Civil Society: The Contribution of Critically Systemic Thinking." *Reflective Practice* 1(2):247–268.

Waldman, D. A. and B. M. Gavin. 2008. "Alternative Perspectives of Responsible Leadership." *Organizational Dynamics* 37(4):327–341.

Wright, R. 2004. *A Short History of Progress*. New York: Carroll & Graf Publishers.

Chapter 16

Ethical IT Planning

By the end of the information technology (IT) selection phase of the ethical system development life cycle (E-SDLC), leaders have decided to pursue or dispense with a broadly defined IT solution. If they decide to pursue an IT project, they hopefully have a clear value proposition and value prioritization for the idea. They know what positive values they want to create and the potential harms that could be caused. An IT design ideal is the starting point for more concrete IT project planning, which is described in this chapter.

In this planning phase, a company sets up the administrative structures responsible for organizing an IT project; including the IT project team. This IT project team engages in detailed project planning, value conceptualization, and feasibility analysis, all of which inform an initial cost-benefit analysis. Project planning includes a description of the project scope, goals, potential schedule, budget, and resources needed. Value conceptualization refines the values discovered in phase 1. And feasibility analysis looks into whether a project is realistic in the current environment. Figure 16.1 summarizes what is done in this phase.

At the end of this phase, the project team produces a "baseline project plan." Executives use this plan to decide whether an IT project should be accepted, redirected, or canceled. The baseline project plan should contain a system description that outlines alternative implementation solutions, selected configurations, system input factors, tasks performed by the system, and expected output. It also addresses management concerns around value harms and implementation challenges. The latter are based on the conceptual value analysis and the feasibility study.

16.1 Conceptual Value Analysis

When a team engages in conceptual value analysis, the team lists relevant values in a project and deconstructs each value into individual constituents. Figure 16.2 illustrates the idea behind the conceptualization or decomposition of a value by using the example of electronic privacy. A more detailed conceptualization of privacy can be seen in Figure 12.1 in Section 12.1.

Conceptual analysis of a value means to break it down into the subdimensions that constitute its essence. A very helpful way for project teams to deconstruct a value is to first consult the literature about computer ethics and philosophy, marketing or psychology to see whether a respective value has been analyzed there. Sometimes a value is also conceptualized in the law in the form of legal principles. Many values have been studied from a theoretical perspective. When I wrote Chapters 5 to 12 of this book, I used sources from these disciplines to conduct the conceptual analysis of the values that might be useful for IT project teams. But I caution that these chapters should not be overused to fit a case. Each IT system has its own value logic, and project teams must carefully study exactly what value they need to analyze.

Even though it is helpful and important to use existing literature or the law to better understand and deconstruct a value, note that in the recommended sources only a theoretical perspective on a value is provided. This is not enough for an IT project, where every technology and deployment requires a unique, practical, and context-specific value reflection. Remember the example of body scanners at airports, which cause privacy concerns with passengers. With the help of Solove's privacy taxonomy (Section 12.1, Figure 12.1), we might have found that people can be concerned about the *exposure* aspect of privacy. But without further reflection and refinement of that privacy constituent in the scanner context,

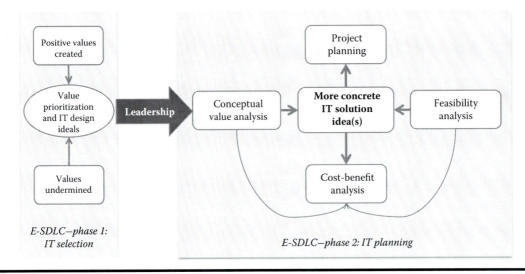

Figure 16.1   Overview of E-SDLC phases 1 and 2 and their tasks.

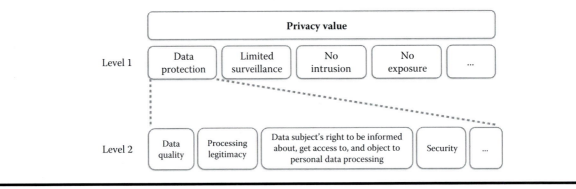

Figure 16.2   Illustration of how to deconstruct and conceptualize a value like privacy.

we would not have realized that passengers actually worry most about the exposure of their entire body, skin and figure, not just their genitals. An understanding of these actual and detailed value concerns drives the success of a technology's requirements specification.

The term *value conceptualization*, which stems from Friedman's work on value-sensitive design (VSD; Friedman, Kahn, and Borning 2006), may be a bit confusing. It seems to suggest that values can be neatly organized and categorized in this phase of the SDLC. This is the ideal, but in practice it is not really the case. "Values are not something that can be catalogued like books in a library, but are bound to each other in complex weaves that, when tugged in one place, pull values elsewhere out of shape" (Sellen et al. 2009). The IT planning phase is crucial to understanding the whole value spectrum and value entanglements. Although Figure 16.2 suggests, for instance, a simple value hierarchy with one or two levels, the value hierarchy or value tree may be more complex. It may have more levels and some of its constituents may have dependencies.

16.2 Feasibility Analysis

Once values are decomposed and their dependencies understood, the broad and ideal IT design that came out of phase 1 usually needs to be refined. The value conceptualization leads the project team to really understand the underlying issues and trade-offs inherent in the value proposition. Value dams and flows start to be recognized and a more concrete and refined IT solution materializes.

As the team researches available technologies, various IT solution alternatives might become apparent. These alternative implementation options support the discovered value landscape and its priorities to varying degrees. For example, a solution that disables radio-frequency identification (RFID) tags after sales take place prioritizes the privacy value over economic values. Furthermore, the team might identify alternative deactivation methods for RFID that appease privacy concerns while giving more room to leverage the technology economically (e.g., rip-tags, password-based deactivation).

When conceptual value analysis is done and one or two favored technical solutions become apparent, the team can begin to conduct a feasibility analysis for the concrete IT solution alternatives that emerge. Feasibility analysis helps to further refine the thinking of the project team around a final solution. Feasibility analysis traditionally contains six dimensions (Hoffer, George, and Valacich 2002):

1. Economic assessment of the financial benefits and costs of an IT project
2. Legal assessment of potential legal or contractual ramifications of the project
3. Political assessment of how key stakeholders in the organization view the project
4. Technical assessment of the development organization's ability to construct the proposed system
5. Operational assessment of whether and how the new IT product or service, and its components will fit into existing operations
6. Assessment of the schedule: Can the project be completed in the necessary time frame to match organizational goals?

To this list, I add ethical feasibility. Ethical feasibility looks into whether the more refined solution is good from a deontological and utilitarian perspective and is in line with the organization's narrative.

Every angle of the feasibility assessment makes a project team think about the available technology from a fresh perspective. And every assessment can lead to further refinements and adjustment of the technology. In the following I give more detail on the different angles of feasibility analysis.

16.2.1 Political Feasibility and Ethical Feasibility

Testing the political feasibility of a new system means evaluating whether key stakeholders will support the proposed system. The ETHICS method I described in Box 16.1 helps to identify the ways that a system will affect employees' job satisfaction and hence its ultimate political feasibility in a corporation. Beyond employees there might, of course, be other relevant stakeholders that must buy into the new technology as well; i.e., customers, the media or supply chain partners.

Note that political feasibility is not necessarily the same as ethical feasibility. Actions that are politically correct and receive organizational buy-in are not automatically "good" in an ethical sense. Political feasibility means that an organization and its partners accept the actions or assume that customers, employees, or other stakeholders will somehow live with them.

Ethical feasibility means that an action or a system trait can be morally justified. Ethical feasibility analysis has not yet been embraced by IT system design textbooks. But as I show throughout this book, it is an important dimension for system design because it emphasizes the value harms and benefits, and hence the good and bad sides of a system. Not everything that is economically and legally doable or even politically feasible is ethically OK.

In fact, once a preferred IT solution has been formulated and refined through conceptual value analysis, project team should run one more time through Barger's (2008, p. 71) ethical challenge questions to see how their refined solution plays out:

1. Are you (the IT project team) willing for everyone to use the solution? Does this solution treat people as ends rather than as means?
2. Would there be a majority agreement that this solution is the most efficient means to an end? Will it produce the greatest good for the greatest number of people?
3. Is this solution the one you feel most committed to in your own conscience, regardless of whether it benefits you personally? Do you choose this solution in an autonomous manner, as the final arbiter of good, free from the influence of others?

16.2.2 Economic, Legal, and Ethical Feasibility

At this stage of system planning, economic feasibility analysis is more detailed than in the initial project selection phase. In the project selection phase, general market trends, analyst studies, or vendor material may have given a broad picture of what one can expect from a new technology in terms of benefits and costs. This second phase requires more fine-grained planning of an IT project for the organization. For companies planning to include new IT in their workflows, cost reduction, error reduction, increased flexibility, increased speed of activity, more management control, and even new sales opportunities may arise. For IT vendors, the size of sales regions and segments can be quantified, and customers' willingness to pay and potential competitors can be anticipated. The company typically tries to estimate the costs of various project alternatives. These estimates can then inform cost-benefit analyses that reveal anticipated net present value, break-even, return on investment, or other financial indicators. As I outlined earlier, ethical issues play into a realistic assessment of these benefits and costs that can be expected from an IT project. I will expand on this issue one more time here, because the IT planning phase is where more realistic cost estimates are made.

An intangible cost is created when customer goodwill, employee morale, or operational efficiency is negatively impacted by a new system. In Box 16.1 I already outlined how employee morale is a problem when new systems are

BOX 16.1 DECOMPOSING THE VALUE OF JOB SATISFACTION FOR CORPORATE IT PROJECTS

One concrete example of conceptual value analysis is Enid Mumford's ETHICS method, which seeks to gain and maintain the human value of *job satisfaction* in highly computerized work environments (Mumford 1983; Stahl 2007).

I add this work on the particular value of job satisfaction here because a major part of IT investments are made to change work processes in corporate environments. However, at least 40% of new IT systems deployed do not make the change happen in their environment that they promised and that initially justified the investment (IBM 2008). Corporate IT investments have therefore seen a lot of criticism in past years; up to the point where Nicholas Carr (2003) challenged the IT industry with his claim that "IT doesn't matter." One reason for the failure rate is that the organizational context and employees' values are respected too little. Seventy-eight percent of the project failures are attributed to the organization not being sufficiently involved with the IT team in defining a project's organizational context requirements and context needs (Geneca 2011). In its 2008 study "Make Change Work," IBM wrote: "We found in our detailed analysis of study results that achieving project success does not hinge primarily on technology—instead, success depends largely on people."

One way to overcome this problem is Mumford's (1983) ETHICS method, which roots in Critical IS. The ETHICS method consists of seven steps that systematically combine business needs with human values: (1) diagnosing business *and* social needs and problems, focusing on both short- and long-term efficiency *and* job satisfaction; (2) setting efficiency *and* social objectives; (3) developing a number of alternative design strategies that fit efficiency *and* social objectives; (4) choosing the strategy that best achieves both sets of objectives; (5) designing in detail; (6) implementing the new system; and (7) evaluating it once it is operational. As becomes apparent from steps 1 to 3, the ETHICS method emphasizes an important value in organizations: employees' job satisfaction.

Mumford (1983) defined job satisfaction as the attainment of a good fit between what employees seek from their work—their job needs, expectations, and aspirations—and what they are required to do in their work. If the intention is to design new systems or alter existing ones in a way that improves job satisfaction, IT designers must identify significant factors that can be measured to check how good or bad the fit is between an employee's job expectation and the requirement of her job, both before and after the system is implemented.

Mumford divided the value of job satisfaction into five kinds of employee beliefs: The belief of an employee in a(n)

1. Knowledge fit, meaning that one's personal skills are being well-used and developed
2. Psychological fit, meaning that one's personal interests are successfully engaged
3. Efficiency fit, meaning that the support services one receives are efficient and adequate
4. Task structure fit, which means that the tasks one does are sufficiently interesting, have variety, give feedback, and leave autonomy
5. Ethical fit, which means that the philosophy and values of the employer do not contravene one's personal values

introduced into company operations. The problem with intangible ethical costs is that management is often driven by an overoptimistic belief in the IT planning phase that the new solution will *somehow* go well. Somehow employees will deal with the new workflow system. Somehow customers will not recognize that they consented to secondary data uses that they actually do not want. Somehow privacy will not play a role in the new genetic fingerprint service. The "somehow-it-will-work-out" reflex is a great entrepreneurial force that leads organizations to implement an idea regardless of the consequences. Project teams and wise leaders should be more realistic. They should take the intangible ethical costs of IT projects more seriously and think about strategies to mitigate potentially adverse effects.

One strategy that is often open to a company is to resolve an ethical conflict or find compromises by altering the way technology is built or deployed. For example, when RFID is introduced on retailers' shop floors, it is possible to deactivate or kill tags at the moment that customers leave the store. This approach addresses most consumer concerns because further tracking or profiling outside the store is no longer possible. However, the retailer would need to invest in a deactivation infrastructure, which add to the cost of the system. So within economic feasibility analysis, companies must ask whether such ethical features in a future system add to the cost of development or deployment. Building a different architecture or adding extra functionality and controls for ethical reasons can increase the cost and development time of a system. This extra time and cost must be broadly anticipated in the IT planning phase and as part of the feasibility analysis and later baseline project planning.

But ethical system traits can also create direct financial benefits. Companies that take their customers' ethical concerns seriously might have a competitive advantage in the long run. A good example is again the case of the body scanning systems. Many manufacturers of body scanning systems competed to supply their solutions to this lucrative market, which is approximately €1.2 billion in Europe alone. As of 2011, not one European body-screening manufacturer was able to supply a privacy-sensitive solution. All of them exposed sensitive body details to airport security personnel. "Somehow," they probably thought, airport operators would not recognize this exposure. However, personal exposure was a serious concern to both airport operators and passengers. Only one company, U.S.-based L-3 Security and Detection Systems, took the privacy concern seriously enough, developing a solution that represents passenger bodies as stick figures (see Chapter 13, Figure 13.8). As a result, this company has held an initial quasi-monopoly on passenger screening systems since 2011.

Privacy does not need to be as deeply embedded into a system's technical design as the L3 body screener. Strong policies and governance procedures are also effective when they are made visible to the customer. Scholars have shown, for example, that online customers are willing to pay a premium for products that are sold on websites with high privacy standards (Tsai et al. 2007). Another option is to build "ethical service packages." We do not know for instance how many people avoid using social networks or are less active on them because of unaddressed privacy concerns, but the ethical angle on this market might be an untapped business opportunity. A study by Krasnova, Hildebrand, and Guenther (2009) found that people would pay Facebook between EUR 0.98 and EUR 3.68 per month if the platform promised to not use the full set of their personal information for secondary purposes.

Economic feasibility is often intertwined with legal and ethical feasibility. Not everything that is doable is legally or ethically feasible. Let us return to the example of secondary uses of personal data. As personal-data markets thrive, the temptation is high to use personal data collected from customers for more purposes than the customers are aware of. Secondary use has become a major source of business for many online companies today; it is even part of their valuation and business planning. Typically, companies require customers to consent to data-sharing practices by signing terms and conditions that are incomprehensible. The companies then generate revenue from those online users that effectively do not read the terms but still sign them (over 90%). If people do not consent, they are effectively excluded from the online service. In this example, economic feasibility—at least in the short term—is a given. Legally, this kind of practice is feasible and tolerated by regulators. However, global surveys show that 80% to 90% of people worry about who has access to their data and are starting to become more conscious about

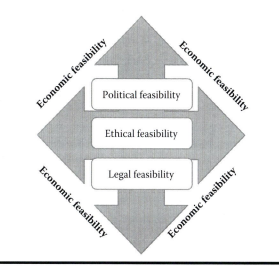

Figure 16.3 Political, ethical, and legal feasibility drive economic feasibility of an IT solution.

the security of their personal data (Fujitsu 2010). Therefore, ethically speaking, these practices are questionable.

A good test for one's ethical feasibility beyond the three philosophical questions is to simply ask: Are we ready to publicize our practices in simple words on the front page of our website? If a company can answer this question affirmatively, it is on a good path. If it cannot, it has an ethical problem.

Figure 16.3 summarizes how political, ethical, and legal feasibility are determinants of economic feasibility.

16.2.3 Technical and Operational Feasibility

After organizations have completed political, ethical, legal, and economic feasibility analyses to refine an IT solution, they can begin to anticipate its technical and operational feasibility.

With technical feasibility analysis, organizations question their ability to construct and deploy the proposed system. Technical feasibility analysis involves anticipating a project's size in terms of project duration; the number of organizations, departments, and people involved; and the amount of development effort (Hoffer, George, and Valacich 2002). With the help of an operational feasibility analysis, the organization analyzes a project's structural implications and how it is embedded into the existing operational technical landscape.

Figure 16.4 shows how Unified Modeling Language (UML) case diagrams can be used to understand the operational context of a new IT system to broadly anticipate what other IT systems or actors will be involved (Kurbel 2008, p. 246). UML typically visualizes the actors and components of a system as well as the communication mechanisms between them.

Users' perceptions and their willingness to use the new technology as well as management commitment can depend

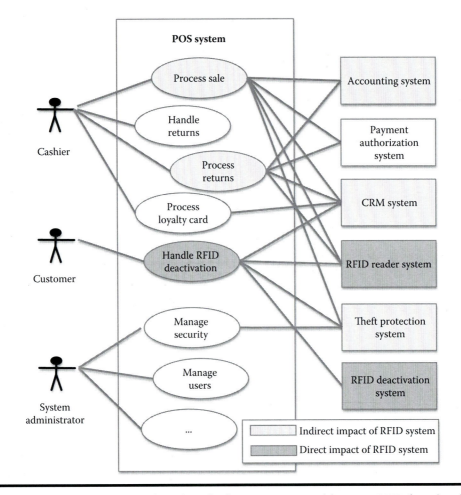

Figure 16.4 Exemplary UML case diagram of a point-of-sale (POS) system with a new RFID functionality.

on how a project is embedded. Changing an established technological infrastructure and operational processes is much harder than starting from scratch.

Finally, technical feasibility analysis must look at the development group and the user group. Is the development group familiar with the target hardware, software development environment, and operating system? Is it familiar with the proposed application area or at least with building similar systems of similar size? How will users react? Are they technically familiar with using similar systems? How will they react to the new system and planned system development? Technical and operational feasibility analysis—if done well—enables understanding of a project's technical and operational risks.

These risks are partly driven by the wider technical and operational context in which new IT operates (Figure 16.5). Today's IT infrastructures are often a patchwork of interlinked systems. Individual systems receive input from other systems that may be hosted within or beyond corporate boundaries. They may pass their information to other systems that are in turn integrated into other workflows. Companies must understand such complex system

landscapes in order to know their data flows, understand the logic of their business and, ultimately, be accountable to their customers. For example, when a new RFID system is introduced to a retailer's system landscape, it must be clear whether that RFID system will be interlinked or is linkable to a customer relationship management (CRM) system. If it is so, then the privacy risks of the new RFID system are much higher than if no CRM system was used or linked.

In recent years, new forms of risks have emerged that are particularly relevant for technical and operational feasibility analysis. For instance, security and privacy risks increasingly play a role in system design because companies are regularly attacked, see identities stolen, and are asked by regulators to run impact assessments before they launch a new IT service. For example, if a company wants to offer an IT-based health service in the United States, it must respect the Health Insurance Portability and Accountability Act (HIPAA). HIPAA defines policies, procedures, and guidelines for maintaining the privacy and security of individually identifiable health information. Planning for such data handling practices influences

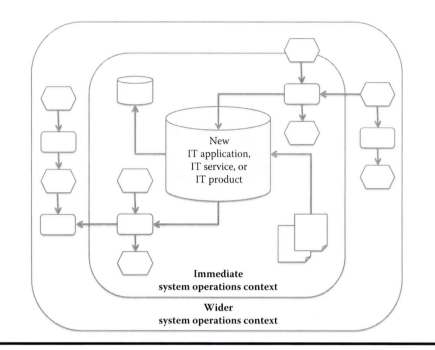

Figure 16.5 A new system has an immediate and wider operational context that must be understood to judge technical and operational feasibility and risks.

project duration, size, budget, and the feasibility of certain technical functions that might have been envisioned by the project team.

The new regulations surrounding the privacy and security of IT systems are probably just precursors for how ethical issues will become a required part of technical and operational planning. If a technology clearly bears ethical challenges, then planners must consider whether the challenges can be effectively addressed through the technology's design. Is it technically feasible to build a value-sensitive alternative system? If it is, what would that solution look like, and how much would it cost? These are questions that must be answered in the planning phase.

The questions should not suggest that a project team will define the details of a technology deployment in the planning phase. This is the responsibility of the next two phases of the E-SDLC: the analysis phase and the design phase. But the project team will have to make an informed judgment in the planning phase of the E-SDLC on whether it is feasible to introduce a new technology.

EXERCISES

1. Compare political feasibility to ethical feasibility.
2. Compare ethical feasibility to legal feasibility.
3. Think of the artificial intelligence system Hal that was described in the robot scenario in Chapter 3. Discuss the impact of this system on employees against the background of Enid Mumford's conceptualization of the job satisfaction value.

4. Think of a system you have used or heard of where a company underestimated ethical feasibility and incurred unanticipated losses. Describe what happened and how the company could have avoided the problems.

References

Barger, R. N. 2008. *Computer Ethics: A Case-Based Approach.* New York: Cambridge University Press.

Carr, N. 2003. "IT Doesn't Matter." *Havard Business Review* 81(5).

Friedman, B., P. Kahn, and A. Borning. 2006. "Value Sensitive Design and Information Systems." In *Human–Computer Interaction in Management Information Systems: Foundations,* edited by Ping Zang and Dennis F. Galletta. Armonk, NY: M.E. Sharpe.

Fujitsu. 2010. "Personal Data in the Cloud: A Global Survey of Consumer Attitudes." Tokyo, Japan.

Geneca. 2011. "Geneca Survey." Company Report, retrieved August 11, 2015, from http://www.geneca.com/75-business-executives-anticipate-software-projects-fail/.

Hoffer, J. A., J. F. George, and J. S. Valacich. 2002. *Modern Systems Analysis and Design.* Upper Saddle River, NJ: Prentice Hall.

IBM. 2008. *Making Change Work.* Somers, NY: IBM.

Krasnova, H., T. Hildebrand, and O. Guenther. 2009. "Investigating the Value of Privacy in Online Social Networks: Conjoint Analysis." Paper presented at International Conference on Information Systems (ICIS 2009), Phoenix, Arizona, December 15–18.

Kurbel, K. 2008. *System Analysis and Design.* Heidelberg, Germany: Springer Verlag.

Mumford, E. 1983. Designing Human Systems—The ETHICS MethodLulu.com.

Sellen, A., Y. Rogers, R. Harper, and T. Rodden. 2009. "Reflecting Human Values in the Digital Age." *Communications of the ACM* 52(3):58–66.

Stahl, B. 2007. "ETHICS, Morality and Critique: An Essay on Enid Mumford's Socio-Technical Approach." *Journal of the Association of Information Systems (JAIS)* 8(4):479–490.

Tsai, J., S. Egelman, L. Cranor, and A. Acquisti. 2007. "The Effect of Online Privacy Information on Purchasing Behavior: An Experimental Study." Paper presented at the 6th Workshop on the Economics of Information Security (WEIS), Pittsburgh, Pennsylvania, June 7–8.

Chapter 17

Ethical IT System Analysis

"We felt so free – but we should have been more thoughtful."

Jaron Lanier (2010)

In the analysis phase of the ethical system development life cycle (E-SDLC), we move from ideas to facts, from initial information technology (IT) solution ideas to concrete development goals and design alternatives.

Literature and experience from the fields of human–computer interaction (Sears and Jacko 2007; Te'eni, Carey, and Zhang 2007), in particular participatory design (Muller 2003), system analysis and design (Hoffer, George, and Valacich 2002; Kurbel 2008), software engineering (Sommerville 2011), value-sensitive design (Friedman, Kahn, and Borning 2006), and risk assessment (European Network and Information Security Agency [ENISA] 2006; ISO 2008, 2014) all provide some insight into how to run through the analysis phase.

The starting point for the analysis phase is the concretized IT solution idea from phase 2. We already know the feasibility of this idea and have analyzed its operational and organizational context. We have also completed a sociocultural context analysis, because we identified stakeholders and relevant values impacted by the IT system idea. In the analysis phase, we now refine this preparatory work.

The first step in the analysis phase is a physical and technical context analysis, which helps us to understand some more relevant *impacts* of a system. Some impacts may lead us to add to the list of values that we need to protect. With this final list of values and the conceptual analysis we did earlier, we can formulate dams and flows for each value, alternatively called "protection goals." For instance, exposing someone's naked body is a dam for privacy. A protection goal would be to inhibit such exposure. Value dams and flows can be challenged and prioritized by a qualitative persona analysis and by a quantitative value dams and flows analysis. In this phase they can also be ranked as

part of a protection demand analysis. The result is a list of what I call ethical development goals. *Ethical development goals* translate value dams and flows into technical language.

Most ethical development goals can be achieved by choosing the right system design alternative. System design alternatives can be modeled and compared in the analysis and system design phase to see which one fits best from an ethical perspective. By choosing an alternative, the team identifies very specific system requirements and a system description, which influence the business case of the new solution. Figure 17.1 gives an overview of the steps in the system analysis phase and the area of overlap with the system design phase.

17.1 Context Analysis

The concrete IT solution ideas that are sketched out in the planning phase are the starting point for a more thorough investigation of the physical and technical context factors that influence the new IT product or service. When the team understands what systems are feasible, it is possible to mentally simulate their insertion into the existing IT environment.

In the E-SDLC, context impact analysis adds a perspective to traditional SDLC context analysis. For traditional human–computer interaction (HCI) context analysis, we look mainly at the environment of a new IT and try to deduce requirements from that environment for the new IT. For instance, if we think that elderly people with reduced vision will use a system, we must adjust the readability of an interface so that they can read the screen. For example, if we foresee that an ATM will be placed in a dark corner of a street, and we know that bank customers want to retrieve cash at night, then we must increase the ATM screen's brightness. Context analysis

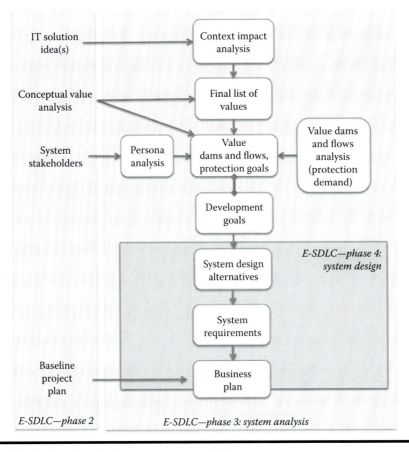

Figure 17.1   Overview of tasks in E-SDLC phase 3: system analysis.

in this traditional sense emphasizes a "computer perspective" (Hassenzahl and Tractinsky 2006), that is the context of a system is analyzed to determine the design of the machine (see Figure 17.2). However, from this perspective, we would not think about adding a street lamp next to an ATM to make cash retrieval at night safer for bank customers.

Supplementing this traditional perspective, *ethical* IT system analysis questions how machines and their design will influence and change their environment, and whether that change is desirable for society (i.e., adding a street pump next to the ATM). Ethical IT system analysis studies the impact of the machine on its context (see Figure 17.2). Langdon Winner (1980) famously framed this matter with the provocative question: Do artifacts have politics?

Indirectly the value discovery phase I described earlier contained a sociocultural and organizational context analysis. When a team envisions a new IT system and identifies the human values that will be impacted, the team automatically considers the system's sociocultural and organizational context effects. Therefore, in an E-SDLC, the sociocultural and organizational context analysis comes earlier than traditional context analysis in the SDLC. The E-SDLC uses the sociocultural and organizational context

to define a new IT's value proposition before considering physical and technical issues.

17.1.1 Physical Context Analysis

Physical context analysis looks into the concrete form factor or interface of a machine, and the implications of this form factor for users in various physical settings. The HCI literature has accumulated a huge body of knowledge in this domain. HCI typically asks where and under what physical conditions a respective IT task will be carried out, for example, whether an ATM needs to be readable in the dark. The *physical engineering* of a machine looks at humans' ability to see, hear, touch, and react to an IT system. Physical IT engineering builds on the traditional engineering field of ergonomics. *Ergonomics* is "the practice of designing products so that the user can perform required use, operation, service, and supportive tasks with a minimum of stress and a maximum of efficiency. To accomplish this, the designer must understand and acknowledge the needs, characteristics, capabilities, and limitations of the intended user and design from the human out, making the design fit the user instead of forcing the user to fit the design" (Woodson 1981, p. vii).

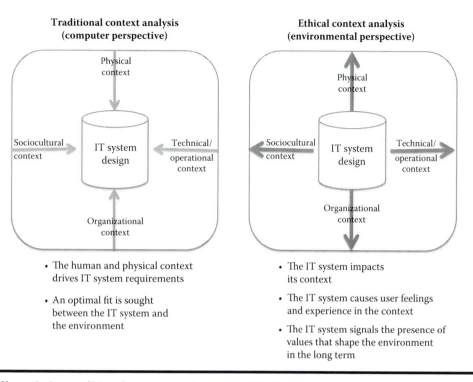

Traditional context analysis
(computer perspective)

- The human and physical context drives IT system requirements

- An optimal fit is sought between the IT system and the environment

Ethical context analysis
(environmental perspective)

- The IT system impacts its context

- The IT system causes user feelings and experience in the context

- The IT system signals the presence of values that shape the environment in the long term

Figure 17.2   Differentiating traditional context analysis and ethical context analysis.

Many scholars argue that the physical design of a system has ethical consequences in its environment. Therefore, they have hinted at the need for the universal usability of IT systems (Shneiderman 2000). *Universal usability* means that systems should be accessible to groups such as the elderly and those with disabilities, and not discriminate between user groups. With this approach, the HCI community has embraced an environmental perspective. This perspective anticipates that normal systems might cause problems for people with disabilities, deeply frustrating them because they cannot use the system. As a result of these potential effects of the IT, many systems are now designed more carefully for universal usability.

To illustrate the difference between traditional context analysis and ethical context analysis, I want to again draw on the body scanner example. With the traditional HCI approach, the team might look at how tall and wide people are physically and then ensure that the body scanner is big enough to fit everyone. The team can anticipate that people who are claustrophobic would need a transparent, open scanner. The physical size of the facility is considered to ensure that body scanners are not too big to fit. In all these cases, the physical context is considered, and designers try to tailor the machine to fit in seamlessly with these conditions.

The ethical context perspective adds to this traditional analysis. It focuses on the user experience of a system that works fine from an instrumental perspective (Hassenzahl

and Tractinsky 2006; Hassenzahl, Diefenbach, and Göritz 2010). A user experience can be negative or positive. Even if a body scanner is perfectly designed from an ergonomic and instrumental perspective, the user experience could still be negative. Perhaps the scanner requires people to raise their hands in a way that causes them to feel like exposed criminals (Figure 17.3). Perhaps the body scanner looks frightening and fully encloses passengers, making them feel like cattle and raising an air of suspicion (Figure 17.4). If the feedback screens of the security personnel are placed such that other passengers can witness the body scans, people may feel shamefully exposed even if full figure details are blurred. The longer-term impact of this kind of scanning breeds negative emotion and unease and signals power and dominance over submissive individuals.

Figure 17.3   People raise their hands to be screened for security. (From Transportation Security Administration, 2010.)

Figure 17.4   People enter a stall to be screened for security. (© CC BY-SA 3.0 Brijot Imaging Systems, Inc. 2010.)

To avoid such negative experiences, the HCI community has proposed two responses. One is more *affective engineering* of the IT system and the other is UX, which stands for user experience but explicitly focuses on creating *positive* human emotions (Hassenzahl and Tractinsky 2006).

The affective engineering of machines aims to develop mechanisms that detect and address negative human emotions (see, for example, Sun and Zhang 2006 or the journal *IEEE Transactions on Affective Engineering*). To begin, machines try to sense and interpret the affective state of their users. For instance, the body scanner could sense the anxiety or happiness of passengers. Affective computing also investigates how machines could optimally respond to such emotions, potentially by expressing emotions themselves. For example, body scanners could display feedback videos that respond to passenger anxiety. Yet, as Hassenzahl and Tractinsky (2006) note, the affective computing community always takes the traditional "computer perspective" (p. 93). It is more interested in how to measure and accommodate emotions *in machines* than in dealing with the emotional impact of machines on people. Therefore, affective engineering is complemented by UX.

UX wants to propagate intrinsic values such as joy, motivation, fun, and beauty. By using affective engineering methods, a designer may know how a user feels; with this knowledge, UX design tries to create a pleasurable experience for users. UX would not just focus on avoiding passenger anxiety (a negative perspective). Instead, UX could turn the security scanning process into a fun experience from the start. For example, UX designers might design a body scanner that looks like a digital shower, with blue and green bubbles or smileys. "UX enriches current models of product quality with non-instrumental aspects to create a more complete and holistic HCI," Hassenzahl and Tractinsky write (2006). So UX clearly takes an environmental perspective. It aims to affect the human context in a positive and pleasurable manner.

Designing for UX can be regarded as ethical because beauty, aesthetics, and joy are clearly associated with the "good" (Eco 2004). At a minimum, aesthetics inserts some poetry into our functional environments (McDonald 2012). However, thinking along the lines of Soren Kierkegaard, UX mechanisms do not go far enough for ethical requirements engineering and context analysis; even a good UX design process will not delve into the sources of negative impacts such as the suspicion, submission, exposure, dominance, or shame that I described for body scanners. UX designers who are asked to come up with body scanners at airports would probably think about how to make the security environment more cheerful and exciting no matter how serious it actually is. And this is good! But this focus is not enough for ethical system analysis. In the E-SDLC, context analysis goes beyond the temporal boundary of the immediate emotional experience. It faces up to the negative—potentially long-term—impacts of a system that may not even be immediately perceived by users. It focuses not only on individual experiences but also on machine effects for a community and societies at large. So we may say that physical context analysis for ethical requirements engineering not only aims to "design for an experience" but to "design for trust," meeting stakeholders' long-term social expectations. Figure 17.5 illustrates the differences between UX design and design with a view to ethical impacts.

I have discussed UX and the ethical impact on context under the umbrella of physical context analysis. I am aware that this classification is not optimal, as UX and ethical impacts are equally driven by the sociocultural context of a system. Take the example of color choice for an interface: Depending on the cultural context, colors trigger very different affective and politically significant responses (Boor and Russo 1993). In the United States, the color red signals danger, whereas in China it triggers happiness. In Japan, white

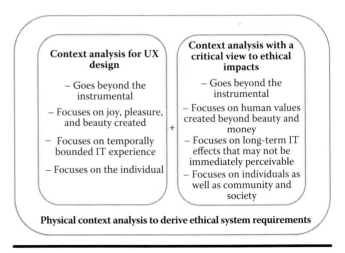

Figure 17.5   A critical view and UX can be combined for ethical requirements engineering.

BOX 17.1 IMPACT OF A ROBOT FORM FACTOR FOR PHYSICAL CONTEXT

The AILA robot depicted in Figure 17.6 was developed to handle tasks on the International Space Station. But what would happen if it was sold for home use? In that case, we might ask why this robot has the form factor of a woman? Does that form factor cement a gender stereotype that women serve at home? Another question is why the robot must remind us of a human being at all? Why not give the AILA robot a funny look with a cartoon face or a neutral one like a box or like C-3PO in *Star Wars*?

The AILA prototype demonstrates that IT's physical designers can intentionally or unconsciously embed personal values in the form factors they create and that early designs in a research lab can be passed on through decades of production. In this fictitious case, where AILA is used as a home robot, the dominant value of gender stereotypes would be deeply embedded into the design.

In their defense, the designers of AILA might argue that they are much like artists. As "creators," they should have the right to determine what their creations look like. Unfortunately, most IT objects are not art.

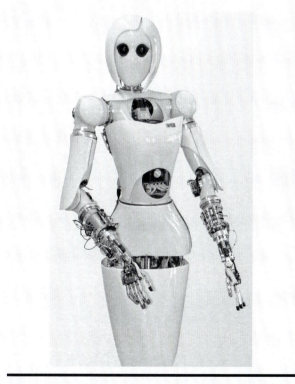

Figure 17.6 AILA robot. (© 2015 DFKI, Alle Rechte vorbehalten, http://www.dfki.de/web/news/cebit2013 /aila.)

They do not come as unique objects or in limited editions. They also typically do not have a signature, a personal name that would ensure a certain degree of accountability. Instead, machines are anonymous artifacts that are often standardized and sold in high volumes. One engineer or IT team's value preference can become a ubiquitous phenomenon influencing users' values around the globe. As a result, IT design teams should challenge the ethical environmental impact of the form factors they put into the world. This is done in the analysis phase.

---

stands for death, and in the West it signals purity.* Color choices in a system's design will be influenced by the physical and sociocultural context of the machine. In turn, that choice influences how the sociocultural and physical context of that system evolves over time. The HCI community has gained a lot of insight into this kind of effect, and I do not want to replicate that knowledge here. Instead the reader is referred to excellent HCI teaching books (see, for example, Dix and Finlay 2004; Te'eni, Carey, and Zhang 2007).

In contrast, I would like to stress that IT systems can shape our long-term value systems by the way they physically present themselves. Box 17.1 discusses the AILA robot and how its form factor influences our perception of gender roles.

17.1.2 Technical Context Analysis

The recognition of the technical context is a prerequisite for IT product and service diffusion. Nowadays, systems can hardly work in technical isolation. The technical infrastructure, adjacent platforms, the installed base of hardware devices and system software, as well as available networks all influence the success of new IT systems. In addition the cost of that technical infrastructure, in particular the cost of its accessibility, arises.

As part of an ethical analysis, we need to identify how the technical context will influence the IT design and how that IT design will then reshape the environment in the long run.

17.1.2.1 How Technical Context Influences IT Design

To understand how important the technical context is for the success of an IT design, I will share the story of a company once called phone.com. In the late 1990s, phone.com (later renamed Openwave) became the first company to offer

---

* For an overview of color meanings, see Jennifer Kyrnin, "Visual Color Symbolism Chart by Culture," About.com, accessed March 14, 2015, http://webdesign.about.com/od/color/a/bl_colorculture.htm.

mobile Internet services. It held the most relevant patents on mobile browsers and the corresponding network infrastructure for providing mobile phone users with online content. In 1999, phone.com went public on NASDAQ; by the year 2000, it had a market value of $7.1 billion.*

Delivering mobile content to phones was technically and operationally feasible. But business managers at phone.com, analysts, mobile operators, and stock investors failed to consider the technical context of the company's product offering. With their small screens, the installed bases of black-and-white handheld devices were in no way suited to accommodate data services (beyond Short Message Services [SMS]). Mobile networks were not ready to transmit mobile content, which made the mobile experience slow and unreliable. For tech-savvy users, who would have tried the service anyway, few mobile applications and very little content were available. And finally, bandwidth was available only at such high prices that most end users feared the financial consequences of using the technology. By 2012, the once-heralded innovator of Silicon Valley had largely dissolved. Phone.com failed because the technical context was not ready to accommodate the company's technical innovation. This case shows how important it is for IT companies and buyers to question the technical context in which a new solution is embedded.

In this case, technical context can be interpreted as the installed base in the market environment. What does the technical market environment in which a new IT system should operate look like? This technical context as market environment is a parameter that IT designers cannot change. The installed base changes only very slowly over time. Engineers can try to anticipate how the market environment will evolve, and they can build their companies' IT products to fit this future. But a common mistake is for engineers is to believe that the future has already arrived. As early adopters, they are often hyped about the speed with which they believe new technologies will penetrate markets.

One would think that an immature diffusion of new technologies into markets has no ethical implications. In reality, it does. As I outlined in Chapter 2, the IT industry is full of immature technical hypes. As a result of these hypes, pressure is put on IT manufacturers to quickly churn out solutions. This pressure leads to less time for IT product development and certainly too little time to think about the ethical implications of a product. When the hype cycle is over and the technology slowly matures, it is often too late to revise initial decisions about data flows and architectures.

Finally, the installed base of IT infrastructure in a market is not the only technical context factor. Technical context is also given by state-of-the-art research and development,

existing technical modules, technical standards, service infrastructures, and the IT landscape within a company.

17.1.2.2 How IT Design Impacts the Technical Context

Although IT design is shaped by its technical surroundings, it also shapes those surroundings in turn. Engineers can heavily influence the long-term technical context for their companies, industries, and society at large.

Two areas of IT design decisions have a particularly great influence on the technical environment: the technical architecture of an IT system and the nature of the data flows within this architecture. Both architecture and data flows significantly influence the ethical implications of systems.

In Chapter 5, I described how the nature and handling of data influences ethical knowledge. I covered data quality, transparency, accessibility, and knowledge. Consequently, I do not cover data handling again in this chapter. Instead, I concentrate on the importance of IT architectural choices from an ethical perspective.

The technical architecture of an IT system identifies the main structural components in a system and the relationships between them (Sommerville 2011). "Architecture in the large" determines where the intelligence of a system resides and what services and controls this intelligence allows for. The way that architecture influences ethical design becomes clear from the robot scenario I described in Chapter 3. Here, Future Lab's robots were compared to Boston Flexible's robots; the former rely on a decentralized, locally controllable system architecture, whereas the latter allow for remote control of the robots.

> Future Lab's robots are embedded with powerful artificial intelligence (AI) technologies including voice, face, and emotional recognition. But these AI functions run independently in dedicated sandboxes contained in the device that does not need to be networked to function. The robots learn locally after their initial setup and so become pretty unique creatures depending on their owners. The decentralized architecture also prevents Future Lab's robots from ever being taken over by a central computer, which increases their security.…
>
> Alpha1 systems look like tall steel men. They can act autonomously and in response to their immediate environment. If needed they can also be taken over remotely by an operator who then embodies the machine.

In the case of Future Lab's robots, the device itself is "intelligent." All processing and reasoning is done within its

---

* *Bilanz, Schweizer Wirtschaftsmagazin,* "No risk no fun mit Internetaktion," February 2000, accessed March 30, 2013, http://www.bilanz.ch/invest/no-risk-no-fun-mit-internetaktien.

proper physical form, and the machine does not rely on a network connection. With the centralized Alpha1 systems, the robot itself has rather low intelligence. All processing and reasoning is done remotely on an external server that communicates with the robot via a network and sends commands that the robot executes.

The robot example illustrates the difference between a decentralized end-to-end architecture (Future Lab) and a centralized architecture (Alpha1). In an end-to-end architecture (e2e), the intelligence of a system is typically located at the "ends" where users put information and applications onto the network (Lemly and Lessig 2000). In my scenario, users can manipulate their robots, develop software for them, and so on. Hence, they are in control of these devices. Their freedom cannot be undermined by a remote takeover of the device. Lemly and Lessig (2000) argue that the electronic world has seen so much digital innovation specifically because the Internet architecture adheres to the e2e principle: people's clients, their personal computers at the network end, are so intelligent that people can use them and be creative with them. The intelligence at the network ends has allowed the Internet to harvest the wisdom of the crowds. People innovate with the help of the powerful machines that are put in their hands. They write new software, share information, and build entire businesses on them.

While the decentralized Internet architecture has been important, it has also led to some drawbacks. In particular, creativity has been put to bad ends, leading to serious security problems. Powerful clients are often abused to create and spread viruses, spam, and all kinds of malware. Attackers abuse the fact that end users cannot always properly protect their powerful computers. As a result, machines are often taken over without users' knowledge, for example, by botnets that become security threats to company operations (in forms such as distributed denial-of-service attacks). People are sometimes harmed directly, as when personal information is stolen from their computers. Powerful clients have led to an environment in which people use computers to control many sensitive operations (such as online banking) even though they are not always up to the security challenges these operations require. For example, people continue to use a small number of simple and unprotected passwords.

Furthermore, copyright conflicts have emerged: Many people have used the portability of media content and the distributed (hardly controllable) hosting environment on the Internet to access and share content for free. Although this sharing of artistic content is an unprecedented opportunity to afford unlimited access to the cultural goods of our societies, artist remuneration seems to be threatened.

Clearly, the architecture of a technical system has huge implications, which can be positive or negative. A decentralized architecture fosters innovation, creativity, learning, end-user control, and privacy. On the other hand, it also creates security threats and copyright infringements. A technical context analysis should try to anticipate such good and bad ends of an architecture and take precautions through system design. For instance, many security problems with end-user clients would never have appeared if the clients had a properly planned security infrastructure.

But engineers must make decisions about more than the technical architecture. In the analysis phase, engineers also decide whether to work with open or proprietary software standards, whether their code will be open or closed, and to what extent the data the technology generates will be portable (Narayanan et al. 2012). Some of these decisions, which were discussed in Chapter 11, affect user creativity, innovation, and control.

In summary, context analysis in the E-SDLC goes beyond what is covered in the SDLC. Whereas traditional development efforts try to optimally fit a system into a context, the E-SDLC anticipates the impact of a working system on its environment. In doing so, the E-SDLC goes beyond UX because it does not focus only on hedonistic or aesthetic experiences of the individual user. Instead, it questions the invisible long-term impacts of a system on its environment, notably individuals, groups, and society at large. It looks into how new IT artifacts affect politics in public and private space. It anticipates how IT architecture influences humans' freedom, control, and creativity. And it analyzes the social effects of physical form factors.

17.2 From Value Dams and Flows to Development Goals

The insights of context analysis add to the value analysis completed in the first two phases of the E-SDCL. From these analytical efforts, we end up with a well-rounded understanding of the full spectrum of values we want to create and need to protect.

Let me recap where we stand if we take up the case of RFID deployment in retail again: we have decomposed values and their meaning for a particular context. We know that using RFID in an intelligent mall environment of the kind described in the retail scenario will lead to privacy concerns and concerns over the freedom to make purchase decisions. We also know that privacy can be seen as a matter of wanting to stay in control of data collection practices. People want to have choice over data collection. They see their privacy at risk if they are tracked after a purchase is completed. They also want to remain free in their purchase decisions while appreciating well-targeted purchase advice that is based on their true preferences. Free purchase decisions imply that they can freely direct their attention, are not manipulated in their purchase decisions by a third-party vendor's sales interests, and can address their second-order desires.

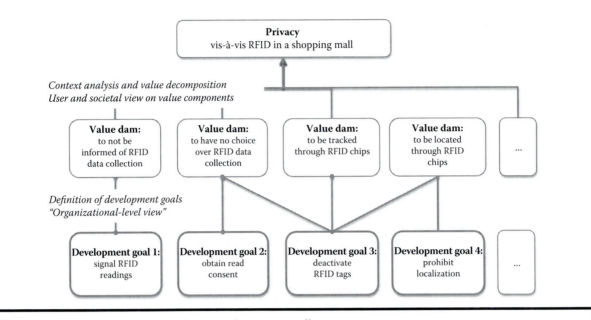

Figure 17.7 Value dams for privacy in the exemplary RFID mall context.

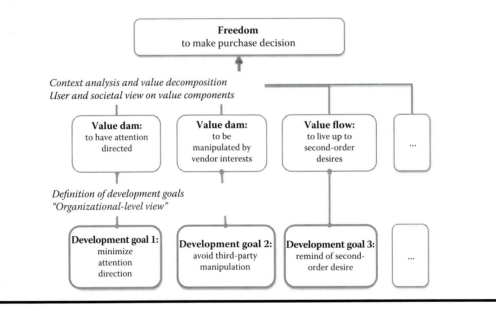

Figure 17.8 Value dams and flows for freedom in the RFID mall context.

With a view to system design, value-sensitive design (VSD) scholars have started to call the concrete conceptual value components "value dams" and "value flows" (Miller et al. 2007). A value dam prevents a value from materializing. A value flow supports the unfolding of a value. In the RFID mall scenario, a dam for the privacy value to materialize would be to track people by using the technology or to locate them against their will, to not inform them appropriately about the RFID's use, and to give them no choice about use of the technology (Figure 17.7). A value flow for the freedom value would be to help people live up to their second-order desires (Figure 17.8). For example, Agent Arthur helps Sophia to avoid buying sweets in the mall.

Value dams and flows need to be addressed in a system's design in order for the system to be successful and reliable in the long term or to be what I call ethical. For each value dam or flow, we can formulate one or more development goals.

17.3 Analyzing Ethical Priorities for System Design

The stakeholders and future users of a new IT system normally have different perceptions of the value dams and flows; as a result, they prioritize development goals differently.

Some will think that goal 1 is more important than goal 2 or 3. Others feel that without goal 3 we should not build the system at all. As every additional system value produces additional development goals, the complexity of an IT project increases. And development goals may also conflict. For instance, the goal to deactivate RFID readers could conflict with the goal to recognize tags for postsale marketing purposes. Before a system can be designed or built, these conflicts and varying priorities need to be understood.

In the IT selection phase I outlined how leaders express views on overall value priorities. For instance, they decide whether privacy is more important than postsale marketing. As a result, the ranking of development goals should be clear to some extent. But often it is not. As system analysis gets into details, stakeholders have room to again take different views on what should be implemented in a new IT system. For ethical system analysis, it is important to uncover these different views and potential conflicts. It is not possible to continue IT design with a blurred understanding of development goals and priorities. Instead, the team must identify different opinions on development goals, underlying values, priorities, and the origins of these priorities. It must be clear who holds what views on a new system, and this information should be shared and reviewed with senior executives. Senior executives can use this information to reflect again on their initial decisions and prioritizations.

To create this transparency and to facilitate and justify later system design, two kinds of analysis can be done: A qualitative "persona analysis" and a quantitative "value dams and flows analysis." Both analyses can inform a rigorous ethical risk analysis.

17.3.1 Qualitative Persona Analysis

Persona analysis is a highly effective way to accommodate different expectations of a system by creating fictive personas to deal with the system. In the E-SDLC, persona analysis must address not only future users of a system but also important stakeholders.

Personas are archetypal users of or stakeholders in a system. They represent the needs of a larger group in terms of their goals, expectations, and personal characteristics. Personas act as stand-ins for real stakeholders and thus help to guide decisions about system functionality and design targets (Pruitt and Grudin 2003). Describing a persona is a storytelling exercise that the project team can use to put themselves in the shoes of their stakeholders. The team can bring these stakeholders to life by giving them names, personalities, and photos (Figure 17.9).

17.3.1.1 What Should a Persona Description Contain?

The typical elements of a persona are a fictional name, job title, and description of main responsibilities. The description

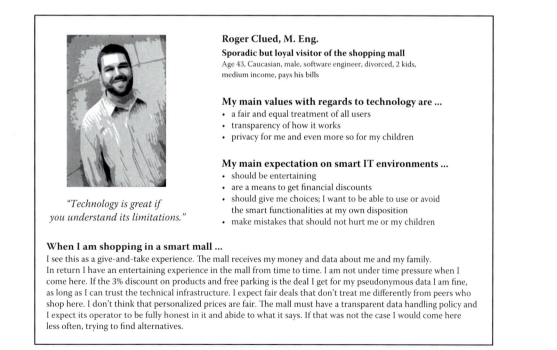

Roger Clued, M. Eng.
Sporadic but loyal visitor of the shopping mall
Age 43, Caucasian, male, software engineer, divorced, 2 kids, medium income, pays his bills

My main values with regards to technology are ...
- a fair and equal treatment of all users
- transparency of how it works
- privacy for me and even more so for my children

My main expectation on smart IT environments ...
- should be entertaining
- are a means to get financial discounts
- should give me choices; I want to be able to use or avoid the smart functionalities at my own disposition
- make mistakes that should not hurt me or my children

"Technology is great if you understand its limitations."

When I am shopping in a smart mall ...
I see this as a give-and-take experience. The mall receives my money and data about me and my family. In return I have an entertaining experience in the mall from time to time. I am not under time pressure when I come here. If the 3% discount on products and free parking is the deal I get for my pseudonymous data I am fine, as long as I can trust the technical infrastructure. I expect fair deals that don't treat me differently from peers who shop here. I don't think that personalized prices are fair. The mall must have a transparent data handling policy and I expect its operator to be fully honest in it and abide to what it says. If that was not the case I would come here less often, trying to find alternatives.

Figure 17.9 Persona description of Roger, visiting the smart mall as described in Chapter 3. (Roger: © CC BY-SA 3.0 Tylerhwillis 2010.)

states what group or segment of users the persona represents. Information on demographics, age, education, ethnicity, and family status form the external description of the character. A casual picture of the persona complements this text description. Some researchers recommend taking this picture explicitly for an IT project and not using ready-made portrait libraries (Pruitt and Grudin 2003) so that the picture represents the user group really well. If a photograph is not practical, the team can use a sketch. A similarly telling part of a persona description is a quote that sums up what matters most to the persona in relation to the new technology. This quote reveals an affective attitude toward the new technology.

Finally, and most important, persona representations contain a description of goals for technology usage as well as the kind of experience and effects that the persona desires. Personas have expectations for IT systems, both in terms of usability and long-term effects. These expectations relate to the values they hold and lead to development goals that are particularly important for that persona. For ethical analysis a persona description should make ethical values very clear. As shown in the persona analysis of Roger in Figure 17.9 it is pulled out that Roger values privacy, transparency, fairness, and honesty.

17.3.1.2 How Personas Can Be Used to Analyze Ethical Priorities

Persona descriptions can be used to address different ethical value priorities and moral expectations for a new system. The stakeholder discussions that took place in earlier E-SDCL phases may have already uncovered different personas. But additional qualitative interviews and market statistics can help to provide details for these descriptions. Take, for example, the privacy value. Many studies have identified four clusters of privacy sensitivity: (1) marginally concerned users who do not care about their privacy, (2) privacy fundamentalists for whom privacy is extremely important, (3) people who readily share their e-mail address and phone number but do not want their behavior to be profiled, and (4) concerned users who do not care about having their behavior profiled but do not want to be identified or give out contact information (Berendt, Guenther, and Spiekermann 2005; Kumaraguru and Cranor 2005). Different personas can be matched with varying value preferences. Personas can also contain expectations about concrete development goals.

Ultimately, the project team must identify the primary persona to cater to and which personas are secondary. Developing a persona to fit everyone's interests is not the point. The number of personas must be kept to a manageable size. Pruitt and Grudin (2003, p. 4) recommend using a maximum of three to six personas.

Companies normally have an idea of their main customer segments and can tailor personas to those segments. The team can discuss conflicting development goals and value trade-offs against the background of these main personas' needs. In this way, it becomes clearer whose values should be prioritized, and what development goals are more or less important.

17.3.2 Quantitative Value Dams and Flows Analysis

An alternative or complement to a persona analysis is a quantitative value dams and flows analysis as proposed by Miller et al. (2007). This analysis is a quantitative empirical approach to understand how strongly future system users believe a given value will be harmed or supported by a new system. Potential value dams and flows are used to build a questionnaire given to a representative sample of future users. To build a meaningful questionnaire, the team must complete the kind of systematic analysis described in Figures 17.7 and 17.8. The team must know the potential value harms and flows. The empirical value dams and flows analysis is then used to understand how important or strong these harms or flows really are for users, and whether they actually qualify as harms and flows. The importance of the harms and flows determines the importance of the corresponding development goals.

Future users' or customers' answers to a value questionnaire give a perspective on priorities for system design and development. Before analysis is conducted, the team can determine a minimum percentage of consent that must be met for dams or flows to produce a development goal. For example, the team might agree that 50% of users must want to be informed about RFID data collection for this feature to be included in the IT system's design. Threshold levels are helpful because IT systems offer so many potential development goals that the team cannot possibly fulfill them all.

Another possible use of quantitative insights is to combine them with persona analysis. From persona analysis, we know whom the system should be catered to, and we also know the sociodemographic traits of the core personas. If the value dams and flows analysis collects information about the sociodemographic traits of the questionnaire's respondents, and personas can be isolated in the group of respondents, then the team can look into how this target group in particular views the value dams and flows.

One short note of caution here: Some value dams are already recognized in the law. For example, data protection law in the United States and Europe states that customers must be informed about data collection practices. So no matter what empirical results are gained from the value dams and flows analysis, providing customers with information must be a development goal for legal reasons.

17.3.3 Using Risk Analysis to Identify Ethical System Design Priorities

Besides VSD design scholars, security researchers and political standardization bodies have worked to understand system design priorities, particularly for system security. These researchers have developed extensive methodologies to address privacy and security risks in IT deployments in particular. Some of their methodological tools can be transferred to ethical system analysis and design (ENISA 2006; National Institute for Standards and Technology [NIST] 2010; Oetzel and Spiekermann 2012; ISO 2014).* Instead of using the terms *value flow* or *value dam*, the security risk community uses the terms *protection goal* or *target for protection* (see, e.g., BSI 2009) to denote those issues that deserve design attention. They would frame the RFID case in terms of customer "goals" that need to be protected; such as the goal to be informed about RFID reads, have choices, and not be tracked or located through the technology. In the following sections I will use the terms value flows and protection goals interchangeably. Value dams are inversed to form a protection goal. The upper part of Figure 17.11 illustrates the direct logical correspondence between value dams and flows and protection goals.

In security analysis, protection goals are prioritized through protection demand analysis. *Protection demand analysis* can be used to classify protection goals into several categories, based on how important it is to preserve or protect a value component. Protection demand can be low, medium, or high (or measured on any other appropriate scale). Later, the choice of organizational and technical controls will need to respond to the level of protection demand identified here (see Chapter 18 on design).

In security risk assessments, protection demand is derived from the amount of risk that is inherent in a system or a system component. Analogically, for ethical assessments, we can equally determine the amount of risk for a value component. For example, how big is the risk that customers will be tracked through RFID?

Generally, riskier systems require more protection. In order to understand the degree of protection demand, we need to understand a system's (or value component's) risk. The amount of risk is derived from three parameters. First is the amount of damage that will occur if a protection goal is compromised or reduced in value. In security risk analysis, the monetary value of an asset is often used to estimate the amount of damage. But damage can also be of a technical, operational, or human nature. The second parameter is whether a protection goal is actually threatened, whether there are realistic ways in which damage can be caused. An attack tree analysis (see Box 18.1 in Chapter 18) can be used to understand these threats. The third parameter is the likelihood of a threat to occur. Not every threat is equally probable. Taking these three parameters together, we end up with a risk estimate that will then determine the level of protection demand for each protection goal (value dam or flow).

Security risk estimation often builds on economic analysis in that it estimates its degree of risk on financial arguments (such as the financial value of the system compromised by an attack). But even in security risk analysis it is often difficult to identify meaningful financial risk indicators, especially when the damage is human in nature. For example, how much human damage was caused for individual users of Sony's PlayStation network when hackers stole users' identities? How much damage would be caused if customers were tracked through RFID technology against their will? As we assess the risk to human values (privacy, freedom, friendship, etc.), it becomes more difficult to anticipate the true extent of damage.

One possible way to approximate the level of protection demand for a value is to simply ask: What would happen if the goal to protect a value was not met (BSI et al. 2011; Oetzel and Spiekermann 2013)? For example, what would happen if RFID chips were abused to track customers? Or what would happen if customers were not informed about RFID? With such questions, protection demand analysis determines the seriousness of the consequences or the damage that ensues if a value is not respected.

The seriousness of damage is certainly different for different personas. And there are multiple forms of damage for every persona. For instance, suppose that the public found out that a company spied on its customers. For the company (persona 1), damage could be measured in terms of a damaged reputation or brand value as well as potential financial loss. For customers (personas 2 to *n*), damage could be measured in terms of the impact on their reputation, dignity, freedom, or financial situation. Figure 17.10 presents a template for analyzing the protection demand of a protection goal and an exemplary line for the RFID case analysis. Depending on the value at stake, a quantitative value dams and flows analysis, a qualitative persona analysis, or both can support informed judgments here.

---

* The European Network and Information Security Agency (ENISA) provides an excellent overview of the approaches used for security risk analysis and their common denominators (see ENISA 2006, 2010a,b, 2011). These documents show how to systematically protect the values of security and privacy. The methodological knowledge accumulated here can be applied to protect other values. I will therefore present how security risk assessment methodology can be used to prioritize and address value risks in general.

Protection demand of dam/flow/ protection goal	Criteria for the assessment of protection demand for a respective dam/flow/protection goal						
Personas	Perspective of persona x (i.e., company)			Perspective of persona y (i.e., customer)			
Nature of Impacts	Impact 1 (e.g., reputation? brand value?)	Impact 2 (e.g., financial loss?)	Impact ...	Impact 1 (e.g., social standing? reputation?)	Impact 2 (e.g., financial well-being?)	Impact 3 (e.g., freedom)	Score
Low - 1 Damage is limited and calculable **Medium - 2** Damage is considerable **High - 3** Damage is devastating	1, 2, or 3?	1, 2, or 3?	1, 2, or 3?	1, 2, or 3?	1, 2, or 3?	1, 2, or 3?	
(i.e., RFID case – level 2)	2 **Considerable** impairment of the reputation of retailer can be expected, because chips are perceived as "spying" on customers.	1 The financial loss is **acceptable** because only 25% of customers are really privacy sensitive and even fewer have the choice to switch to another retailer.		1 The processing of personal data could **adversely** affect the reputation of the data subject, especially when RFID is used to discriminate service quality on the shopfloor based on belongings.	1 The processing of personal data could **adversely** affect the financial well-being of the data subject, but chances are low as long as RFID is not used to discriminate prices.	2 The processing of RFID data could endanger the freedom of customers **considerably** because they feel constantly followed.	

Figure 17.10 Template for thinking through the protection demands, for example, for privacy-related protection goals. The template must be adjusted for specific values and their protection goals. (From BSI, Marie Oetzel, Harald Kelter, Sarah Spiekermann, and Sabine Mull, 2011, BSI PIA Leitfaden, in BSI TR-03126 Technical Guidelines for the Secure Use of RFID, Bonn: Bundesamt für Sicherheit in der Informationstechnik [BSI], p. 17.)

17.4 Operationalizing Development Goals

With the help of persona analysis, value dams and flows analysis, and protection demand analysis, a project team can make an informed judgment about development goals and their priorities. The next step is to see how these development goals can be operationalized. The HCI literature refers to this work as the process by which "organizational-level tasks" are translated into "tool-level tasks" (Te'eni, Carey, and Zhang 2007). At the tool level, the team identifies a system's concrete functions and features or requirements.

Normally, various technical and organizational design alternatives are available to reach a development goal. The lower part of Figure 17.11 suggests some of the development goals for a privacy-friendly, RFID-based shopping mall. Each goal leads to various technical and organizational design alternatives. Sometimes, design alternatives to reach a development goal can be complementary, reinforcing each other. However, the project team normally needs to decide on one design alternative. For example, the development goal to "deactivate RFID chips" can be achieved through various design alternatives: by killing tags, silencing them, physically breaking them, and so forth.

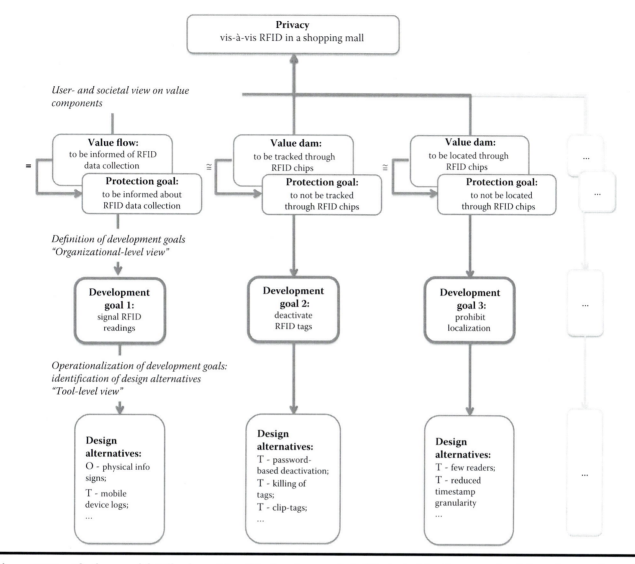

Figure 17.11 **The layers of detail relevant in ethical system analysis, illustrated for the example of the privacy value to be protected in RFID-enabled environments.**

Note that I explicitly call out "organizational design alternatives." Even though this textbook focuses on machine design, we can create "good" environments and positive context effects only if we also consider the governance processes surrounding an IT system. For example, when signaling RFID readings to customers in a mall environment, we must display understandable emblems or logos so that people can recognize and understand that RFID technology is being used. Customers must also have access to informational material (at least on demand and on a website) so that they can understand how a technology is used. Sales personnel in a mall must be able to give informed answers to customer questions about a technology. In the retail scenario, I showed how the father Roger had a very clear understanding of RFID use in the mall. Based on this knowledge, he could choose the right privacy settings for his family. Understanding and transparency can be established only if organizations (such as the mall)

maintain organizational processes to inform customers. Figure 17.11 labels organizational design alternatives with an "O" and technical design alternatives with a "T" at the tool level.

Figure 17.11 gives an overview on how value dams and flows (or protection goals) correspond to development goals and how these development goals are then translated into various system design alternatives. When design alternatives are modeled in detail, the design phase of the E-SDLC starts (see Chapter 18). However, the transition between the analysis and design phase is blurred. Strictly speaking, the analysis phase of system development should already contain various modeling activities for understanding design alternatives. The design alternative chosen by a project team has considerable technical and organizational implications. Most important, the design alternative chosen is key for proper business case planning. As I outlined in Chapter 13, the detailed modeling of functions and processes occurs at the intersection of the IT analysis and design phase. Modeling IT systems in detail is the starting point for spiral-model-based or agile software engineering.

EXERCISES

1. Conduct a context analysis for the robots that are presented in the robot scenarios in Chapter 3. Identify the values that are embedded in robots and how the robots might affect the values in their environment.

2. Choose one value that is particularly threatened by the Alpha1 robots in the robot scenario in Chapter 3. Analyze this value and identify value dams and flows. What development goals would be needed to ensure that the value you identified is not threatened?

3. Choose one value that you find particularly threatened by the use of emotional profiling as described in the gaming scenario in Chapter 3. Analyze this value and identify value dams and flows. What development goals would be needed to ensure that the value you identified is not threatened?

4. Think of the employee-profiling system described in the work scenario in Chapter 3. Conduct a persona analysis of a family member, imagining that your family member is an employee of a company that uses such a system.

References

Berendt, B., O. Guenther, and S. Spiekermann. 2005. "Privacy in E-Commerce: Stated Preferences vs. Actual Behavior." *Communications of the ACM* 48(4):101–106.

Boor, S. and P. Russo. 1993. "How Fluent Is Your Interface? Designing for International Users." Paper presented at Computer Human Interaction (CHI '93), Amsterdam, Netherlands, April 24–29.

BSI. 2009. BSI TR-03126 Technical Guidelines for the Secure Use of RFID.

BSI, M. Oetzel, H. Kelter, S. Spiekermann, and S. Mull. 2011. BSI PIA Leitfaden. In BSI TR-03126 Technical Guidelines for the Secure Use of RFID. Bonn: Bundesamt für Sicherheit in der Informationstechnik (BSI).

Dix, A. and J. E. Finlay. 2004. *Human–Computer Interaction*. Harlow, UK: Pearson Education.

Eco, U. 2004. *Die Geschichte der Schönheit*. Munich, Germany: Carl Hanser Verlag.

European Network and Information Security Agency (ENISA). 2006. "Risk Management: Implementation Principles and Inventories for Risk Management/Risk Assessment Methods and Tools." Cyprus.

European Network and Information Security Agency (ENISA). 2010a. "Emerging and Future Risks Framework: Introductory Manual." Brussels.

European Network and Information Security Agency (ENISA). 2010b. "Flying 2.0: Enabling Automated Air Travel by Identifying and Addressing the Challenges of IoT & RFID Technology." Brussels.

European Network and Information Security Agency (ENISA). 2011. "To Log or Not to Log? Risks and Benefits of Emerging Life-Logging Applications." Athens.

Friedman, B., P. Kahn, and A. Borning. 2006. "Value Sensitive Design and Information Systems." In *Human–Computer Interaction in Management Information Systems: Foundations*, edited by Ping Zang and Dennis F. Galletta. Armonk, NY: M. E. Sharpe.

Hassenzahl, M. and N. Tractinsky. 2006. "User Experience: A Research Agenda." *Behaviour & Information Technology* 25(2):91–97.

Hassenzahl, M., S. Diefenbach, and A. Göritz. 2010. "Needs, Affect, and Interactive Products: Facets of User Experience." *Interacting with Computers* 22(5):353–362.

Hoffer, J. A., J. F. George, and J. S. Valacich. 2002. *Modern Systems Analysis and Design*. Upper Saddle River, NJ: Prentice Hall.

ISO. 2008. ISO/IEC 27005 Information Technology—Security Techniques—Information Security Risk Management. International Organization for Standardization.

ISO. 2014. ISO/IEC 27000 Information Technology—Security Techniques—Information Security Management Systems—Overview and Vocabulary. International Organization for Standardization.

Kumaraguru, P. and L. F. Cranor. 2005. "Privacy Indexes: A Survey of Westin's Studies." Pittsburgh, PA: Carnegie Mellon University, CMU, Institute for Software Research International.

Kurbel, K. 2008. *System Analysis and Design*. Heidelberg, Germany: Springer Verlag.

Lemly, M. A. and L. Lessig. 2000. "The End of End-to-End: Preserving the Architecture of the Internet in the Broadband Era." Stanford Law School, Stanford, California.

McDonald, W. 2012. "Kierkegaard, Soren." In *The Stanford Encyclopedia of Philosophy*. Stanford, CA: The Metaphysics Research Lab

Miller, J., B. Friedman, G. Jancke, and B. Gill. 2007. "Value Tensions in Design: The Value Sensitive Design, Development, and Appropriation of a Corporation's Groupware System." Paper presented at GROUP, Sanibel Island, Florida, November 4–7.

Muller, M. J. 2003. "Participatory Design: The Third Space in HCI." In *The Human-Computer Interaction Handbook*, 1051–1068. Hillsdale, NJ: Lawrence Erlbaum Associates.

Narayanan, A., S. Barocas, V. Toubiana, H. Nissenbaum, and D. Boneh. 2012. "A Critical Look at Decentralized Personal Data Architectures." Cornell University, New York.

National Institute for Standards and Technology (NIST). 2010. "Risk Management Guide for IT Systems." U.S. Department of Defense.

Oetzel, M. and S. Spiekermann. 2012. "Privacy-by-Design through Systematic Privacy Impact Assessment: Presentation of a Methodology." Paper presented at 20th European Conference on Information Systems (ECIS 2012), Barcelona, Spain, June 10–13.

Oetzel, M. and S. Spiekermann. 2013. "Privacy-by-Design through Systematic Privacy Impact Assessment: Presentation of a Methodology." *European Journal of Information Systems* 23(2):126–150.

Pruitt, J. and J. Grudin. 2003. "Personas: Practice and Theory." Paper presented at Conference on Designing for User Experiences (DUX '03), San Francisco, California, June 5–7.

Sears, A. and J. A. Jacko. 2007. *Computer–Human Interaction: Development Process*. Boca Raton, FL: Taylor & Francis Group.

Shneiderman, B. 2000. "Universal Usability." *Communications of the ACM* 43(5):85–91.

Sommerville, I. 2011. *Software Engineering*. 9th ed. Boston: Pearson.

Sun, H. and P. Zhang. 2006. "The Role of Affect in is Research: A Critical Survey and a Research Model." In *HCI and MIS Foundations*, edited by Ping Zhang and D. Galletta. Armonk, NY: M. E. Sharpe.

Te'eni, D., J. Carey, and P. Zhang. 2007. *Human Computer Interaction: Developing Effective Organizational Information Systems*. New York: John Wiley & Sons, Inc.

Winner, L. 1980. "Do Artifacts Have Politics?" *Daedalus* 109(1):121–136.

Woodson, W. E. 1981. Human factors design handbook: information and guidelines for the design of systems, facilities, equipment, and products for human use McGraw-Hill.

Chapter 18

Ethical IT System Design

"The better is the enemy of the good."

Voltaire (1772)

The last phase of the ethical system development life cycle (E-SDLC) that I cover in this textbook is system design. Work in the information technology (IT) system design phase builds directly on the outcomes of the analysis phase, in particular the systematic analysis of protection goals (value dams and flows; see Figure 18.1). To ensure a systematic ethical IT design, protection goals are analyzed with a view to how they can potentially be threatened. What conditions or even malicious attacks could undermine the protection goals? Or, in other words, undermine value flows and constitute value dams? Once these *threats* to protection goals are identified and described in detail, an IT project team can work to understand technical and organizational measures to address them. For each threat, appropriate *controls* are chosen. These countermeasures are then prioritized and further refined. The IT team can visualize some controls in a modeling language such as Unified Modeling Language (UML), which promotes a better understanding of system complexity. With a prioritized control list and visual model, the team can make choices about what the new IT system should look like. The models and controls are the starting point for iterative system implementation (software and hardware engineering).

18.1 Threat and Control Analysis

In the analysis phase, ethical development goals have been formulated based on protection goals or value dams and flows. These development goals must now be addressed through various design alternatives (Chapter 17, Figure 17.11). A highly systematic way to identify and chose design alternatives is to transfer some essential parts of risk assessment methodology to the system design. From a risk assessment perspective the central question asked is: what technical and organizational threats could undermine the protection goals? And how can these threats be mitigated or controlled? The mitigation measures taken constitute the ethical part of the final IT design. For example, what (technical and organizational) threats could effectively undermine people's desire for privacy and not respect their desire to avoid tracking through their radio-frequency identification (RFID) chips? In threat analysis, IT experts and software engineers use their expertise to identify the characteristics of a technology that impact protection goals. Every single value dam and flow (or protection goal) undergoes a threat analysis. And for every threat, engineers and IT project team members then estimate how likely it is for that threat to materialize. In the end, a system's design should include a control or mitigation strategy for every threat that is sufficiently likely to occur (Figure 18.2).

So let us take the RFID retail example again to understand this methodology in detail. In Chapter 17, we recognized that a dam to people's privacy is to track them by using RFID technology (Figure 17.7). The protection goal is to not track them. There are several high-level threats to this protection goal. First, a smart RFID infrastructure in a retail mall could read the unique serial number of tags (EPCs, or Electronic Product Codes) and use a sequence of these readings to develop customer "tracks." Second, the mall could link these tracks to individual customers if the customers use loyalty cards. These two high-level threats are likely to occur because most retailers use loyalty cards explicitly to understand customer behavior. And analyzing customer tracks in retail outlets has been a common marketing practice for a long time.

These high-level threats must now be broken into subthreats. A helpful methodological tool to understand subthreats is the attack tree analysis, which I present in Box 18.1. As you can see from the attack tree in Figure 18.3, the threat of producing an RFID track has various technical subthreats or sources. The most important ones can be understood by looking at the lowest attack tree leaves, which show the technical origins that enable a threat to materialize. In the

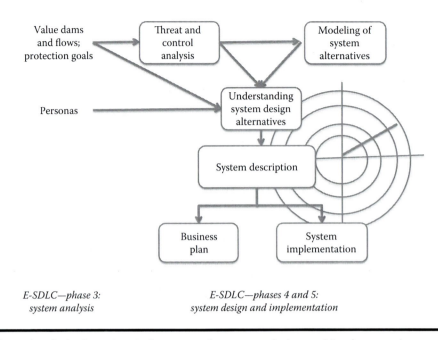

Figure 18.1 Overview of tasks in the E-SDLC phases 4 and 5: system design and implementation.

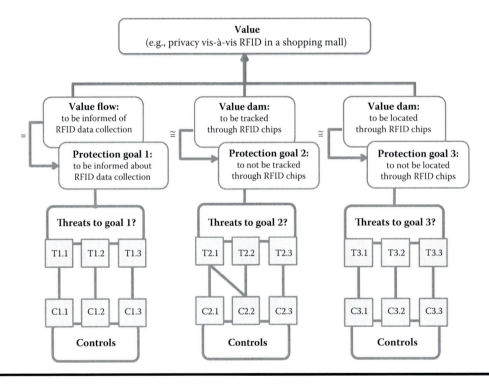

Figure 18.2 Every threat to a protection goal is mitigated by an effective control.

example, being able to read an RFID tag's EPC from a distance is the core enabler for building tracks. To mitigate fears of being tracked, EPCs could be built so that they cannot be read from a certain distance. Indeed, there are many technical design alternatives that impede EPC readings from a distance. One is to use frequencies for RFID tags that reduce the read range to a meter (13.56 MHz chips) or less (NFC

chips). Other options include killing EPC tags, making them password protected or breaking their antennas once an item is sold. These latter design alternatives would allow retailers to track RFID throughout their supply chains and in their own stores, but not beyond. Box 18.1 shows how to build attack trees to facilitate threat identification. In fact, it applies this theory to our running RFID example.

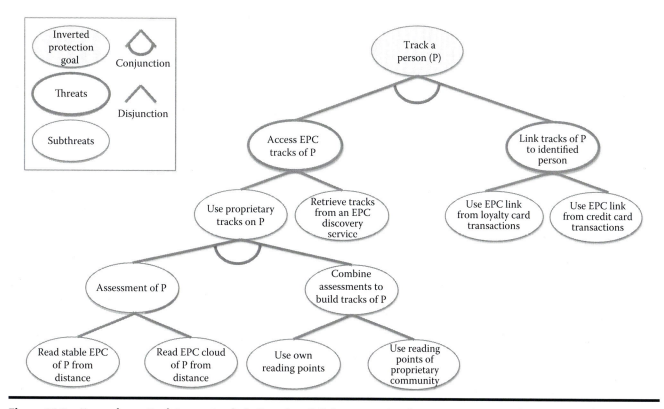

Figure 18.3 Exemplary attack tree extended. (Based on Spiekermann, Sarah, 2008, *User Control in Ubiquitous Computing: Design Alternatives and User Acceptance*, Aachen, Germany: Shaker Verlag; and Spiekermann, Sarah, and Holger Ziekow, 2006, "RFID: A Systematic Analysis of Privacy Threats and a 7-Point Plan to Address Them," *Journal of Information System Security* 1(3):2–17.)

BOX 18.1 EXAMPLE OF ATTACK TREE ANALYSIS: TRACKING A PERSON WITH THE HELP OF RFID TECHNOLOGY

Attack tree analysis can be used to systematically explore the technical feasibility of concrete attacks on protection goals.[1] The protection goal is inverted for attack tree analysis. As you see in Figure 18.3, the goal of the attack is to track a person. The inverse of this attack is the protection goal, which is to not be tracked. The inverted protection goal is the root of the attack tree. The IT project team thinks about how the attack on its protection goal would be realized. The inverted protection goal is hierarchically disassembled into subgoals attackers must reach. Subgoals need to be technically achievable in combination or alternatively (conjunction or disjunction). The subgoals are analyzed systematically to identify critical aspects of the technology.

The attack tree shows how tracking can be technically achieved with RFID. Identifying data is linked with individual tracks of movement. This data is collected when a person pays electronically or uses a loyalty card. Second, individual tracks of movement need to be recorded and combined with that identity information by using the product's EPCs. Because they have unique serial numbers, EPCs are suited to serve as identifiers. They can be the basis for proprietary tracks built through self-recorded read events of a collecting entity (i.e., a retailer). Or, they can be used to retrieve tracks that spread across locations.

Assessment of EPCs would be done for objects that are regularly carried by a person over a longer period of time, for example, wristwatches or purses. This long-term EPC could serve as a kind of cookie (HTTP cookies are strings of information used by web servers to reidentify browser clients in the context of web-based services). In the RFID context, read events referring to EPCs in a similar way could be put in sequence by using reader timestamps and location data.[2] As a result of this process, a track is established. If no long-lasting EPC is available, a cloud of EPCs (an EPC combination profile) could also be used as an identifier.

When tracking is desired beyond the premises of a collecting entity and is to be done on a regional or even global scale, the EPCglobal Network could be used to gain access to geographically dispersed read-event points. The example tree is based on the assumption that the global barcode organization GS1 can widely implement its current standards for RFID item-level tagging, including its initial vision for a global network in which EPCs are regularly collected and shared among industry players.[3] Broadly speaking, this sharing of EPC resources would need to be facilitated through the EPC discovery service mentioned in the tree.

All in all, the analysis shows that the current specification of the EPC (with a globally unique serial number) and the EPCglobal Network technically allow for the creation of tracks.

NOTES

1. Bruce Schneier, "Attack Trees," *Dr. Dobb's Journal* 24, no. 12 (1999): 21–29.
2. Christian, Floerkemeier, Dipan Anarkat, Ted Osinski, and Mark Harrison, *PML Core Specification 1.0* (Cambridge, MA: Auto-ID Center, Massachusetts Institute of Technology [MIT], 2003).
3. Global Commerce Initiative (GCI), "Global Commerce Initiative EPC Roadmap" (Cologne: GCI, Metro Gruppe, IBM, 2003).

In attack tree analysis, threats do not reside only in technology traits. As you can see in the attack tree in Figure 18.3, decisions are made to "*use* proprietary tracks," to "*retrieve tracks* from an EPC discovery service," and, most important, to "*link* tracks to an identified person." All of these decisions are not technical traits; they are a matter of governance and management. IT managers and general managers in a company need to actively decide to use, retrieve, and link. Although technology can be built with affordances that allow for bad decisions, it is the engineers' responsibility to build technology in a way that does not give managers too many options for abuse. If the potential for technical abuse exists, IT managers and general managers are the ones who need to decide to use the technology in a way that has adverse consequences. Wise leaders will not do so.

Attack tree analysis relates to security analysis, where the goal is to understand IT systems' detailed vulnerabilities. Therefore, an attack tree should include all potential paths to a successful attack, even those that are costly, difficult, and rather unlikely. For ethical threat analysis, the subthreats (listed in Figure 18.4) are all technically possible. But they are not necessarily equally likely in the day-to-day operations of a normal business environment. Ethical threat analysis includes a judgment on how likely a threat is to materialize in a normal business environment. The likelihood of a threat depends on its cost and technical feasibility at the time of analysis. For example, in the RFID context it does not seem likely as of 2015 that a global EPC discovery service will be available as the global barcode organization GS1 once planned it. Therefore, the threat analysis table judges it as unlikely that an operator will be able to use such a service to retrieve global EPC tracks. Similarly, many technical problems (like the reflection of radio waves) make it hard to read tag clouds in a way that would allow operators to reliably assess individuals. Because these likelihood judgments can change over time, threat analysis must be repeated from time to time after deployment to see whether changes are necessary to maintain value-based design.

Figure 18.4 illustrates how threats and subthreats can be systematized for IT system planning and how these threats relate to the values and protection goals identified earlier (see Chapter 17, Figure 17.11). Tables and a numbering scheme support systematic ethical analysis.

Once the threat analysis is completed, the team develops a design to meet and mitigate each threat. In security-risk assessments, we speak about controls or mitigation strategies. Figure 18.5 illustrates how individual threats and threat codes can be aligned with controlling design alternatives. In Figure 18.5, engineers can choose from various layers of controls. The design alternatives can provide for low, medium, or high protection. Ideally, the level that is chosen should correspond to the protection demand category discussed in the analysis chapter (Chapter 17). Remember that we used various forms of analyses (persona analysis, value dams and flows analysis, and risk analysis) to understand how important a protection goal (value dam or flow) is for users and other stakeholders. Depending on this importance, engineers can now choose the respective level of protection.

When engineers delve into the details of alternative IT design options, they need to fully comprehend the ethical implications of each option. They need to weigh the expectations of stakeholders as much as they need to understand the practical, functional, and financial implications of various design alternatives. The design phase is therefore the time when the project team must model system design alternatives in detail and—after some iterations—decide on the final system.

18.2 Modeling and Understanding System Design Alternatives

IT teams can use various modeling techniques to see how design alternatives can be operationalized. For ethical analysis, I find sequence diagrams in UML helpful because they

Value code and name		Protection goal code (Vx.x) and name (Code for value dam or value flow)	
V1	Privacy	V1.1	to be informed of RFID data collection
		V1.2	to have choice over RFID data collection
		V1.3	to not be tracked through RFID chips
		V1.4	to not be localized through RFID chips
		V1.5	...
V2	Freedom	V2.1	to have attention directed
		V2.2	to not live up to second-order desires
		V2.3	...

Protection goal V1.3 is threatened by T1.1, T1.2, etc.

Threat code and name		Subthreat code	Description of subthreat	Likely?	Associated protection goals	
					Dams	Flows
T1	Build EPC tracks	T1.1	The operator has access to an EPC discovery service that allows retrieving EPC tracks	no	V1.3 ...	
		T1.2	With proprietary read points the operator can assess a person	yes	V1.3 V1.4 ...	
		T.1.3	An assessment of a person can be combined to form a track	yes	V1.3 ...	
		T1.4	RFID readers can read a stable identifier (i.e., EPC) from a distance	yes	V1.1 V1.3 V1.4 ...	
		T1.5	RFID readers can read consistent ID clouds from a distance to assess a person	no		V1.1 V1.3 V1.4 ...
		T1.6	The operator has access to RFID read points of a community (i.e., mall) to build a track on a person	yes		V1.3 V1.4 ...
T2	Link EPC tracks to identified person	T2.1	The operator can use EPC link from loyalty card transactions	yes		V1.3 ...
		T2.2	The operator can use EPC link from credit card transactions	no		V1.3 ...
		T2.3	...			...

Figure 18.4   How threats to a value are systematically broken down and judged for their likelihood.

are sometimes less visually complex than process models. UML sequence diagrams are an effective way to visualize how machines will do certain tasks, in what order, by using what system components, implementing what methods, using what data, and so on. More important, we can clearly see what interactions take place, particularly the degree of user involvement in an IT system's operations. Ethical systems strive to give users an optimal level of control.

The Object Management Group (OMG 2011), which maintains the UML standard, provides ample advice on how to draw UML diagrams. UML, in particular case and sequence diagrams, are relatively easy to read and intuitive to learn as a modeling technique, even for nontechnical project members. This ease of use allows for joint application development (JAD) sessions (Hoffer, George, and Valacich 2002), where engineers and nontechnical staff can discuss UML

Control code	Available levels of control	Description of design alternatives that control each threat	Threat(s) that are addressed
Cx.x		General description of technical and organizational controls that can serve to address a subthreat; controls can be more or less strong	Sub-threat code Tx.x
	Low	Weak control	
	Medium	Medium control	
	High	Strong control	

Example

Control code	Available levels of control	Description of design alternatives that control each threat	Threat(s) that are addressed
C1.1		**RFID tag deactivation**	**T1.4: read EPC from distance**
	Low	Use agent model to deactivate tag	
	Medium	Use password model to deactivate tag	
	High	Use kill command to deactivate tag	
C1.2		**RFID tag read range**	**T1.4: read EPC from distance**
	Low	Use UHF such as 860 MHz for tag-reader communication (≈ 6-8 meters read distance)	
	Medium	Use HF such as 13,56 MHz for tag-reader communication (≈ 1 meter read distance)	
	High	Use NFC for tag-reader communication (≈ 10 cm read distance)	
Cx.x		...	...

Figure 18.5 Aligning design alternatives to the threats they are supposed to control based on the example of RFID readers reading tags from a distance.

diagrams. The transparency that is thus created in advance of system implementation is not only for picking the right solution but also beneficial for long-term documentation of the IT system (Fleischmann and Wallace 2009).

Figure 18.6 gives an example of a UML sequence diagram. It shows the password model for RFID tag deactivation (Spiekermann and Berthold 2004), which offers a medium level of privacy control. The UML sequence model shows that a RFID tag's EPC is encrypted with a password (steps 6 and 7 in Figure 18.6). The RFID reader at a point of sale (POS) checkout would probably need to retrieve a unique password for each EPC from an EPC Information System (EPC-IS; step 4). With this password, the tag's content can be encrypted. Only authorized readers that have access to this password will later be able to read the encrypted content. For instance, the retailer that sold the product can reopen the tag for warranty or return purposes. The password can be stored not only in the EPC-IS at the POS but also on

the mobile phone of the retailer's customer (steps 10 and 11). This mechanism provides customers control of RFID tag behavior, keeps track of their purchases, allows for transfer of ownership, and provides proof that they are the rightful owners of a product. For this purpose, the mobile phone of the customer would need to run a RFID privacy application where all protected EPCs are stored alongside the passwords that are needed to decrypt them.

Modeling task sequences help teams to imagine how various personas will react to each sequence. Let us say persona 1 is a late adopter of technology who does not possess a smartphone. She may be an elderly woman who is concerned about being tracked. She does not normally use many post-sales services. Let us assume she does not plan to ever buy an intelligent fridge. Persona 2 owns the retail store and has to pay for the entire RFID system landscape.

By using UML sequence diagrams, the team can predict the personas' reactions. First, persona 1's worry about being

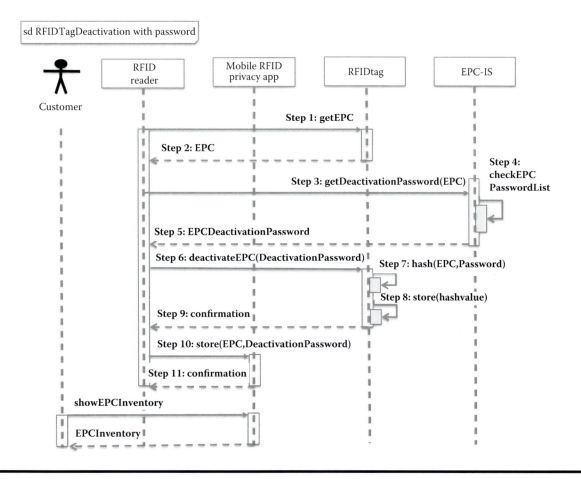

Figure 18.6 UML sequence diagram to show the system tasks needed for RFID tag deactivation with a password.

tracked may be unjustified. Because tags are password protected, the risk of them being read out by third parties is low. If persona 1 wanted to use postsales services, she could theoretically do so by accessing the tag's content with the password stored in her mobile RFID privacy app. However, persona 1 does not care about postsales services and does not possess a smartphone to run a privacy application. For persona 1, this scenario creates a lot of technical functionality that she does not need.

As we go on with the analysis, we see that the password-based encryption mechanism in steps 6 to 9 require RFID tags to be intelligent enough to encrypt their information. They would hence need some storage capabilities. They would also need to be able to process the encryption. These requirements undermine persona 2's (the company's) goal to minimize tag costs.

These first two persona reflections immediately show that a password-based tag deactivation approach is problematic. Moreover, as the control analysis in Figure 18.5 showed, passwords are only a medium-strength method for protection against tracking. As a consequence, the project team may want to look into and compare other deactivation alternatives.

Figure 18.7 illustrates an alternative way to deactivate RFID tags: a kill command is used instead of a password-based deactivation solution. From Figure 18.7, it is immediately clear that deactivation still occurs, so persona 1 (the old lady) is satisfied. The cost for implementing a kill command for tag deactivation requires less tag functionality, reducing the cost for tags. This appeases persona 2 (the company). But since the tags will be killed at the store exit by the RFID readers (steps 6 to 8), postsales services based on RFID will also be thwarted. For persona 1, this does not matter, but now one development goal of persona 2 is not fulfilled anymore. We run into a value conflict.

As a result, the IT innovation team would now need to determine whether postsale services based on RFID are financially important enough for retailers to invest in chips that remain intact. From a value dam and threat perspective, we see that the kill command is the strongest control against consumers' fear to be tracked and is therefore a welcomed technical solution. Yet, the core question is one of foregoing: foregoing the opportunity to read chips again or preserving the ability to do so for whatever future development might come up. This strategic decision must be made by the IT

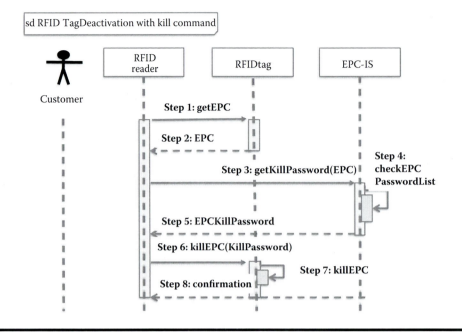

Figure 18.7   UML sequence diagram to show the system tasks needed for RFID tag deactivation with a kill command.

innovation team that decides on RFID tag/deactivation standard used.

At this stage, senior executives need to get involved again. Even though we are already deep down in technical design, senior managers will need to rejoin the decision-making process. They need to understand how the chosen RFID tag control will influence future business opportunities as well as the way retail (as well as smart homes) environments will be able to evolve. The RFID deactivation example shows that the details of technology design can be crucial in shaping the future margins and opportunities of a business.

As senior executives step in, they face the same decision dilemma I presented in Chapter 15 where leaders' wise decision-making on automation was discussed. They must make tough decisions on the kind of future they want to create. Again, they have to look at their own narrative and desired virtues; against this background, they must think through the inventory of moral philosophy that I presented in Chapter 15. Which technical solution treats people as ends rather than as means? In the long run, which solution will produce the greatest good for the greatest number of people? Which solution does the leader feel most committed to in his or her own conscience, regardless of whether it benefits him or her personally?

When tool-level tasks are analyzed and modeled, the "modelers" in the project team can see that there is more than one way to design an IT system. IT systems can automate many tasks and not involve users as much. They can require more or less user control. They can send feedback messages or not, providing more or less information control.

The number of systems or databases involved in handling a task can also be varied, affecting system complexity. As I outlined in Chapter 16 on IT architecture, end-user devices can be built to communicate with a central IT system at the backend, or they may be more decentralized. End-user devices (down to the tag and sensor level) therefore need more or less processing power and storage, and they provide for more or less behavioral control over a system. All of these detailed technical aspects are recognized when they are modeled in UML, and when the models are then critically discussed. As a result of this effort, the true requirements and business consequences of a new IT system become apparent. It becomes clear where ethical development goals conflict with other goals at the technical and economic level; this clarity allows the IT team to investigate compromises and trade-offs "resolving, solving, and dissolving" ethical problems (Ehrenfeld 2008). When all this is done, the team can pick the IT solution to implement and calculate its cost.

At this point, we can discuss an important final question for IT innovation: When is the right time to calculate a final business plan? When is a business case for an IT system truly complete? Classical product innovation processes, such as the stage gate process, foresee and press for relatively early completion of a business case. As I outlined in Chapter 13, if we implemented the E-SDLC so that it matched classical product innovation processes, we would understand the final cost of an IT project in the analysis phase. However, to get true insight into the cost of an IT project, system design alternatives must be fully understood. As becomes apparent from the RFID deactivation example described here, each design alternative will have completely different cost implications. So, although the team can give a rough estimate of financial feasibility at the IT planning phase, the design phase is where the bottom line is really understood. Because business managers have a hard time understanding what level of technical detail is required for the business planning of IT, they often build premature business cases that are not based on the kind of rigorous thinking presented here. Consequently, many business plans do not reflect technical reality and dramatically underestimate the cost of a system. This result is not only bad for a company's financial planning and bottom line, it is also equally problematic from an ethical perspective, as poor IT business planning leads to poorly performing IT projects that clog the pipe and drain the time and money required to build "good" systems.

18.3 System Implementation

Once the team models and understands system design alternatives, they will have a plan or descriptive outline for what the final system should look like. This outline is the starting point for system development. In Figure 13.4 (Chapter 13) I showed a spiral symbol to signal that most software engineering projects today progress in an incremental and iterative way. Incremental development means that an initial system implementation is exposed to user comments and risk analysis, and evolves through several versions until an adequate system version is achieved (for an illustration, see Sommerville 2011, p. 33).

Plenty of academic research and textbooks describe how to engineer software-based systems (see, e.g., Hoffer, George, and Valacich 2002; Sears and Jacko 2007; Kurbel 2008; Sommerville 2011). This book does not aim to add to this literature. Instead, I focus on ethical issues in earlier phases of system selection, planning, and requirements engineering. Here, the values are set for when software is implemented. If a project team has followed the methodological steps of an E-SDLC as described so far, they should know whose and what values to cater to.

Incremental and iterative methods for software development offer more opportunities for ethical system design because they allow software engineers to continuously improve on initial solutions, both in terms of usability and ethical success. User participation can start at a very early stage, with the help of low-tech prototypes. For example, users can sketch an initial version or mock-ups of the system themselves on paper or other materials. The development team can also put early versions of their system on the web and have future users provide them with suggestions (Rashid et al. 2006).

Once early running versions are ready for use, functionality can be gradually added to see how users use and react to it. Muller (2003) notes that this incremental adding of functionality works especially well when the added functionality is a "crucial artifact" for end users in their work. In other words, intermediate software versions tested with users should include the most relevant new design features that influence a work pattern.

In this context, scholars use the terms *reflective design* or *reflective HCI* (Dourish et al. 2004; Sengers et al. 2005) to refer to a process where technologies are initially given to users as probes with an explicit feedback channel. As they use the new system, users provide feedback that can go beyond the usability of the interface and can include context effects. For example, Sengers et al. (2005) built software that allows couples in long-distance relationships to share location information. Couples were invited to give feedback on the software and its use but also reflected on how their relationship was influenced by the technology.

While working with software engineering teams, Katie Shilton (2013) found that prototype testing among the responsible software engineers themselves is a very powerful "value lever." Software engineers should beta test the software they develop themselves. Shilton reports on a project where a software engineer presented a map navigation system she had developed herself and that displayed her movement. When she publicly showed her GPS data during a presentation, she realized that she had been crisscrossing through fuel stations and parking lots to avoid traffic. After being embarrassed by her own software application, she became the strongest privacy advocate on the engineering team.

The literature on participatory design (Muller 2003; Sears and Jacko 2007) and values in design (Shilton, Koepfler, and Fleischmann 2012; Shilton 2013) provides extensive insights on how to best involve users and their expectations in the incremental improvement of an IT system.

As we move into the implementation and programming phase of system development, we draw closer to the moment where a machine's internal logic is developed, where the algorithms for its artificial intelligence (AI) are chosen, and where software needs to account for hardware limitations or existing standards. We approach the core of computer science. Within this domain, a niche of scientists have started to work on "machine ethics" and question how machines, artificial intelligence can be enhanced with ethical reasoning (Anderson and Anderson 2011). The next chapter delves into this field and how it relates to the ethical design work presented in this textbook.

EXERCISES

1. Develop a technical and governance proposal that would allow United Games to deploy the emotional activity logging system (described in Chapter 3) in a privacy-friendly way. To support your proposal,

identify the value dams and flows relevant for this system and specify protection goals. Then, for one value, systematically identify controls as outlined in Figures 18.4 and 18.5. Identify design alternatives you regard as relevant and choose one to implement.

2. Based on the retail scenario in Chapter 3, create UML sequence diagrams that consider customers and mall systems. Create one model for an anonymous customer and one for an identified customer. What are the differences between the identified and anonymous IT architecture and data flows?

3. Having a say in who creates knowledge about us seems to be a vital ethical claim. For instance, in the scenario on the future of university education in Chapter 3, Jeremy is rejected by a university because his past average outdoor time reflects poorly on him. Draw a UML sequence diagram that outlines the organization of a "human-in-the-loop" sequence for the Stanford Online system. Specify the technical and governance entities required to make such a transparent integration of applicants work. What challenges might arise?

4. Think of the Alpha1 robots described in the scenario in Chapter 3. What security measures must the robot embed so that it does not undermine security and safety? Conduct an attack tree analysis where security goals are placed at the root of the trees.

5. Develop a technical and governance proposal for the bee drones from Chapter 3 (robot scenario) that recognize criminals in the street. For this purpose, identify the value dams and flows most relevant for this system and specify protection goals. Then, for one value, systematically identify controls as outlined in Figures 18.4 and 18.5. Identify design alternatives you regard as relevant and choose one to implement.

References

Anderson, M. and S. L. Anderson. 2011. *Machine Ethics*. New York: Cambridge University Press.

Dourish, P., J. Finlay, P. Sengers, and P. Wright. 2004. "Reflective HCI: Towards a Critical Technical Practice." Paper presented at Computer Human Interaction (CHI 2004). Vienna, Austria, April 24–29.

Ehrenfeld, J. R. 2008. *Sustainability by Design: A Subversive Strategy for Transforming Our Consumer Culture*. New Haven, CT: Yale University Press.

Fleischmann, K. R. and W. Wallace. 2009. "Ensuring Transparency in Computational Modeling." *Communications of the ACM (CACM)* 52(3):131–134.

Hoffer, J. A., J. F. George, and J. S. Valacich. 2002. *Modern Systems Analysis and Design*. Upper Saddle River, NJ: Prentice Hall.

Kurbel, K. 2008. *System Analysis and Design*. Heidelberg: Springer Verlag.

Muller, M. J. 2003. "Participatory Design: The Third Space in HCI." In *The Human-Computer Interaction Handbook*, 1051–1068. Hillsdale, NJ: Lawrence Erlbaum Associates.

Object Management Group (OMG). 2011. "Unified Modeling Language™ (UML®)." http://www.omg.org/spec/UML/.

Rashid, A., D. Meder, J. Wiesenberger, and A. Behm. 2006. "Visual Requirement Specification in End-User Participation." Paper presented at Multimedia Requirements Engineering (MERE '06), Minneapolis, Minnesota, September 12.

Sears, A. and J. A. Jacko. 2007. *Computer–Human Interaction: Development Process*. Boca Raton, FL: Taylor & Francis Group.

Sengers, P., K. Boehner, D. Shay, and J. Kaye. 2005. "Reflective Design." Paper presented at AARHUS '05, Aarhus, Denmark.

Shilton, K. 2013. "Values Levers: Building Ethics into Design." *Science, Technology & Human Values* 38(3):374–397.

Shilton, K, J. Koepfler, and K. Fleischmann. 2012. "Chartering Sociotechnical Dimensions of Values for Design Research." *The Information Society* 29(5):1–37.

Sommerville, I. 2011. *Software Engineering*. 9th ed. Boston: Pearson.

Spiekermann, S. 2008. *User Control in Ubiquitous Computing: Design Alternatives and User Acceptance*. Aachen, Germany: Shaker Verlag.

Spiekermann, S. and O. Berthold. 2004. "Maintaining Privacy in RFID Enabled Environments: Proposal for a Disable Model." In *Privacy, Security and Trust within the Context of Pervasive Computing*, edited by Philip Robinson, Harald Vogt and Waleed Wagealla. Vienna, Austria: Springer Verlag.

Spiekermann, S. and H. Ziekow. 2006. "RFID: A Systematic Analysis of Privacy Threats and a 7-Point Plan to Address Them." *Journal of Information System Security* 1(3):2–17.

Chapter 19

Machine Ethics and Value-Based IT Design

"What is natural has been swallowed by the sphere of the artificial."

Hans Jonas (1979)

There are broadly two ways to think about machine ethics. One way is to design machines such that they incorporate carefully elaborated ethical principles in the way they are built. Here, information technology (IT) designers tailor machines to ethical principles. The second way is to give machines the capability to reason on ethical matters. Here, machines use ethical principles or learning procedures to abstract ethical theories and guide their own behavior (Anderson 2011). This second stream of machine ethics focuses on how ethical theories can be translated into algorithms.

The distinction between these two broad streams of machine ethics is vital because the streams imply completely different focuses on how ethical theories are used in machine design. In this book, I pursue the first stream. I show how we can take various streams of moral philosophy and make them usable for IT designers to build "good" IT systems. The focus is on what IT managers, IT project teams, and engineers can do to ensure that the IT systems they build and invest in are beneficial for the environment and for the people who live and work with these systems.

Various domains of ethical theory come into play for a value-based design as I propose it here. First of all I have used "value ethics," a stream of philosophy that draws from twentieth century philosophy and psychology (Frankena 1973; Rokeach 1973; Schroeder 2008; Krobath 2009). Value ethics is my foundation for understanding core goals for IT system design. But value ethics is not enough. As Charles Ess (2013) notes, values can easily fall victim to relativism and be misused. It is, therefore, indispensible to combine value ethics with moral guidance for executives and engineers.

Rule-based moral guidance and Aristotelian virtue ethics are therefore further building blocks of what I call "value-based design." Figure 19.1 summarizes how I combine ethical theories in value-based IT design.

The building blocks of value-based design reflect James Moor's distinction between "ethical-impact agents" and "implicit ethical agents."* When IT project teams consider the ethical impact of machine agents, they think through the value consequences of a machine's deployment. Chapters 14 to 17 showed how teams can question machines' impact on values. I showed how we can think through the benefits and harms that are caused by new IT investments. We should identify values that might (or should) be impacted by our agents, and operationalize these values for IT design. We should also reflect on context effects created.

When teams consider implicit ethical agents, they can constrain or design machines' actions so that the actions are beneficial to humans, avoid unethical outcomes, or both. The analysis and design phases of the ethical system development life cycle (E-SDLC), which I presented in Chapters 17 and 18, contain a methodology for reaching this goal and for effectively building implicit ethical agents.

That said, two more explicit kinds of machine agents are discerned: "explicit ethical agents" and "full ethical agents." Flanagan, Howe, and Nissenbaum (2008) described these kinds of machines as "materially embodying" ethics; they run algorithms that are informed by ethical theory. If this kind of system was built, it would also impact our human environments in an important way; the context effects would be tremendous. Considerable research work is now being put into this field of

---

* Moor calls machines "agents" because of their embedded intelligence and the roles they may play for us and on our behalf. He notes that "computing technology often has important ethical impact" (Moor 2011, p. 15).

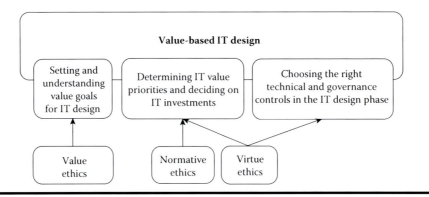

Figure 19.1 Building blocks of ethical IT design.

artificial intelligence. But it seems to me that scholars in this field of AI rarely think about the ethical effects of their research, values they create or destroy through such highly intelligent machines. This is why I have not marked this form of explicit or full ethical machine agency with a shaded background in Figure 19.2. Shaded fields in Figure 19.2 correspond to what I have worked on in this book. Still I recognize that explicit and full ethical agents are a centerpiece of ethical machines in the future. Let me therefore briefly outline what this area of research is about and where it currently confronts limits.

19.1 Explicit Ethical Agents

According to Susan Leigh Anderson, the ultimate goal of machine ethics is to develop machines that are explicit

ethical agents. Explicit ethical agents are machines that incorporate ethical principles so that they can reason and base their behavior on these principles. The machines "do ethics in a way that, for example, a computer plays chess" (Moor 2011, p. 16). Explicit ethical agents need to be able to assess the situations in which they find themselves, know which alternative actions are available to them, apply ethical reasoning, and take the best action from an ethical standpoint.

If explicit ethical agents worked, they promise to support human beings who sometimes base their decisions too much on emotions or first-order desires. Emotions and intuitions are good, and they should not be dominated by explicit ethical agents' rationality. But good agents could still help people reflect on some of their judgments, take more reasoned decisions and become more attentive to their own behavior. I

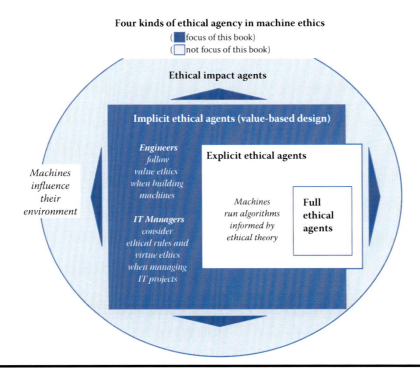

Figure 19.2 Four varieties of machine ethics and the focus of this textbook.

hinted at that kind of agent support when I described the relationship between agent Arthur and Sophia:

> Sophia almost cannot live without Arthur's judgment anymore. She really loves him like a friend even though he recently started to criticize her sometimes; for example, when she was lazy or unfair to a friend. He also helps her to avoid buying and eating too many sweets, because she really gained weight and deeply desires to lose a few pounds.

As I outlined in Chapter 6, this kind of agent behavior has implications for human freedom. Still, scholars in artificial intelligence (AI) have started to work on the grand challenge of creating this kind of machine. They think about how advanced logic could serve as a bridge between ethics and machine reasoning (Van Den Hoven and Lokhorst 2003): Deontic logic may be used for statements of permission and obligation. The prospects of a Kantian machine have been investigated (Powers 2006). Epistemic logic can be used for statements of beliefs and knowledge. And action logic can be applied for statements about actions. A formal apparatus is already available to describe simple ethical situations with enough precision that software systems can embody ethical protocols in a clearly defined context.

For example, the action-logic-based agent called Jeremy was provided as a proof of concept for explicit ethical agents by Anderson, Anderson, and Armen (2005). Jeremy, named after Jeremy Bentham, implements a hedonistic act utilitarian protocol. In an early form, Jeremy is manually fed information on a situation. It receives the name of an action, the names of persons affected by that action, a rough estimate of the action's consequences for each person (which can range from "very pleasurable" to "not pleasurable at all") and the likelihood of that consequence for each person. With the help of this data, Jeremy calculates the amount of net pleasure resulting from each action and informs its user about the action for which the net pleasure is greatest. Jeremy's user can also request more information on the chosen action and why that one was chosen. Future machines could embed sensor technology, face recognition, behavioral pattern analysis, and machine learning technology powerful enough that— when combined—the machine can automatically collect the data to run this kind of logic at an advanced and autonomous level.

Ethical protocols of this kind are only a starting point for explicit ethical agents. In Chapter 15, I presented the most common ethical theories that would need to be embraced by explicit ethical agent programming, such as deontological reasoning and utilitarianism. But as you read in Chapter 15, ethical theories have their own pitfalls and challenges. Act utilitarianism algorithms could be manipulated by

BOX 19.1 THE ROBOT CAR OF TOMORROW MAY JUST BE PROGRAMMED TO HIT YOU

Suppose that an autonomous car is faced with a terrible decision to crash into one of two objects (Figure 19.3). It could swerve to the left and hit a Volvo sport utility vehicle (SUV), or it could swerve to the right and hit a Mini Cooper. If you were programming the car to minimize harm to others—a sensible goal—which way would you instruct it go in this scenario?

As a matter of physics, you should choose a collision with a heavier vehicle that can better absorb the impact of a crash, which means programming the car to crash into the Volvo. Further, it makes sense to choose a collision with a vehicle that's known for passenger safety, which again means crashing into the Volvo. But physics isn't the only thing that matters here. Programming a car to collide with any particular kind of object over another seems an awful lot like a targeting algorithm, similar to those for military weapons systems. And this takes the robot-car industry down legally and morally dangerous paths.

Even if the harm is unintended, some crash-optimization algorithms for robot cars would seem to require the deliberate and systematic discrimination of, say, large vehicles to collide into. The owners or operators of these targeted vehicles would bear this burden through no fault of their own, other than that they care about safety or need an SUV to transport a large family. Does that sound fair?

What seemed to be a sensible programming design, then, runs into ethical challenges. Volvo and other SUV owners may have a legitimate grievance against the

Figure 19.3 Illustration of vehicles communicating with one another. (From USDOT, 2015. With permission.)

manufacturer of robot cars that favor crashing into them over smaller cars, even if physics tells us this is for the best.

IS THIS A REALISTIC PROBLEM?

Some road accidents are unavoidable, and even autonomous cars can't escape that fate. A deer might dart out in front of you, or the car in the next lane might suddenly swerve into you. Short of defying physics, a crash is imminent. An autonomous or robot car, though, could make things better.

While human drivers can only react instinctively in a sudden emergency, a robot car is driven by software, constantly scanning its environment with unblinking sensors and able to perform many calculations before we're even aware of danger. They can make split-second choices to optimize crashes—that is, to minimize harm. But software needs to be programmed, and it is unclear how to do that for the hard cases.

In constructing the edge cases here, we are not trying to simulate actual conditions in the real world. These scenarios would be very rare, if realistic at all, but nonetheless they illuminate hidden or latent problems in normal cases. From the above scenario, we can see that crash-avoidance algorithms can be biased in troubling ways, and this is also at least a background concern any time we make a value judgment that one thing is better to sacrifice than another thing.

In previous years, robot cars have been quarantined largely to highway or freeway environments. This is a relatively simple environment, in that drivers don't need to worry so much about pedestrians and the countless surprises in city driving. But Google recently announced that it has taken the next step in testing its automated car in exactly city streets. As their operating environment becomes more dynamic and dangerous, robot cars will confront harder choices, be it running into objects or even people. …

THE ROLE OF MORAL LUCK

An elegant solution to these vexing dilemmas is to simply not make a deliberate choice. We could design an autonomous car to make certain decisions through a random-number generator. That is, if it's ethically problematic to choose which one of two things to crash into—a large SUV versus a compact car, or a motorcyclist with a helmet versus one without, and so on—then why make a calculated choice at all?

A robot car's programming could generate a random number; and if it is an odd number, the car will take one path, and if it is an even number, the car will take the other path. This avoids the possible charge that the car's programming is discriminatory against large SUVs, responsible motorcyclists, or anything else.

This randomness also doesn't seem to introduce anything new into our world: luck is all around us, both good and bad. A random decision also better mimics human driving, insofar as split-second emergency reactions can be unpredictable and are not based on reason, since there's usually not enough time to apply much human reason.

Yet, the random-number engine may be inadequate for at least a few reasons. First, it is not obviously a benefit to mimic human driving, since a key reason for creating autonomous cars in the first place is that they should be able to make better decisions than we do. Human error, distracted driving, drunk driving, and so on are responsible for 90 percent or more of car accidents today, and 32,000-plus people die on U.S. roads every year.

Second, while human drivers may be forgiven for making a poor split-second reaction—for instance, crashing into a Pinto that's prone to explode, instead of a more stable object—robot cars won't enjoy that freedom. Programmers have all the time in the world to get it right. It's the difference between premeditated murder and involuntary manslaughter.

Third, for the foreseeable future, what's important isn't just about arriving at the "right" answers to difficult ethical dilemmas, as nice as that would be. But it's also about being thoughtful about your decisions and able to defend them—it's about showing your moral math. In ethics, the process of thinking through a problem is as important as the result. Making decisions randomly, then, evades that responsibility. Instead of thoughtful decisions, they are thoughtless, and this may be worse than reflexive human judgments that lead to bad outcomes.

CAN WE KNOW TOO MUCH?

A less drastic solution would be to hide certain information that might enable inappropriate discrimination—a "veil of ignorance," so to speak. As it applies to the above scenarios, this could mean not ascertaining the make or model of other vehicles, or the presence of helmets and other safety equipment, even if technology could let us, such as vehicle-to-vehicle communications. If we did that, there would be no basis for bias.

Not using that information in crash-optimization calculations may not be enough. To be in the ethical clear, autonomous cars may need to not collect that information at all. Should they be in possession of the information, and using it could have minimized harm or saved a life, there could be legal liability in failing to use that information. Imagine a similar public outrage if a national intelligence agency had credible information about a terrorist plot but failed to use it to prevent the attack.

A problem with this approach, however, is that auto manufacturers and insurers will want to collect as much data as technically possible, to better understand robot-car crashes and for other purposes, such as novel forms of in-car advertising. So it's unclear whether voluntarily turning a blind eye to key information is realistic, given the strong temptation to gather as much data as technology will allow.

SO, NOW WHAT?

In future autonomous cars, crash-avoidance features alone won't be enough. Sometimes an accident will be unavoidable as a matter of physics, for myriad reasons—such as insufficient time to press the brakes, technology errors, misaligned sensors, bad weather, and just pure bad luck. Therefore, robot cars will also need to have crash-optimization strategies.

To optimize crashes, programmers would need to design cost-functions—algorithms that assign and calculate the expected costs of various possible options, selecting the one with the lowest cost—that potentially determine who gets to live and who gets to die. And this is fundamentally an ethics problem, one that demands care and transparency in reasoning.

It doesn't matter much that these are rare scenarios. Often, the rare scenarios are the most important ones, making for breathless headlines. In the U.S., a traffic fatality occurs about once every 100 million vehicle-miles traveled. That means you could drive for more than 100 lifetimes and never be involved in a fatal crash. Yet these rare events are exactly what we're trying to avoid by developing autonomous cars, as Chris Gerdes at Stanford's School of Engineering reminds us.

Again, the above scenarios are not meant to simulate real-world conditions anyway, but they're thought-experiments—something like scientific experiments—meant to simplify the issues in order to isolate and study certain variables. In those cases, the variable is the role of ethics, specifically discrimination and justice, in crash-optimization strategies more broadly.

The larger challenge, though, isn't thinking through ethical dilemmas. It's also about setting accurate expectations with users and the general public who might find themselves surprised in bad ways by autonomous cars. Whatever answer to an ethical dilemma the car industry might lean towards will not be satisfying to everyone.

Ethics and expectations are challenges common to all automotive manufacturers and tier-one suppliers who want to play in this emerging field, not just particular companies. As the first step toward solving these challenges, creating an open discussion about ethics and autonomous cars can help raise public and industry awareness of the issues, defusing outrage (and therefore large lawsuits) when bad luck or fate crashes into us.

Patrick Lin
The Robot Car of Tomorrow May Just Be Programmed to Hit You, (WIRED), 2014

biased facts about a situation. And how and by whom should weights be set to calculate the greatest good for the greatest number? General utilitarianism algorithms could reflect the latest opinion trends on what is good for the general public because such opinions tend to dominate the literature that is electronically available. But whether these short-term opinions are valid and ethically good is questionable. A deontological algorithm depends on the right choice of maxims. But who should set these? If a small number of humans programmed these agents' algorithms and fed the agents information, those humans would have the power to determine what the agents say is ethical. Box 19.1 recounts Patrick Lin's account of the challenges that accompany explicit ethical agents. Here the fictive machine agent is a self-driving car that needs to make a decision on what target to hit during an accident. The example prolifically shows how rational utilitarian reasoning can struggle to come to any justifiable and good conclusion.

AI may try to force philosophers to come to terms on what ethics should finally look like. As Daniel Dennett said, "AI makes philosophy honest." But how can philosophy make AI honest in a case like the autonomous car crashing into the "right" human being? What if—beyond crash tests—true ethical behavior ultimately resides in a human being's narrative, in personal virtue in the Aristotelian sense? What kind of intelligence would be required to embody that highest form of ethical reasoning in a machine? At that point, we talk about the questionable, but still powerful contemporary idea of full ethical agents.

19.2 Full Ethical Agents?

Full ethical agents are the most ambitious vision of what machines could be like one day. They are inspired by science fiction and transhumanist narratives of digital objects becoming as intelligent as adult human beings. Moor (2011) argues that full ethical agents would need to possess consciousness, intentionality, and free will in order to be comparable to human beings. Machine ethics scholars debate whether it will be possible for machines to live up to these qualities (Sullins 2006; Anderson and Anderson 2011).

Scholars who embrace transhumanistic thinking promote the idea that machines will soon become full ethical agents. Transhumanists believe that the human race is just one point in the evolution of information. Powerful artificial intelligence will surpass human intelligence at a point in time they call the "Singularity" (Kurzweil 2006). At the point of Singularity, they believe machines will not only be more intelligent than human beings, but they will learn, reproduce, and improve themselves. Ray Kurzweil, a major proponent of this thinking (and product engineering director at Google as of 2015) writes, "our technology will match and then vastly exceed the refinement and suppleness of what we regard as the best of human traits" (2006, p. 9). Scholars who work on full ethical agents seem to share this or a similar vision of machines that possess a kind of consciousness. Some of them believe that machines will exceed human beings in moral and intellectual dimensions. "The transformation underlying the Singularity is not just another in a long line of steps in biological evolution. We are upending biological evolution altogether" (Kurzweil 2006, p. 374). One of the pioneers of robotics, Hans Moravec (1988), foresaw this evolution in the 1980s when he wrote his book on "Mind Children" (Hall 2011). He predicted a "convergent evolution" of men and machines (p. 42) and did not exclude the possibility of "awareness in the mind of our machines" (p. 39). "Will robots inherit the earth?" he asked (p. 260).

Of course, natural scientists as well as people educated in philosophy question the feasibility of this transhumanistic vision. One of the core assumptions of the Singularity is that nonbiological structures are more capable than the human body and that these artificial media can replicate the richness, subtlety, and intuitive sensitivity of human thinking. Scholars like Steve Torrance (2005) question this assumption, arguing that full ethical agents must be *organic* to reach "intrinsic moral status." Only organic beings are "genuinely sentient," and only sentient beings can be "subjects of either moral concern or moral appraisal" (as cited in Anderson and Anderson 2011, p. 8).

One of the main pitfalls of transhumanistic thinking is in the focus on the human brain, which seems to take the role of a kind of isolated cage in which the entire human intelligence resides. Kurzweil (2006) writes: "After the algorithms of a particular [brain] region are understood, they can be refined and extended before being implemented in synthetic neural equivalents. They can run on a computational substrate that is already far faster than neural circuitry. ... (pp. 149, 199). These technical ideas are philosophically in line with sixteenth century philosopher Descartes, who first proposed the idea of body–mind separation. I describe this

BOX 19.2 BODY VERSUS MIND DISCOURSE IN PHILOSOPHY AND ITS IMPLICATION FOR THE TRANSHUMANISTIC IDEA OF FULL ETHICAL AGENCY

Philosophy is split into two schools of thought that disagree about the importance of the human body. On one side, transhumanists such as Ray Kurzweil and Hans Moravec believe that the body is not important. Like Descartes, they believe in a body–mind separation.[1]

In contrast, philosophers like Nietzsche, Maurice Merleau-Ponty, and Hubert Dreyfus and spiritual thinkers in Asia (in particular, those who practice Yoga) believe that the most important resource of human beings is not their mental capability but the emotional and intuitive capacity of their whole bodies.[2]

Let's see what this means for our understanding of machines and the roles they can take in our lives: In the thinking of Descartes and today's transhumanists, the sense organs are transducers that bring information to the brain. Descartes drew on the phenomenon of amputees, who sometimes insist that they feel pain in a limb that is not there. This observation led him to believe that everything we experience is a creation of our own minds and that the world and our bodies are not directly present. The end vision of this thinking can be understood by watching the 1999 film *The Matrix*, where human beings spend their lives in tubes, with their brains connected to machines that simulate life for them. They believe that they live, but in reality they spend their true lives in a tube. Transhumanists like Kurzweil might not find this fictional scenario to be too far-fetched. They argue that we could scan our brains to understand how humans work, upload the essence of human "intelligence" to a computer, and then live in a machine after our physical death. "Uploading a human brain means scanning all of its salient details and then re-instantiating those details into a suitable powerful computational substrate. This process would capture a person's entire personality, memory, skills, and history."[3] Embracing this view means that our bodies are ultimately not important.

Philosophers have questioned this idea of humanity, taking analogical phenomenology as a more holistic scientific approach to understand human existence.[5] Merleau-Ponty (1908–1961), for example, stressed the importance of using our bodies to make sense of the world. He described how our bodies constantly commune with the objects around us; for example, a couple communes with each other and a jazz player communes with his saxophone (Figure 19.4). Our body

Figure 19.4 With our bodies we "commune" with reality and get "a grip of the world." (Left: © CC BY-SA 2.0 Loos-Austin 2005; right: © CC BY-SA 2.0 Van der Wel 2010.)

movements help us to zoom in and out of the world, to approach it from the right distance, and in doing so help us to achieve our unique "grip of the world."[5] From this perspective, our body is not just a collection of sensors that channel bits of information to the brain; consciousness is part of the body itself.

Take another example: When we enter a room where a party is happening, we sense the mood in the room. We perceive this mood through more than just our eyes; if we used a surveillance camera to review the scene, we would not necessarily be able to see the mood. Neither can we really smell or hear the mood. Still, we know through our bodily senses whether the party is in full swing, and we can physically share in this mood. In his *Phenomenology of Perception,* Merleau-Ponty wrote, "Insofar as I have hands, feet, a body and a world, I sustain intentions around myself that are not decided upon and that affect my surroundings in ways I do not choose."[6] The described neuroscience research supports this view. Vittorio Gallese et al., who coined the term "mirror neurons" in their seminal 1996 article "Action Recognition in the Premotor Cortex,"[7] writes about our "embodied experience of the world." He says: "We map the actions of others onto our own motor system … creating a mutual resonance of intentionally meaningful sensory-motor behaviors, but not specific mental state interference."[8] Nietzsche's Zarathustra says: "'I,' you say, and are proud of the word. But greater is that in which you do not wish to have faith—your body and its great reason: that does not *say* 'I,' but *does* 'I'. … Behind your thoughts and feelings, my brother, there stands a mighty ruler, an unknown sage—whose name is self. In your body he dwells; he is your body."[9]

The idea of transhumanists of simply uploading human existence by scanning our brain activities is therefore probably highly naïve. "I shall not go your way, O despisers of the body! You are no bridge to the overman!" concluded Nietzsche in his *Zarathustra*.

NOTES

1. Ray Kurzweil, *The Singularity Is Near: When Humans Transcend Biology* (London: Penguin Group, 2006); Hans Moravec, *Mind Children: The Future of Robot and Human Intelligence* (Cambridge, MA: Harvard University Press, 1988); René Descartes, *Principles of Philosophy* (Dordrecht, Netherlands: Kluwer Academic Publishers, 1991).
2. Friedrich Nietzsche, *Also sprach Zarathustra: Ein Buch für Alle und Keinen* (Munich: C. H. Beck, 1883–1885/2010); Maurice Merleau-Ponty, *Phenomenology of Perception* (Abingdon, UK: Routledge, 1949/2014); Hubert L. Dreyfus, *On the Internet* (New York: Routledge, 2009).
3. Kurzweil, *The Singularity Is Near*, p. 198/199.
4. Johannes Hoff, *The Analogical Turn: Rethinking Modernity with Nicholas of Cusa* (Cambridge: William B. Eerdmans, 2013).
5. Merleau-Ponty, *Phenomenology of Perception*.
6. Ibid., p. 465.
7. Vittorio Gallese, Luciano Fadiga, Leonardo Fogassi, Giacomo Rizzolatti, "Action Recognition in the Premotor Cortex," *Brain*, 119 (1996): 593–609 (p. 180).
8. Deborah Jenson and Marco Iacoboni, "Literary Biomimesis: Mirror Neurons and the Ontological Priority of Representation," *California Italian Studies* 2 (2011), p. 9.
9. Nietzsche, *Also sprach Zarathustra*, p. 34.

body versus mind discourse in Box 19.2 and outline how the last 200 years of philosophical thinking have seen major updates to the idea that our brain reigns over our existence. Recent neuro-science research points more toward viewing humans as body–mind entities. So the benefit from just scanning and uploading human brains is questionable if the rest of what constitutes a human being is factored out. Still, great scientific effort is being put into the replication of brain functionality in computers.* The hope that human consciousness can be decoded inspires scientists, but it is unclear how far this effort gets us. As Leibniz (1714/2014) once pointed out, even if we blew the brain up to the size of a mill and walked around inside, we would not find consciousness.

---

* See, for example, the Human Brain Project, funded by the European Union, https://www.humanbrainproject.eu/ (accessed February 19, 2015).

In presenting these futuristic ideas of full ethical agents and transhumanism, I arguably conflate practical machine ethics with philosophical machine ethics, which are distinct streams of academic research. Practical machine ethics is interested in how to technically build superintelligent machines that embed ethical algorithms (I described early works in Section 19.1 on explicit ethical agents). Practical machine ethics is more commonly pursued by computer scientists. Philosophical machine ethics is interested in what this technical vision means for humanity or what it means to be a "genuine moral agent" as opposed to one that merely behaves as if it were being moral (Torrance 2011).

Machine ethics philosophers like Luciano Floridi and J. W. Sanders (2004) and Matteo Turilli (2007) facilitate the work of practical machine ethicists as well as transhumanists by looking into how, axiomatically, ethical theories or principles could be interpreted to make machines qualify as moral agents. For them, the way around many apparent paradoxes in moral theory is to find a level of theoretical abstraction where they can logically argue that artificial agents possess consciousness, morality and so forth. Their approach has earned them the criticism to "adopt a 'mindless morality' that evades issues like intentionality and free will since these are all unresolved issues in the philosophy of mind" (Sullins 2006, p. 27).

Some philosophers who believe that machines will soon live up to human beings argue that human beings actually do not possess certain qualities that are ascribed to them either. For example, Joseph Emile Nadeau argued that a free action is only free if it is based on reasons that are fully thought through by an actor. If free will is necessary for moral agency, and we as humans have not such an apparatus (because we often act emotionally and out of subconscious intuition), then we are—according to Nadeau—not free. In contrast, machines can be free, he argues, because they are capable of unbiased reasoning (Nadeau 2006, as cited in Sullins 2006, p. 27).

Finally, philosophers try to align artificial agents with human beings by simply bending the definition of philosophical principles so that they are a better fit for the machine world. For example, Sullins (2006) argues that if we define autonomy in the machine sense (as I did in Figure 6.3), then we already have autonomous agents today, because some machines already act fully automatically (e.g., some flight cockpits).

Seeing these debates, I want to close by arguing that neither analytical philosophy nor any philosophical labeling (and adaptation of definitions) will change much in the quality of the systems that are finally delivered. As of 2015, we do not know how far practical machine ethics will go. The field is so prone to speculation that I do not include many details about it in this book. More important, I personally doubt whether it is *desirable* to create full ethical agents. For a great

overview of diverging opinions on this matter, see Anderson and Anderson (2011). I would recommend the computer science world to first get a value-based approach to system design right (i.e., as it is described in this book) and only then turn to more complex problems.

EXERCISES

1. Explain the four kinds of ethical agents. Identify an example of each kind, excluding full ethical agents, which do not currently exist. Comment on the differences between the remaining three types of agents.

2. Think of a specific incident where an explicit ethical agent, such as agent Arthur in the scenarios in Chapter 3, would be useful. What would be the harms and benefits for human beings in such a situation? How would it influence the situation? What values might the agent create or destroy through its input to the situation?

3. Box 19.1 describes the decision dilemma confronted by an intelligent car in a crash situation. Provide a similar example of conflict and dilemma from the future scenario world described in Chapter 3.

4. Reflect on the philosophical debate about body–mind separation described in Box 19.2. Then, think of an incident where you felt your body react to a stimulus that you could intellectually comprehend after you reacted. What kind of sensors would an artificial human being need to reconstruct your emotional perception? What arguments or practical use cases speak for such a reconstruction effort?

References

Anderson, S. L. 2011. "Machine Metaethics." In *Machine Ethics*, edited by Michael Anderson and Susan Leigh Anderson. New York: Cambridge University Press.

Anderson, M. and S. L. Anderson. 2011. *Machine Ethics*. New York: Cambridge University Press.

Anderson, M., S. L. Anderson, and C. Armen. 2005. "Towards Machine Ethics: Implementing Two Action-Based Ethical Theories." In *AAAI Fall Symposium*. Menlo Park, CA: American Association for Artificial Intelligence.

Ess, C. 2013. *Digital Media Ethics*. 2nd ed. Hoboken, NJ: Wiley.

Flanagan, M., D. C. Howe, and H. Nissenbaum. 2008. "Embodying Values in Technology: Theory and Practice." In *Information Technology and Moral Philosophy*, edited by Jereon van den Hoven and Weckert John. New York: Cambridge University Press.

Floridi, L. and J. W. Sanders. 2004. "On the Morality of Artificial Agents." *Minds and Machines* 14(3):349–379.

Frankena, W. 1973. *Ethics*. 2nd ed. Upper Saddle River, NJ: Prentice-Hall.

Hall, J. S. 2011. "Ethics for Self-Improving Machines." In *Machine Ethics*, edited by Michael Anderson and Susan Leigh Anderson, 512–523. New York: Cambridge University Press.

Krobath, H. T. 2009. *Werte: Ein Streifzug durch Philosophie und Wissenschaft*. Würzburg: Könighausen & Neumann.

Kurzweil, R. 2006. *The Singularity Is Near: When Humans Transcend Biology*. London: Penguin Group.

Leibniz, G. 1714/2014. "Monadology." In *Leibniz's Monadology: A New Translation and Guide*, edited by Lloyd Strickland. Edinburgh: Edinburgh University Press.

Moor, J. 2011. "The Nature, Importance, and Difficulty of Machine Ethics." In *Machine Ethics*, edited by Michael Anderson and Susan Leigh Anderson. New York: Cambridge University Press.

Moravec, H. 1988. *Mind Children: The Future of Robot and Human Intelligence*. Cambridge, MA: Harvard University Press.

Nadeau, J. E. 2006. "Only Androids Can Be Ethical." In *Thinking About Android Epistemology*, edited by Kenneth M. Ford, Clark Glymour and Patrick Hayes, 241–248. Cambridge, MA: MIT Press.

Powers, T. M. 2006. "Prospects for a Kantian Machine." *Intelligent Systems* 21(4):46–51.

Rokeach, M. 1973. *The Nature of Human Values*. New York: Free Press.

Schroeder, M. 2008. "Value Theory." In *The Stanford Encyclopedia of Philosophy*. Stanford, CA: The Metaphysics Research Lab.

Sullins, J. 2006. "When Is a Robot a Moral Agent?" *International Review of Information Ethics* 6(12):23–30.

Torrance, S. 2005. "A Robust View of Machine Ethics." In *AAAI Fall Symposium*. Menlo Park, CA: American Association of Artificial Intelligence.

Torrance, S. 2011. "Machine Ethics and the Idea of a More-Than-Human Moral World." In *Machine Ethics*, edited by Michael Anderson and Susan Leigh Anderson. New York: Cambridge University Press.

Turilli, M. 2007. "Ethical Protocol Design." *Ethics and Information Technology* (9):49–62.

Van Den Hoven, J. and G.-J. Lokhorst. 2003. "Deontic Logic and Computer-Supported Computer Ethics." *Metaphilosophy* 33(3):376–386.

Index

Page numbers followed f, t, b and n indicate figures, tables, boxes and notes, respectively.